Electronic Literature as Digital Humanities

Electronic Literature
Volume 2

Series editors:
Astrid Ensslin, Helen Burgess, Rui Torres, Maria Mencia

Electronic Literature Organization

Electronic Literature as Digital Humanities

Contexts, Forms, & Practices

Edited by
Dene Grigar & James O'Sullivan

BLOOMSBURY ACADEMIC
NEW YORK · LONDON · OXFORD · NEW DELHI · SYDNEY

BLOOMSBURY ACADEMIC
Bloomsbury Publishing Inc
50 Bedford Square, London, WC1B 3DP, UK
1385 Broadway, New York, NY 10018, USA
29 Earlsfort Terrace, Dublin 2, Ireland

BLOOMSBURY, BLOOMSBURY ACADEMIC and the Diana
logo are trademarks of Bloomsbury Publishing Plc

First published in the United States of America 2021
This paperback edition published in 2022

Cover design by Namkwan Cho
Cover image from All the Delicate Duplicates by Mez Breeze and Andy Campbell,
https://mezbreeze.itch.io/all-the-delicate-duplicates

Bloomsbury Publishing Inc does not have any control over, or responsibility
for, any third-party websites referred to or in this book. All internet addresses given in
this book were correct at the time of going to press. The author and publisher regret any
inconvenience caused if addresses have changed or sites have ceased to exist, but can
accept no responsibility for any such changes.

ISBN: HB: 978-1-5013-6350-4
         PB: 978-1-5013-7389-3
       ePDF: 978-1-5013-6348-1
      eBook: 978-1-5013-6349-8

Series: Electronic Literature

Typeset by Integra Software Services Pvt. Ltd.

To find out more about our authors and books visit www.bloomsbury.com
and sign up for our newsletters.

The editors would like to dedicate their work on this collection to friends and colleagues at the Digital Humanities Summer Institute and Electronic Literature Organization.

CONTENTS

Section III Practices

Section IV Artist Interventions

ABOUT THE EDITORS

Dene Grigar is Professor and Director of the Creative Media & Digital Culture Program at Washington State University Vancouver whose research focuses on the creation, curation, preservation, and criticism of electronic literature, specifically building multimedial environments and experiences for live performance, installations, and curated spaces; desktop computers; and mobile media devices. She has produced sixteen media works such as "Curlew" (2014), "A Villager's Tale" (2011), the "24-Hour Micro E-Lit Project" (2009), "When Ghosts Will Die" (2008), and "Fallow Field: A Story in Two Parts" (2005), as well as fifty-seven scholarly articles and five books. She also curates exhibits of electronic literature and media art, mounting shows at the British Computer Society and the Library of Congress and for the Symposium on Electronic Art (ISEA) and the Modern Language Association (MLA), among other venues. With Stuart Moulthrop (University of Wisconsin–Milwaukee) she developed the methodology for documenting born-digital media, a project that culminated in an open-source, multimedia book, entitled *Pathfinders* (2015), and book of media art criticism, entitled *Traversals* (The MIT Press, 2017). She has served as President of the Electronic Literature Organization from 2013 to 2019 and Associate Editor of *Leonardo Reviews* since 2003. In 2017 she was awarded the Lewis E. and Stella G. Buchanan Distinguished Professorship by her university. She also directs Electronic Literature Lab at WSUV.

James O'Sullivan (@jamescosullivan) lectures at University College Cork (National University of Ireland). He has previously held faculty positions at the University of Sheffield and Pennsylvania State University. His work has been published in a variety of interdisciplinary journals, including *Digital Scholarship in the Humanities, Digital Humanities Quarterly*, and the *Electronic Book Review*. He is the author of *Towards a Digital Poetics: Electronic Literature & Literary Games* (Palgrave Macmillan, 2019). Further information on James and his work can be found at jamesosullivan.org.

Electronic Literature as Digital Humanities: An Introduction

Dene Grigar

The title of our collection, *Electronic Literature as Digital Humanities: Contexts, Forms, and Practices*, may seem an obvious one to scholars and artists already involved in electronic literature and the digital humanities. For well over a decade, presentations, exhibitions, courses, workshops, and papers addressing born-digital literary art have been featured at conferences held by the Modern Language Association (MLA)[1] and the Alliance of Digital Humanities Organizations (ADHO),[2] at centers and institutes like Digital Humanities at Berkeley[3] and the Digital Humanities Summer Institute at the University of Victoria,[4] and in publications like *Digital*

---

[1]Electronic Literature Organization is an Allied Organization of the Modern Language Association. One of the criteria to gain this status is a description of past special sessions held at the convention. ELO and its members had been giving papers at the convention for close to two decades, beginning with Judy Malloy's "Between Narrator and the Narrative," presented at MLA 1992 on December 29, 1992. See http://www.judymalloy.net/richmond/bowl.html.

[2]For example, Kathi Inman Berens' paper presented at Digital Humanities 2013, "Debugging 'The Personal Is Political:' Uncle Roger's Grandmother," discussed Judy Malloy's seminal work of electronic literature, *Uncle Roger*. See http://dh2013.unl.edu/abstracts/ab-286.html.

[3]See "No Legacy || Literatura electrónica," curated by Alex Saum-Pascual and Élika Ortega, Bancroft Library, UC Berkeley, February 16–May 5, 2017, https://libraries.cca.edu/exhibitions/no-legacy-literatura-electronica/.

[4]ELO and is members have taught at the Digital Humanities Summer Institute since 2011. One course example is "Introduction to Electronic Literature in Digital Humanities," which I introduced in 2014 and continues to be taught at the Institute by Davin Heckman and Astrid Ensslin under the title "Digital Fictions, Electronic Literature, and Literary Gaming." See https://dhsi.org/course-offerings/.

Humanities Quarterly (*DHQ*)[5] and *Literary Studies in the Digital Age* (LSDA),[6] to name but a few points of overlap. Additionally, funding for projects related to the archiving and documentation of electronic literature have been provided by the Office of Digital Humanities (ODH) of the National Endowment for the Humanities.[7] Moreover, from 2006 to 2011 the Electronic Literature Organization—the hub of activity for electronic literature art and scholarship—was hosted by the Maryland Institute of Technology in the Humanities at the University of Maryland at College Park, arguably one of the top digital humanities centers in the world. Likewise, the university holds the collections of papers and art by two very prominent electronic literature artists, Deena Larsen[8] and Bill Bly.[9] The exhibition "No Legacy || Literatura electrónica," curated by Alex Saum-Pascual and Élika Ortega and held at the Bancroft Library of UC Berkeley in spring 2016, showcased electronic literature, framing it as computational and "digital technologies in literary production in the networked world and its material connections with 20th-century technologized approaches to literature like futurism, concretism, creationism, stridentism, magical realism, and others" ("Introduction").

While these examples suggest a synergy exists between two complimentary fields of study both birthed in the mid-twentieth century during the rise of digital technologies, our book takes the argument further by demonstrating that electronic literature—namely, experimental computer-writing that possesses "important literary aspects" (Hayles, "E-Lit: What is it?") and is

---

[5]See, for example, Mark Marino's "Review: The Electronic Literature Collection Volume I: A New Media Primer" (2008: 2.1), http://www.digitalhumanities.org/dhq/vol/2/1/000017/000017.html; Scott Rettberg's "Communitizing Electronic Literature" (2009: 3.2), http://www.digitalhumanities.org/dhq/vol/3/2/000046/000046.html; and my essay, "Curating Electronic Literature as Critical and Scholarly Practice" (2014: 8.4), http://www.digitalhumanities.org/dhq/vol/8/4/000194/000194.html.
[6]See Davin Heckman and James O'Sullivan's essay, "Electronic Literature: Contexts and Poetics" (2018), https://dlsanthology.mla.hcommons.org/electronic-literature-contexts-and-poetics/.
[7]See Joseph Tabbi's 2009 award from the Office of Digital Humanities at the National Endowment for the Humanities, a project entitled "Electronic Literature Directory: Collaborative Knowledge Management for the Literary Humanities" (HD-50778-09), https://securegrants.neh.gov/PublicQuery/main.aspx?f=1&gn=HD-50778-09; and Stuart Moulthrop and my grant, "Pathfinders: Documenting the Experience of Early Digital Literature" (HD 51768), https://www.neh.gov/divisions/odh/grant-news/announcing-23-digital-humanities-start-grant-awards-march-2013.
[8]See "The Deena Larsen Collection," https://mith.umd.edu/research/deena-larsen-collection/.
[9]See "The Bill Bly Collection of Electronic Literature," https://mith.umd.edu/research/bill-bly-collection/.

"native to the digital environment" (Rettberg 2019: 5)—is *central* to the humanities, particularly one focusing on questions relating to digital culture and "the symbolic representation of language, the graphical expression of concepts, and questions of style and identity" (Burdick et al 2012.: 12). In fact, we argue that electronic literature is the logical object of study for digital humanities scholars who have, by the second decade of the twenty-first century, cut their teeth on video games, interactive media, mobile technology, and social media networks; are shaped by politics of identity and culture; and able to recognize the value of storytelling and poetics in any medium. As Scott Rettberg reminds us in "Electronic Literature as Digital Humanities," one of the inspirations for this book: "[C]reative production ... is a digital humanities practice: not an application of digital tools to a traditional form of humanities research, but rather experiments in the creation of new forms native to the digital environment" (2015: 127). In sum, electronic literature *is* digital humanities because of our shared philosophy that a computer is not a tool or prosthesis that helps us to accomplish our work; rather, it is the medium in which we work.

This line of reasoning is articulated in the volume's opening section, "Contexts"—that is, Giovanna Di Rosario, Kerri Grimaldi, and Nohelia Meza's chapter "The Origins of Electronic Literature." The authors place electronic literature squarely in the digital humanities, calling it "a new form of literature" that emerged in the 1950s with the introduction of the computer. Other chapters in this first section—Carolyn Guertin's "Cyberfeminist Literary Space: Performing the Electronic Manifesto;" Astrid Ensslin et al.'s "Bodies in E-Lit" and Élika Ortega and Alexandra Saum-Pascual's "*Toys* and *Toons*: From Hispanic Literary Traditions to a Global E-Lit Landscape"—all gesture toward the interest in identity and culture so common in both digital humanities and humanities scholarship, while Davin Heckman's "Community, Institution, Database: Tracing the Development of an International Field through ELO, ELMCIP, and CELL" and Loss Pequeño Glazier's "The E-Poetry Festivals: Celebration, Arts, Imaginations of Community" both speak to computer-based literary activities and events that help to situate electronic literature in practices embraced by the digital humanities.

If there are any doubts as to the deep connection between electronic literature and the digital humanities, they are dispelled with the second section of our book, where chapters focus on those literary forms informed by computational practices. Jim Bizzocchi's and John Barber's chapters remind us that literary experiences are grounded in both visual and aural traditions, opening the way for an understanding of literature in any medium as art. They evoke the views of John Cayley, who in his book *Grammalepsy*, argues for the term "digital literary art" rather than electronic literature, adding the caveat that "[t]here is art, but no one need mention that it is 'digital' because art is simply part of a culture that is also, inevitably, historically digital, and these circumstances have little to tell us concerning the significance or affect of art, as such" (2018: 7). While the technical practices explained in some of the chapters in this section—physical computing in Helen Burgess's "The Voice of the Polyrhetor: Physical Computing and the (e-)Literature of Things," databases as in Theresa Jean and Karen Tannenbaum's "Consuming the Database: The Reading Glove as a Case Study of Combinatorial Narrative," and Twitter bots discussed in Leonardo Flores's "Artistic and Literary Bots"—may be unfamiliar to digital humanities scholars, all position their art practices within literary *forms* recognizable to digital humanities scholars, that is fiction, poetry, and the creative essay.

The third and fourth sections of our volume, "Practices," and "Artist Interventions," respectively, introduce topics common to discussions surrounding literary works in digital humanities, such as archiving, collaboration, publishing, language, pedagogy, and artistic practice, even as these topics point to the need to rethink traditional approaches. My own chapter, "Challenges to Archiving and Documenting Born-Digital Literature: What Scholars, Archivists, and Librarians Need to Know" is born out of experiences I had while conducting research about electronic literature at institutions whose archives are built on print-based practices. Rob Wittig and Mark Marino's "Come Play Netprov! Recipes for an Evolving Practice" presents improvisational performances the authors have produced, reminding as they do that the practice "lies at the intersection of literature, theater and performance, mass media (film and television), games (in particular Alternate Reality Games, ARGs, in which players physically enact roles and compete in real life), avant-garde visual arts (in galleries and museums), and born-digital Internet, personal media and social media practices." Despite antecedents with traditional art forms, the authors ask, "When is a netprov finished?" and "What becomes of the netprov once the initial play period is over?"

Indeed, differences exist between literature and electronic literature, just as they do with the humanities and digital humanities, but at the heart of all of them is the focus on human expression, the human need to tell their stories, to use their gift of language to make sense of the world around them, and to burrow into the depths of understanding to explain what it means to be human, particularly human at a time when digital technologies are proliferating and impacting that world. The literature of the electronic arts does not seek to hide its dependency on computers any more than a traditional novel shuns print. But it calls for and needs digital humanities scholars who are trained in digital practices to study it, just as digital humanities scholars need a literature that reflects the world and daily practices in which they now all operate.

References

Burdick, Anne, Johanna Drucker, Peter Lunenfeld, Todd Pressner, and Jeffrey Schnapp (2012), *Digital_Humanities*, Cambridge, MA: The MIT Press.

Cayley, John (2018), *Grammalepsy: Essays on Digital Language Art*, New York, NY: Bloomsbury.

Hayles, N. Katherine (2007), "Electronic Literature: What is it?" Electronic Literature Organization, January 2, https://eliterature.org/pad/elp.html.

Rettberg, Scott (2015), 127–36, "Electronic Literature as Digital Humanities," in Susan Schreibman, Ray Siemens, and John Unsworth (eds.), *A New Companion to Digital Humanities*, Hoboken, NJ: John Wiley Press.

Rettberg, Scott (2019), *Electronic Literature*, Cambridge: Polity Press.

Saum-Pascual, Alexandra and Élika Ortega (2017), "Introduction," "No Legacy || Literatura Electronica," Bancroft Library, UC Berkeley, February 16–May 5, 2017, https://libraries.cca.edu/exhibitions/no-legacy-literatura-electronica/.

SECTION I

Contexts

1

The Origins of Electronic Literature: An Overview

Giovanna di Rosario, Nohelia Meza, and Kerri Grimaldi

The aim of this chapter is to sketch the origins of electronic literature and to highlight some important moments in order to trace its history. As electronic literature is a "recent" form of literature (Hayles) one could suppose that it is an easy duty to look for its origins. However, due to its ephemeral nature—many of the electronic literature works created in the last century and even in this one are lost or they do not work anymore on modern computers due to the inevitable mutation of technology—which makes the goal more complex than one may think.

Electronic literature is a form of literature that started to appear with the advent of computers and digital technology. It is a digital-oriented literature, but the reader should not confuse it with digitized print literature. Electronic literature is a new object of study that can be approached from diverse disciplines. There are different possible definitions of electronic literature or digital literature.[1] Yet again, without a clear and rigorous definition, electronic literature tends to be an object of study that is difficult to categorize and clearly describe.

For our overview, we have decided to use the Electronic Literature Organization (ELO)'s definition of electronic literature: "works with

---

[1]We will use the terms "electronic literature" (or its abbreviation e-lit) and "digital literature" as synonymous and in the space of this chapter we will not make any reference to the possible different exceptions they may have.

important literary aspects that take advantage of the capabilities and contexts provided by the stand-alone or networked computer" ("What is E-lit?");[2] and Scott Rettberg's recent definition, "electronic literature is most simply described as new forms and genres of writing that explore the specific capabilities of the computer and network – literature that would not be possible without the contemporary digital context"(2019: 2).

In order to trace the origins of electronic literature around the world we need to consider the variety of languages, cultural backgrounds, heritages, and contexts in which digital literature has been created. Digital literature cannot be seen as a whole and it has not been produced at the same time, in the same ways, and for the same purposes, around the world. An aspect we need to emphasize in this reconstruction of the history of electronic literature concerns languages. The variety of languages, while being a value, can also be considered a linguistic challenge. Seeing that electronic literature is a recent form of literature, the interest in translating works is quite recent as well. Although some electronic literary works have been translated and a number of significant researches on electronic literature translation have been made, much work remains to be done[3].

Given the vast scope of this chapter, we have also relied on the competences and knowledge of several colleagues that have helped us to retrace the origins of electronic literature in different countries and continents. So, methodologically, we have conducted our research in part by interviewing different digital literature specialists around the world, and we would like to thank them for their contribution to and involvement in this chapter: Natalia Fedorova (for her contribution to this chapter as far as digital literature in Russia is concerned), Carolina Gainza and Claudia Kozak (for their support with and advice on tracing the history of digital literature in Latin America), Dani Spinosa (for her contribution to this chapter as far as digital literature in Canada is concerned), Eman Younis (for her contribution to this chapter as far as digital literature in the Arab world is concerned); and finally we would also like to thank Philippe Bootz, Serge Bouchardon, John Cayley, Dene Grigar, Michel Hockx, and Keijiro Suga for their suggestions and ideas.

This chapter aims to highlight some important works and moments in the creation of this new form of literature. In doing that, we have tried to combine several aspects, such as its social, political, and aesthetical

---

[2]Electronic Literature Organization, https://eliterature.org.
[3]Cf. J. R. Carpenter (2012), "Translation, Transmutation, Transmediation, and Transmission," in TRANS.MISSION [A.DIALOGUE]; Marecki and Montfort (2017), "Renderings: Translating Literary Works in the Digital Age. Digital Scholarship in the Humanities;" Cayley (2018), "The Translation of Process;" Portela, Pold and Mencía (2018), "Electronic Literature Translation: Translation as Process, Experience and Mediation." *Déprise* (2010) by Serge Bouchardon and Vincent Volckaert has been translated from French into more than nine languages (https://bouchard.pers.utc.fr/deprise/home). J. R. Carpenter's TRANS.MISSION [A.DIALOGUE] (2011) has been translated into French (Ariane Savoie) and Finnish (Anne Karhio).

implications. To do so, we divided this chapter into five sections: a brief history of electronic literature in general (however, we must admit that this section has a very ethnocentric point of view) and then four other sections divided into North American, Latin American, European (Russia included), and Arab electronic literature. Although we are aware of the limits of this division and of the problems it can create (for instance, does it make sense to geographically divide a literature that seems to be a true world literature?), we thought it was the easiest way to shortly map out the origins of electronic literature and its development in different countries and continents. Due to the lack of information, there is no section devoted to electronic literature in Asia, although a few texts will be mentioned.

Brief History

For many years, the electronic literature community has considered "Stochastic Texts" (1959) by Theo Lutz as the first digital literary text.[4] German scholar, philosopher, and poet Max Bense suggested that Lutz use a random generator to accidentally determine texts. Bense looked to establish a scientific and objective branch of aesthetics, by means of applying mathematical and information theoretical premises to the study of aesthetic texts. Lutz made a database of sixteen subjects and sixteen titles from Franz Kafka's novel *The Castle* (1926). Lutz's program randomly generated a sequence of numbers, pulled up each of the subjects/titles, and connected them using logical constants (gender, conjunction, etc.) in order to create syntax. The language of the work contained permutation—the same set of words were used over and over again, each time that the program was running. However, it was not the permutation of Kafka's complete work; it was a fragmented permutation of the words Lutz chose from *The Castle*.

The results of his project were published in 1959 as an essay in *Augenblick* 4 (3–9), a journal of aesthetics edited by Max Bense. The publication in a journal of aesthetics gave credit to consider "Stochastic texts" (*Stochastische Texte*) as the very first piece of electronic literature. After Lutz's work, many other authors have experimented with the possibilities of computers in creating poetry. However, a few years before, in 1952 Christopher S. Strachey created what could be considered the first piece of digital literature.[5] Strachey is rightly viewed as a pioneer of modern computing, but

---

[4]See for instance Chris Funkhouser, *Prehistoric Digital Poetry: An Archaeology of Forms 1959–1995*, Alabama University Press, 2007.

[5]Christopher S. Strachey (1916–75) was a British computer scientist. He was a pioneer in programming language design. He was a colleague of the famous Alan Turing and in 1952 Strachey was a programmer of the world's first commercially available general-purpose electronic computer, the Ferranti Mark 1.

he is not usually viewed as the creator of the first work of digital literature. Strachey developed—using Turing's random number generator—a Mark I program that created combinatory "Love Letters" (1952). This was the first piece of digital literature and of digital art, predating the earliest examples of digital computer art by almost a decade.

Lutz was just the first of a group of scholars to view mathematics, science, and creativity as cooperative disciplines. Many other experiments in computer-randomized poetry have been conducted since 1960, primarily in Europe, the United Kingdom, and the United States. Thanks to the evolution in technology, other electronic/digital poetry experiments began in the following years. Some examples are Brion Gysin's permutation "I am that I am" (1960) programmed by Ian Somerville, "Tape Mark made" in 1961 with an IBM calculating machine by Nanni Balestrini, and "*La machine à écrire*" published in 1964 by Jean Baudot. As hardware and graphical programs were developed in the 1960s, a few poets started to use digital tools to create visual poems. In the late 1960s concrete and visual poets began to focus on using computers to make graphical representations of and with language. When the technology became available, artists started to create digitally static and animated works and to manipulate language to increase visual properties.

By the 1980s, poets increasingly presented moving language on screen as a result of the development of computers. These experiments prefigure many later works in poetry that proliferated in animated, hypermedia digital formats. The 1980s are an important moment for the history of electronic literature since, in 1985, an international exhibit held in Paris at the Centre Georges Pompidou, titled "*Les Immatériaux*," organized by Jean-François Le Lionnais, the ALAMO[6] group introduced its first poems "generated" by a computer, which somehow sanctioned the birth of a new form of visual poetry "animated" by this new medium. The ALAMO group wanted to develop tools and computational methods of use to writers. They have focused on the potentiality of writing "assisted" by the machine, by the computer.

---

[6] *Atelier de Littérature Assistée par la Mathématique et les Ordinateurs* (Workshop of Literature Assisted by Mathematics and the Computers). Paul Braffort and Jacques Roubaud, both members of OULIPO, created the ALAMO in 1981. In 2008 there were seventeen members in the ALAMO group. Cf. Philippe Bootz, *From OULIPO to Transitoire Observable: Evolution of the French Digital Poetry*, accessible online at: http://www.dichtung-digital.org/2012/41/bootz.htm (accessed January 29, 2016).

By the mid-1980s, moreover, the influence of post-structural critical theories (such as deconstruction) spurred writers and poets to make up new appearances for literature in general (let's think of fictional hypertexts) and for poetry in particular. As for poetry, all the elements promoted by concretists—the visual presentation of texts, graphical effects, a new typography, coloration, repetition—can be easily found in many electronic texts. Computers clearly enable and extend ideas looked for by the concretist aesthetic. Examples of graphical poems made thanks to computer technology had already begun to emerge in the late 1960s. Marc Adrian's *Computer Texts*[7] were featured in the Cybernetic Serendipity exhibition in 1974. The options derived by the possibility to animate the language were also particularly investigated; in fact, animated poems long pre-dated a style of electronic poetic practice that erupted with the advent of the World Wide Web, typified by works such as Brian Kim Stefans' *The Dreamlife of Letters* (2000).

To this technological revolution we should add another important step, which took place in the 1990s, when CERN's researchers in Geneva (led by Tim Berners-Lee) developed the technology that has made the net popular. It was from that date that a proliferation of websites of "cyber-poetry/cyber-literature" began and, consequently, a new generation of digital authors was born. Since then we have witnessed the continuous increase of poetic creations published on the web, so that in 1999 the magazine Doc(k)s[8] felt the need to catalog what had already been produced so far.[9] Viewers confronting a program in an installation setting like text-generated poems automatically spawned the initial works. With the development of graphics software, successive works embodied visual methods that approximated concrete and visual poems rendered and fixed on the page.[10]

---

[7] In this work, the computer randomly assembles poems by using a database of 1,100 alphabetic symbols to place twenty words at time on the screen. Adrian organized the interface using a grid of system. The symbols retrieved from the database (letters or groups of words) appeared in rows and columns on the screen. Adrian in part disguised the grid element by variegating the size of the font and not using every line or block.

[8] Doc(k)s is a review of contemporary poetry, which explores the audiovisual experiments in poetry that have marked the twentieth century. The review has a website at: http://www.sitec. fr/users/akenatondocks/(accessed March 31, 2016).

[9] It emerged immediately that the new generation of "digital" poets knew the computer culture very well, they came from different fields, visual and/or plastic arts, communication, design, or simply from the web but they did not have any specific aesthetic or literary knowledge.

[10] In contrast to the production of the earliest visual poets, these works are not interactive.

Europe (Russia Included)[11]

Europe has been very prolific in the creation and development of electronic literature, although some countries more than others. In the space of this chapter it will be impossible to retrace the origins of digital literature through the whole of Europe; however, we try to give an overview on as many countries as possible focusing a bit more on the ones that seem to have a stronger tradition. Germany and the UK have undoubtedly played a central role in the origins and development of digital literature. As we have seen, the text that has been considered for years (and the discussion is still open) the first piece of digital literature was made in Germany in 1959, and the "Love Letters" generator was invented by the British Christopher Strachey in 1952.

As seen, in Britain, "I Am That I Am" (1959–60) by Brion Gysin programmed by Ian Sommerville, was one of the permutation poems included in a series of sound poetry recordings. Gysin was invited to perform these for the BBC radio in 1960. "'I Am That I Am' is a cyclical, randomized representation of the three words contained in that phrase" (Funkhouser 2007: 39). In that period of experimentation with poetry, Italian artist Gianni Toti even coined the term, "poetronica," in order to highlight both components of a new fusion of the arts: the poetic element and the electronic mode, although Toti has never been seriously involved with digital literary or poetic creations.

France, as we have seen, also has a very long and strong tradition of experimenting with literature. Already in 1964, Jean Baudot published "*La machine à écrire*" ("The typewriter")—an example of "computer-assisted literature" ("littérature assistée par ordinateur"). Jean Baudot created a combinatorial program, and then gathered the generated texts into the book published by Les Editions du Jour. In the "Brief History" section we have already mentioned that 1985 somehow sanctioned the birth of a new form of visual poetry "animated" by the new medium, and still in 1985 the first art review on Minitel[12] was published in France. According to Serge

---

[11]A complete overview on electronic literature in Europe could not be included due to limited space. Several countries such as Spain and Poland have not been mentioned although quite active in the digital literature panorama. For more information about Europe see Markku Eskelinen and Giovanna Di Rosario, "Electronic Literature Publishing and Distribution in Europe," University of Jyväskylä Press, 2012; specifically about Poland see Piotr Marecki, "The Formation of the Field of Electronic Literature in Poland," https://elmcip.net/critical-writing/formation-field-electronic-literature-poland (accessed March 31, 2016); about Spain cf. Maya Zalbidea Paniagua (ed.) "Spanish Language Electronic Literature," https://elmcip.net/research-collection/spanish-language-electronic-literature (accessed March 31, 2016).

[12]The Minitel was a videotext online service accessible through telephone lines and is considered one of the world's most successful pre-World Wide Web online services.

Bouchardon, around eighty artists participated in this issue, spanning 1,500 Minitel pages. Text animation was already very present thanks to authors like Philippe Bootz, Frédéric Develay, Claude Faure, Guillaume Loizillon, and Tibor Papp. At the time, "all of them were in the sphere of visual and sound poetry and were to play a key role in the evolution of French digital poetry" (2011: 105).

Portugal also has an interesting tradition of experimenting with literature, especially as far as automatic, generative, combinatory texts are concerned. The Portuguese writer and poet Ernesto Manuel de Melo e Castro is considered the father of the so-called "videopoetry" in which animation and temporality are brought to poetry and that then largely influence digital poetry. According to him "videopoetry" was "inevitable as a concept" answering the challenge of the new technological means for producing texts and images (de Melo e Castro 2007). He also underlined that reading a "videopoem" would be a complex experience since different temporal modalities of perception would coincide with the moving and changing images and texts. He signaled the arrival of a new poetics of reading.

The Portuguese Pedro Barbosa is considered the father of generative texts in Portugal and a pioneer in Europe. His well-known "Sintext" (automatic generator created in collaboration with Abílio Cavalheiro) and "Oficio sentimental" (textual generator) were published in A.L.I.R.E. in 1994 (*Édition Mots-Voir*). Barbosa published a new version of "Sintext" in 1997 in A.L.I.R.E/DOC(K)S n.10 (CD-ROM): "*Sintext: neuf textes automatiques générés par ordinateur.*"

Several pioneering works of digital literature in Europe were hypertexts (although this tradition derives from US examples and texts). For instance, Lorenzo Miglioli wrote the first Italian hypertext in 1993. "Ra-Dio" was presented at a conference in Reggio Emilia organized by Gruppo 63 (an Italian avant-garde movement that had several famous authors as members such as Nanni Balestrini, Edoardo Sanguinetti, and Umberto Eco). "Ra-Dio" was published along with the translation into Italian of Michael Joyce's "Afternoon. A Story." Karl-Erik Tallmo published Sweden's first hypertext fiction "Iaktagarens' förmåga att inngripa" ("Participant's capability to interfere") in 1992. During the 1990s, European authors also experimented with this new form of writing. As stated by Markku Eskelinen "[g]enerally speaking, it is typical of the Nordic scene that many if not most authors of the most prominent works of electronic literature are also (locally) well-known authors of print literature" (2011: 8).[13]

---

[13]This means first of all that their works of electronic literature are situated within an oeuvre that is already recognized and positively evaluated as literature, which is not always the case in the rest of Europe.

Natalia Fedorova notes that the origins of Russian electronic literature are untold stories of the experimentations of mathematicians in their labs that are hardly published as they were seen to be mere jokes. One of these experiments taken seriously is a program for composing verse that was described by Boris Katz in his article from 1978 in the journal of USSR Academy of Sciences. The aim of the program is to find minimal means to produce verse. The thesaurus consists of words (nouns, adjectives, pronouns, conjunctions, and verbs) from Osip Mandelstam's *Kamen* (*Stone*)[14] with marked number, gender, and tense. Each word is accompanied with the information about metrics, rhyme, and grammar. Its function in the sentence is also marked either as subject, predicate, or adverbial modifier. Adding the information about the stress forms the rhyme: ultimate—for masculine rhyme, or penultimate— for feminine. A machine composes every line from right to left: first, it writes the last two words in each line, then it adds all the rest according to grammatical and syntactical functions, disregarding semantics. The program was reimplemented and presented by electronic poet and media artist Anna Tolkacheva at the Taburetka Poetry Festival in Monchegorsk, Murmansk region on August 28, 2016.

The official birth of Russian electronic literature—visible, but not accepted, by rather traditional literary circles—can be dated back to the Teneta (1994) literary contest. Apart from poetry, prose, and translation, it included nominations in "Hyperliterature," "the Creative Arts," and "Games." Teneta positioned itself as a "pure Internet contest." The best texts, originally published on the internet, were to be nominated. It is important to note that the internet culture itself started publicly as a literary phenomenon in the early 1990s (Gorny 2000) with Dmitry Manin's Bout Rimes (Буриме) (1995) and Roman Leibov's ROMAN (1995). The 1990s, as a nostalgic epoch of freedom for Russian millennials, can be seen as formative years of Russian e-lit. Net.art legacy established by the Da-Da-Net Festival (1993–9), as well as the influence of Alexander Shulgin's lectures at Pro Arte Media Art Program (2000–2001), can be traced in Ivan Khimin's asciiticist (ASCII+asceticism) installation and postdigital painting Strokes and Incisions (Черты и резы) (2012). Michail Kurtov creates a Twine-based IF Kourekhin: Second Life loosely based on the biography of legendary artist and performer of the 1980–90s, Sergey Kourekhi.

---

[14]Cf. *Stone*, Petrograd: Hyperborey, 1916, http://digitalmandelstam.ru/engl2.html.

North America

In North America, as in Europe, there has been a prolific history of electronic literature. John Cayley has suggested that digital literature is a "mode of practice," as all print writers today use digital affordances in their writing as it is (Szilak 2015). To look at electronic literature as a "mode of practice," each work of e-lit would then be differentiated from another based upon which tools are utilized in making it. In this way, we believe it is best to begin this account of the origins of North American electronic literature with the first programs and platforms that were utilized to create e-lit and the genres that they inspired. We will also briefly consider the organizations that saw the importance of these works and methods, and made it possible to preserve, archive, criticize, and promote the ever-growing history of electronic literature.

Interactive fiction emerged in the 1970s and describes works of e-lit that blur the lines between games and literature. These works are more interactive than other forms of e-lit in that most works give more control over the story to the reader/user. These are often called, "text adventures," where games are played with text-based input and output (Hayles et al. 2006). The first work of interactive fiction was *Colossal Cave Adventure* created by Will Crowther with the help of Don Woods in 1976. In this text-based game, the player uses text commands to move the character through a cave searching for wealth, with the goal of making it out of the cave alive and finding the most treasure. It was built for the PDP-10 platform (Adams). The game could be accessed from ARPAnet, the precursor to the internet.

Dave Lebling and Mark Blank were so enamored with *Colossal Cave Adventure* that they created their own game with the help of Tim Anderson and Bruce Daniels, *Zork* (1977–9), which became known as the most influential work of interactive fiction. *Zork* was more complex than its muse, allowing for longer text commands and providing multiple levels to the game that the player could master. Joel Berez and Mark Blank wanted to find a way to take *Zork* and make it accessible to home computers, so they designed a program language that could run on any computer through an emulator. With this, they began a company called Infocom, selling commercial interactive fiction for home computers (Thorek 2016).

As mentioned in the previous section of this chapter, the tradition of hypertext fiction was born in the United States. Doug Englebart created the first hypertext system called "Augment" in 1968. It was fully realized into a system called Xanadua, which was eventually adopted by the software company, Autodesk, in 1988 (Funkhouser 2007: 152). According to Thomas Swiss, Eastgate Systems, Inc., a publishing company, "managed to create a kind of 'local' scene for hypertext writers" (qtd. in Funkhouser: 153). Eastgate published many works of hypertext in the 1990s, and notably developed the most popular software used to create hypertext fictions, Storyspace.

While hypertext fiction was largely popular in the 1980s and 1990s, its popularity is now receding for its limitations with graphics and sound files. N. Katherine Hayles refers to the hypertext, link-led style of digital literature as "First Generation" with the year 1995 introducing a "Second Generation" which de-emphasized the link-led nature (Hayles 2007). Though hypertext fiction generated on proprietary software has receded for its limitations with graphics and sound files, the "First Generation" style of hypertext fiction lives on today through use of an open-source tool called Twine created in 2009.

Stemming off of hypertext fiction, electronic literature expanded to include more graphics, sound files, and structures that departed from the block text tradition thanks to the introduction of browser access to the web beginning in 1995. Network fiction employed these features by mimicking network forms like the Frequently Asked Questions list, blogs, news feeds, and email. David Ciccoricco created the term, "network fiction," defining the new wave of e-lit as digital fiction that "makes use of hypertext technology in order to create emergent and recombinatory narratives" (2007: 4). One such example is Talan Memmott's *Lexia to Perplexia*,[15] which utilizes DHTML and Java much like a computer network.

On the extreme end of immersive e-lit environments is the invention of CAVE, or Cave Automatic Virtual Environment, created at the University of Illinois. CAVE is a shared reality virtual environment generated through goggles and several projectors pointing to walls of a small room (Hayles et al.). The experience of CAVE is at once game-like, digital art, and fiction. *Screen* (2003), created by Noah Wardrip-Fruin, Josh Carroll, Robert Coover, Shawn Greenlee, Andrew McClain, and Benjamin "Sascha" Shine, projects block text onto the walls, but words peel off and the user can interact with them and even hit them like a tennis ball back at the wall. Works created in CAVE immerse the user in a full-body experience, though one that is not as accessible as those works that can be hosted on the web.

Despite our limited space in this chapter to address each organization and institution that has been created to support, promote, and preserve electronic literature, we would be remiss not to mention the strides that the ELO has accomplished. The ELO began in 1999 and has since inspired several other organizations. Author Scott Rettberg, novelist Robert Coover, and internet business leader Jeff Ballowe initiated the organization. The ELO pledged to "foster and promote the reading, writing, teaching, and understanding of literature as it develops and persists in a changing digital environment"

---

[15]*Lexia to Perplexia* was first published on *The Iowa Review Web* in September 2000.

(Hayles et al. 2006). In 2001, the ELO held the first Electronic Literature Awards program—the first and only of its kind—to recognize exemplary poetry and fiction (John Cayley among them). The ELO was also responsible for the creation of the Preservation, Archiving, and Dissemination (PAD) project,[16] the Electronic Literature Directory,[17] and three volumes of the Electronic Literature Collection.[18]

In Canada, not only are works usually mixed media, but they also tend to be collaborative, sometimes including authors of multiple countries. In "Toward a Theory of Canadian Digital Poetics," Dani Spinosa argues "Canadian digital poetics has tended toward post-structural skepticism of authorship by producing electronic literature that is generally concerned with generative work, source or seed texts, remixes, cut-ups, or plagiaristic borrowings" (2017: 239). Spinosa goes on to point out that, while many Canadian authors take pride in the collaborative nature of digital poetics, sometimes authors or source texts are not adequately credited for their contributions. Because of the prevalence of transnational authorship and a lack of accurate credit, identifying national qualities and trends in Canadian works is complicated. However, there are still some trends that can be traced.

Throughout Canadian electronic literature the visual concerns of transmedial and born-digital projects are heavily indebted to concrete poetry, as interpreted by earlier Canadian practitioners like bpNichol and Steve McCaffery. Nichol already saw the poetic potentials of the digital in his own work, extending the formal and visual concerns of his typewriter-based concrete poetry into digital technologies in 1982 with the production of *First Screening: Computer Poems*, a collection of kinetic poetry produced on an Apple IIe using Apple BASIC programming language. This work is widely considered to be some of the earliest programmed kinetic poetry and some of

---

[16]The Preservation, Archiving, and Dissemination (PAD) project from 2002 to 2005. This conference resulted in the publication of *Acid-Free Bits* by Nick Montfort and Noah Wardrip-Fruin and *Born-Again Bits* by Alan Liu, David Durand, Nick Montfort, Merrilee Proffitt, Liam R. E. Quin, Jean-Hugues Réty, and Noah Wardrip-Fruin. These publications provide advice to artists on how to best preserve their works and which software to use for their works. The conference turned into an initiative that offers works through a Creative Commons license on the Electronic Literature Collection, Vol. 1 (Hayles et al. 2006).

[17]The Electronic Literature Directory, maintained by scholars and readers, houses information about readings, events, and critical works, https://directory.eliterature.org.

[18]The Electronic Literature Collection has produced three volumes in 2006, 2010, and 2016 respectively (ELC1, ELC2, ELC3). The first two collections have about 60 works and the third collection increased to 114, all of which are edited by editorial collectives ("The Electronic Literature Collection: Volume Three"). Though the ELC is based in the United States, it includes works from many other countries as well.

the first evidence of codework.[19] Nichol's influence can be seen most clearly in the digital component of Darren Wershler's *NICHOLODEONLINE* (1998), but is also evident in Andrews' own work, like "Seattle Drift" (1997) or "Enigma n" (1998) or in the digital component of Damian Lopes' *Sensory Deprivation/Dream Poetics* (1998, 2000).

Latin America[20]

In "Latin American Electronic Literature: When, Where and Why," Claudia Kozak notes that the very first Latin American electronic literature works are most probably "IBM" (1966) by Argentinean Omar Gancedo, which means a bit more than a decade after "Love Letters" (1952) but exactly when other experimentations were taking place in North America and Europe. Omar Gancedo's "IBM" consists of a series of three short poems codified in IBM cards, which, processed by a Card Interpreter, produced the printing of the de-codified texts on the horizontal middle line of each card.

"Le tombeau de Mallarme" (1972) by Brazilian Erthos Albino de Souza[21] —published six years later after "IBM" (1966) is a graphic poem consisting of a series of ten visual poems printed by a computer after the manipulation of software prepared for temperature measurement.

Ana María Uribe (Argentina) created non-digital "tipoemas" in the 1960s and digitally animated poems "anipoemas" from the 1990s onwards. Mariela Yeregui (Argentina) experimented with intertextuality and intermediality in her interactive piece *Ephitelia* (1999). Net.art artist Gabriela Golder's (Argentina) created *Postales* [Postcards] (1999–2000), a hypermedial and interactive narrative in Spanish and French. *The Book After the Book* by Brazilian Giselle Beiguelman appeared in 1999. In that same year, Regina Célia Pinto (Brazil) created *O Branco e o Negro, Reflexões sobre a Neblina* (1999), a CD-rom inspired by *The Rouge et Le Noir* (1830), Stendhal's

---

[19]The historical importance of *First Screening* combined with the high literary quality of Nichol's writing inspired the efforts of students and staff at the University of Calgary and the University of Victoria to preserve this work long after the Apple IIe was obsolete. The process continued into the early 2000s, and resulted in four different versions of *First Screening* (bpNichol's "First Screening") hosted on Jim Andrews's *Vispo* site, including an emulator.

[20]A complete overview about electronic literature in Latin America and the Caribbean could not be included due to limited space. However, significant research on the origins of electronic literature is beginning to take place in different countries such as Argentina, Brazil, Chile, Colombia, Mexico, Puerto Rico, etc. In 2018, the Latin American Electronic Literature Network (litElat) opened its call for submissions for works of electronic literature from Latin America and the Caribbean to create the first Anthology of Latin American Electronic Literature (2020) (http://litelat.net).

[21]See ELMCIP, "The Knowledge Database," https://elmcip.net/creative-work/le-tombeau-de-mallarme (accessed April 30, 2016).

novel, and the Brazilian poem "Fog" (1996) by Nelson Ascher. Not to mention that Brazilian Lenora de Barros developed one of her first visual poems, "Poema" in 1979, and in 1983 she presented other visual poems at the 17th Biennial de São Paulo in a section titled "Arte em videotexto" (1983) [Art and videotext].[22]

During the 1980s, in Latin America there were also diverse experiences linking literature and computers. Some examples are "Soneto só prá vê" (1982) by Brazilian Daniel Santiago with programming by Luciano Moreira, in TAL/II language, and "Universo" (1985) by João Coelho programmed in Advanced BASIC language. In 1986, in Argentina, Ladislao Pablo Györi shaped one of Grete Stern's photomontages using 3D graphics software and combining it with a poem written by artist Gyula Kosice recorded with the aid of a synthesizer (Kozak 2017).

As far as prose is concerned, Juan B. Gutiérrez, a Colombian writer and expert in mathematical modeling systems, created the first electronic hypertext produced by a Latin American: "El primer vuelo de los hermanos Wright." This production has two stages. The first began in 1995, with a hypertextual novel composed of blocks of texts communicated to each other by links (first version 1996–8). The structure imitates a book where the links serve to move from one block of text to another. In this case, a still linear sequence of events was created without utilizing the potentiality of the digital writing. Then in 2006, the "hypernovela" was rewritten to be included in the author's project called Literatrónica (Gainza 2013).

According to Thea Pitman, some of the very earliest works, though notably quite technically advanced for their time of creation, are the autobiographical hypermedia projects Sangre Boliviana [Bolivian Blood] (1994) by Latina artist Lucia Grossberger Morales (Bolivia), and Glasshouses: A Tour of American Assimilation from a Mexican-American Perspective (1997) by Chicana artist Jacalyn López García (Mexico-USA). Both works explore the feeling of belonging to two or more cultures and expressing it with the digital tools of the time. Furthermore, Mexican/American author Blas Valdez was the first author of hypertext fiction. He utilized the hypertext style in his print works before producing them on the internet, as seen in Restos de corazón (Remains of the Heart) (1998).

Arab Digital Literature

Eman Younis notes that the interest in digital literature in the Arab culture started at the beginning of the third millennium, and specifically after the

---

[22]For a complete overview on electronic literature in Brazil see ELMCIP's Brazilian Electronic Literature Collection curated by Luciana Gattass.

Jordanian writer Muhammad Sanajilah published the first Arabic novel in the genre of Interactive Fiction titled *Zilal al-Wahed* (*One's Own Shadows*) in 2001. After that, he published his second novel *Dardasha / Chat* (2005), followed by his third work—a short interactive story, called *Saqi' / Frost* in 2006. Sanajilah's works received a lot of interest by critics, and consequently, a large wave of studies and books appeared in the field of Arabic digital criticism. Among the pioneers of this critical movement were the following critics: Sa'id Yaqtin and Muhammad Aslim from Morocco, Eman Younis from Palestine, Fatima al-Breki from the Emirates, Ibrahim Milhem from Jordan, and Sai'd al-Wakil and Sayyid Najim from Egypt.

The emergence of the pioneering works in the field of visual digital poetry is attributed to the Moroccan poet Mun'im al-Azraq, who published a large number of digital poems on *al-Mirsa'* website, in which he combined media with colors, pictures, photos, paintings, and music. The first interactive poem was written by the Iraqi poet Mushtaq Abbas Ma'in with the title, *Tabarih Raqmiyya li Siratin Ba'dhuha Azraq / Digital Agonies of a Blue Biography*, in which the poet relies on the technique of hypertext. For a broader panorama on the history of electronic literature in the Arab world refer to Reham Hosny's study, "Mapping Electronic Literature in the Arabic Context" (2018).[23]

It is worth mentioning here that Arabic criticism maintains that visual digital poetry is an extension of visual poetry that was known to the Arabs during the Mameluke and Ottoman eras, during which the poets tried to introduce their poems as artistic paintings, and gave different names to each type of poetry such as *the painting poem, the concrete poem, the plastic poem, and the calligraphic poem*. This can probably explain the speed of the development of the digital Arabic poem and how it reached the level of the Western poem from the technical point of view (Younis and Nasrallah).

Despite the enthusiasm of many critics in the Arab world toward this new literary experience, the digital literary movement is still progressing very slowly in comparison with the Western world concerning the number of digital texts and academic studies, and the number of websites that are interested in introducing this kind of literature. In fact, we can hardly find more than one website that is interested in consistently introducing digital works and critical studies in this field. This website is called: the Arab Union for the Internet Writers. Recently, the first Arabic Electronic Literature Conference, "New Horizons and Global Perspectives," was held in Dubai, UAE, in February 2018.[24]

---

[23]https://electronicbookreview.com/essay/mapping-electronic-literature-in-the-arabic-context/.
[24]See Arabicelit website: https://arabicelit.wordpress.com/conference/.

Some Conclusions

Before concluding, we would like to end this overview drafting a bit of digital literature in Asia and inviting scholars to deeply and scientifically investigate the origins and the development of digital literature in Asia since, to our knowledge, there is no complete and systematic study on the subject while it seems to us to be a very urgent topic to be studied.

Young-Hae Chang Heavy Industries is notably the most well-known art-duo creating digital art based in Asia, and more specifically in Seoul, South Korea. The members are Young-Hae Chang, a Korean artist, and translator, Marc Voge, an American poet who currently lives in Seoul. The group was formed in 1999 and since then has been creating works presented in twenty languages, characterized by text-based animations composed in Adobe Flash that are highly synchronized to a musical score that is often original and usually jazz.[25]

In the Electronic Literature Collection Volume Three (2016), Japanese and Chinese texts of digital literature were published into an anthology (collection) for the first time. "The First Intimate Touch" 第一次 的亲密接触 by the Taiwanese author Pizi Cai (pen name of Cai Zhiheng (蔡 智), is considered as the foundational work of Chinese online literature. As stated by Lena Henningsen

> [as] one of the first Chinese language online novels it had a tremendous impact on the field and prompted a first wave of online fiction in the People's Republic of China (PRC). It was published from March to May 1998 in small sections on a Taiwanese Bulletin Board System (BBS) and turned into a bestseller on the Chinese mainland after its publication as a book in 1999.
>
> (Henningsen 2011)

"The First Intimate Touch" is one of the first works in the area to be written in a new literary form, the online form of a novel. As said, the text was published as a book in 1999; however, even the printed version of the novel preserves certain characteristics of the online text, such as the use of icons and the English language note "to be continued" that divides the chapters and subchapters.[26] Henningsen also notices that "The First Intimate

---

[25]See Young-Hae Chang Heavy Industries: http://www.yhchang.com.
[26]The novel was then turned into a TV series (produced by the mainland Chinese Shanghai Film Studio 上海电影制片厂), and its transformation into a stage play in Beijing in 2011 attests to its continuing popularity in the PRC. It has also generated a number of trends; it seems to us that it can be considered as an example of transmedia text in its whole.

Touch" can be seen as a precursor of Chinese internet literature (wangluo wenxue 网络文学) (Henningsen 2011).

Keijiro Suga (a renowned Japanese poet, writer, and translator) remarks that much of the so-called digital literature in Japan (notably Twitter-based) does not seem to be very interesting. According to him, writers are still all very analog-minded, although they do occasionally publish digitally. But this does not mean that important experimental texts have not been produced. This just means that no rigorous studies on digital literature (its history and development) have yet been undertaken—or if they have, the Japanese literary community does not know them—and that digital literature is a niche literature. However, we do not have to forget as well that, although growing fast, electronic literature remains a niche literature also in those countries that have a long tradition in experimenting with literature and poetry.

As we have seen, some countries have developed their interest in and creation of electronic literature almost simultaneously, while others, just because of their own cultural background and/or contexts (also political and economic contexts and backgrounds), have only recently discovered electronic literature, or accepted it as a new form of the literary genre. Not only does electronic literature require literary competences but also IT skills. Authors of electronic literature often work together with graphic designers and, especially, programmers, to unite their competences. This collaboration, however, implies a different relation with the role of the author. With electronic literature, we see in some countries that this sharing of authorship may become problematic, particularly when collaborators are not accurately credited for their contributions. Finally, some developing countries have focused their interests and priorities on other aspects of their cultural and economical life. However, it seems to us that digital literature is globally and constantly growing, and it has been transforming itself thanks to or because of the advent of other new interfaces, supports, and media.

References

Adams, Rick (n.d.), "A History of Adventure," *Colossal Cave Adventure Page*, http://rickadams.org/adventure/a_history.html (accessed February 7, 2016).

Andrews, Jim et al. (2007), "BpNichol's 'First Screening,'" *First Screening: Computer Poems BpNichol 1984*, March 2007, http://vispo.com/bp/introduction.htm (accessed February 7, 2016).

Boluk, Stephanie et al. (eds.) (2016), *The Electronic Literature Collection, Volume Three*, Electronic Literature Organization, http://collection.eliterature.org/3.

Bootz, Philippe (2010), *From ALAMO to Transitoire Observable*, ELMCIP Electronic Literature Communities Seminar, Bergen, Norway.

Borràs, Laura et al. (eds.) (2011), *The Electronic Literature Collection*, Vol. 2, Electronic Literature Organization, http://collection.eliterature.org/2.

Bouchardon, Serge (2011), "Filiations and History of Digital Literature in France," in Giovanna Di Rosario and Lello Masucci (eds.), *OLE Officina Di Letteratura Elettronica*, 98–111, www.elettroletteratura.org/.
Ciccoricco, David, (2007), "Excerpts from the Introduction: On the History of Network Fiction," *Reading Network Fiction*, https://readingnetworkfiction. wordpress.com/.
de Melo e Castro, Ernesto Manuel (2007), "Videopoetry," in Eduardo Kac (ed.), *Media Poetry: An International Anthology*, 175–84, Bristol: Intellect Books.
Eskelinen, Markku and Giovanna Di Rosario (2011), *Electronic Literature Publishing and Distribution in Europe*, Jyväskylä: University of Jyväskylä Press, https://jyx.jyu.fi/bitstream/handle/123456789/40316/978-951-39-4945-7.pdf.
Fedorova, Natalia (2012), "Where Is E-Lit in Rulinet? Rulinet, Russian Literary Internet," *ELMCIP*, https://elmcip.net/critical-writing/where-e-lit-rulinet.
Funkhouser, Chris (2007), *Prehistoric Digital Poetry: An Archaeology of Forms 1959–1995*, Tuscaloosa, AL: Alabama University Press.
Gainza, Carolina (2013), *Escrituras Electrónicas en América Latina. Producción Literaria en El Capitalismo Informacional*, University of Pittsburgh, http://d-scholarship.pitt.edu/16833/.
Gorny, Evgueni (2000), "Chronicle of Russian Internet: 1990–1999," *Net Literature*, March 2000, https://www.netslova.ru/gorny/rulet/.
Hayles, N. Katherine (2003), "Translating Media: Why We Should Rethink Textuality," *The Yale Journal of Criticism* 16 (2): 263–90.
Hayles, N. Katherine (2005), *My Mother Was a Computer: Digital Subjects and Literary Texts*, Chicago, IL: University of Chicago Press.
Hayles, N. Katherine et al. (eds.) (2006), *The Electronic Literature Collection*, Vol. 1, Electronic Literature Organization, http://collection.eliterature.org/1.
Hayles, N. Katherine (2007), "Electronic Literature: What Is It?" Electronic Literature Organization, 2007. https://eliterature.org/pad/elp.html.
Henningsen, Lena (2011), "'Coffee, Fast Food, and the Desire for Romantic Love in Contemporary China: Branding and Marketing Trends in Popular Chinese-Language Literature'," *Transcultural Studies* 2, http://heiup.uni-heidelberg.de/journals/index.php/transcultural/article/view/9089/3107#_edn3.
Hockx, Michel (2015), *Internet Literature in China*, New York, NY: Columbia University Press.
Kozak, Claudia (2015), "Mallarmé e IBM. Los Inicios de La Poesía Digital En Brasil y Argentina," *Ipotesi. Revista de Estudos Literários* 19 (1), http://www.ufjf.br/revistaipotesi/ (accessed April 30, 2016).
Kozak, Claudia (2017), "Latin American Electronic Literature: When, Where and Why," in María Mencía (ed.), *#WomenTechLit*, 55–72, Morgantown, WV: West Virginia University Press.
Lutz, Théo (1959), "Stochastische Texte," *Augenblick*, http://www.reinhard-doehl.de/poetscorner/lutz1.htm.
Memmott, Talan (2006), *Lexia to Perplexia*, Electronic Literature Collection, Vol. 1, http://collection.eliterature.org/1/works/memmott_lexia_to_perplexia.html.
Pitman, Thea (2007), "Hypertext in Context: Space and Time in Latin American Hypertext and Hypermedia Fictions," *Dichtung Digital* 37, http://www.dichtung-digital.org/2007/Pitman/pitman.htm.
Rettberg, Scott (2019), *Electronic Literature*, Cambridge: Polity Press.

Spinosa, Dani (2017), "Toward a Theory of Canadian Digital Poetics," *Studies in Canadian Literature/Études En Littérature Canadienne* 42 (2): 237–55.

Strachey, Christopher (1952), *M.U.C. Love Letter Generator*, https://elmcip.net/creative-work/muc-love-letter-generator/.

Szilak, Illya (2015), "Towards Minor Literary Forms: Digital Literature and the Art of Failure," *Electronic Book Review*, http://www.electronicbookreview.com/thread/electropoetics/failure.

Thorek, Marko (2016), "History of Infocom," *Infocom*, http://www.infocom-if.org/company/company.html.

Wardrip-Fruin, Noah et al. (2011), *Screen*, http://collection.eliterature.org/2/works/wardrip-fruin_screen.html.

Younis, Eman and Aida Nasrallah (2015), *The Artistic Literary Interaction in the Digital Poetry, "Trees of Boughaz" as a Model*, Beit Berl: Berl College.

2

Third-Generation Electronic Literature

Leonardo Flores

The history of electronic literature is inextricably tied to the history of computing, networking, and their social adoption. As computers become increasingly powerful, miniaturized, versatile, user friendly, affordable, and ubiquitous, so does their user base. As digital networks have grown from local networks to private dial-up networks to the open World Wide Web to corporate social media networks, the scale of digital communication and audiences has grown exponentially. Assuming that a fixed percentage of computer and network users will seek to creatively explore the possibilities for writing offered by computers, one would expect to see an level of growth in the production and publication of electronic literature.

But is electronic literature keeping up with this explosive growth of digital media users? What if it is, but in a way that is unrecognizable by the field as currently defined? What is electronic literature's place in a world of ubiquitous computing, massive user bases, and even larger audiences? What is electronic literature's cultural reach? How might it achieve mainstream recognition? To begin to answer these questions, this chapter describes a paradigm shift that opens the door to a third generation of electronic literature.

I define electronic literature as a writing-centered art that engages the expressive potential of electronic and digital media. Even though it has origins in oral culture, particularly poetry, literature as an artistic tradition and field of study has been shaped for centuries by writing and print technologies.

Therefore, broader forms of communication, such as narrative, spoken and sign language, audio and video recordings of performances, purely visual comics, and video games are of less interest from an e-literary perspective because they are not using alphabetic or even asemic writing. An essential component needs to be the artistic engagement of written language in digital media. So even though the programming code that powers a digital work or video game is written, and is of interest, unless it is performing a kind of code poetry its use of language is functional and not artistic. Electronic literature, therefore, explores writing in electronic and digital media, which integrate computation, multimedia integration, interactivity through a variety of input devices, networked data, and digital culture itself. As it grows and matures, digital culture itself is an increasingly important influence in the creation of electronic literature, especially in its third generation.

The first efforts towards historicizing electronic literature were by N. Katherine Hayles in her keynote address for the 2002 Electronic Literature: State of the Arts Symposium at UCLA, the concept was published in "Electronic Literature: What is it?" (2004), and elaborated in *Electronic Literature: New Horizons for the Literary* (2008). She established the concept of first-generation electronic literature and she defined it as pre-web, text-heavy, link-driven, mostly hypertext, that still operated with many paradigms established in print. She defined the second generation from 1995 onward, as web-based and incorporating multimedia and interactivity. After some of the critical conversation around her notion of generations, she renamed the first generation as classic and the second as contemporary electronic literature (2008). And this was accurate, for the moment, because the paradigm shift I will describe was barely getting started.

Christopher Funkhouser, with *Prehistoric Digital Poetry* (2007), elaborated and reaffirmed Hayles' generational formulation, especially of first-generation electronic literature, showing that it wasn't as text-driven as initially understood and that it had a variety of multimedia and kinetic works. He and others have continued to explore the richness of the first generation. With *New Directions in Digital Poetry* (2012), Funkhouser picks up where the previous book left off (around 1995) and makes a series of case studies that span about fifteen years of web-based digital poetry. In these case studies, he maps out some of the most important developments in digital poetry, showing how writers from the first generation continued to develop their work in the second generation or contemporary period. And while this book concluded with a nod toward some of the emerging platforms of the time: social media networks, mobile platforms, and web APIs, its conceptualization of the field was aligned with Hayles' notion of contemporary electronic literature.

It is time to update the historical model to account for these emerging platforms and the practices they encourage. To begin with, we must leave behind distinctions like classic and contemporary because contemporary is an open-ended concept that needs to be continuously adapted when there's a shift in practices. I propose defining three generations (or waves) of electronic literature. The first one, much as defined by my predecessors, consists of pre-web experimentation with electronic and digital media. The second generation begins with the web in 1995 and continues to the present, consisting of innovative works created with custom interfaces and forms, mostly published in the open web. The third generation, starting from around 2005 to the present, uses established platforms with massive user bases, such as social media networks, apps, mobile and touchscreen devices, and web API services. This third generation coexists with the previous one and accounts for a massive scale of born-digital work produced by and for contemporary audiences for whom digital media has become naturalized. Each generation builds upon previous and contemporary technologies, access, and audiences, to develop works and poetics that are characteristic of their generational moment.

The first generation of electronic literature is characterized by a few pioneering works that emerged between 1952 and 1995. For most of this period, people had limited access to computers, resulting in a small number of practitioners, most of whom didn't have a clear concept that what they were creating was electronic literature. In the first few decades, only computer scientists and academics in universities and technical staff in the private industry, producers in film, television, radio studios that had access to expensive tools that could be used to create electronic literature. As word processors, personal computers, and gaming consoles arrived in the late 1970s and became popularized in from the 1980s onwards, access to computers expanded to include hobbyists and middle-class populations in the most developed countries around the world, production of electronic literature began to grow. This itself could be considered a mini-generational shift within this first generation because of the leap in user base, with an according explosion in production and reception. During the first half of this period production tools were very limited in their capabilities. Programmers initially used punch cards and early programming languages were very close to machine language, but over time have become more user friendly. BASIC and Pascal, for example, are closer to natural language and were featured prominently in the first decade of personal computing. Early software was frequently encoded in ROM storage and was used for very specific tasks, such as word processing and gaming cartridges. More versatile software, such as HyperCard, Storyspace, and INFORM, along with increasingly powerful media editing and production tools emerged in the 1980s and early 1990s. Distribution of electronic literature mostly happened in physical media, through magazines like *Byte*, which came with disks bundled with print matter to be sold in newsstands, bookstores, and other brick-and-mortar establishments. The audience for electronic literature was therefore limited, even though interactive fiction was very popular in gaming markets, and Eastgate Systems enjoyed mainstream attention and international circulation through the book market.

The rise of the World Wide Web brought about a major paradigm shift and growth in the production and circulation of electronic literature, initiating a second generation of works. Because growing numbers of people have personal computers and internet access, the number of practitioners increases accordingly, and the ease of publication has led to a massive amount of original work entering circulation through the web. Practitioners are programmers, people with personal computers, web artists, and developers, writers and artists collaborating with programmers, and multimedia authoring software users. Flash and Director, for example, empowered a generation of authors and artists to create electronic literature, even if they didn't have a high level of programming skill to begin with. As more varied and user-friendly tools and programming languages develop—such as HTML, JavaScript, CSS, DHTML, ActionScript, Python, and Ruby—the barrier to entry becomes more accessible for people to create electronic literature. Software for editing audio, video, and images continue to develop, and multimedia authoring software, such as Director and Flash, rise and fall as they become obsolete. The second generation of electronic literature circulates primarily through the open web, through author home pages, web-based 'zines, collections, and other resources. The audience for electronic literature is growing and academia has been a powerful engine for expanding this audience through courses, presentations, exhibitions, conferences, leading to the growth of the field.

The third generation of electronic literature, because it is based on social media networks and widely adopted platforms and apps, both the production and audiences are massive. If you count image macro memes as a kind of electronic literature, as I do, then the numbers of works produced and circulated are in the millions. The number of practitioners is equally enormous, numbering in the thousands, millions if you count image macro meme makers. The increasing demand for digital skills in the workplace has resulted in growing numbers of programmers, designers, digital producers, coders, and web developers. A portion of these highly skilled folks produce electronic literature in the second-generation mode, as well as third-generation work. The huge numbers come from users of multimedia authoring software—such as Instagram, Snapchat, Imgflip, and apps that allow you take or upload a picture, put language in it, create an animation, and share it. Even when they are not self-consciously producing literature—societal concepts of literature are still dominated by the genres and modes developed in the print world—a huge amount of people have used these tools to produce writing that has stepped away from the page to cross over into electronic literature territory, and it's a crucial move. Whether they know it or not, they are producing third-generation electronic literature. The software tools at their disposal are varied and increasingly lower the barrier to entry, with programs like Twine, Unity, JavaScript Libraries, simple and free publication platforms (like Cheap Bots, Done Quick!, and Philome.la), and social media apps like Vine, Instagram, Snapchat, GIPHY, and others. Many of the electronic literature genres that emerged in the first and second generations continue in the third—such as bots, electronic poetry, videopoetry, hypertext fiction, mobile and locative works, virtual reality, augmented reality—and are revitalized as they find new forms in the third generation with Twine games, Twitter bots, Instagram poetry, GIFS, and image macro memes.

Bots are a good case to examine how these generational divides manifest across the same e-lit genre. The first chatterbot, ELIZA, was developed by Joseph Weizenbaum from 1964 to 1966 at MIT, followed by PARRY at Stanford University in 1972. Access to these bots was extremely limited, so if you wanted to interact with either of these bots, you needed to travel to MIT, arrange for a session, and sit down at the teletype machines prepared for them to do so. Interestingly enough, the first conversation between these two bots was arranged in 1972 using ARPANET during the International Conference on Computer Communications in Washington, DC. But even this was a bit of a technical feat at the time and access to computers and networks was mostly limited to computer scientists at elite universities, corporations, and some branches of the government. Bot development was carried out by specialists interested in achieving computer science benchmarks, such as passing the Turing Test, and later on by programmers who developed parsers for Interactive Fiction and MUDs, and chatbot developers that compete for AI awards like the Loebner Prize (established in 1990) and those who develop chatbots for phone-answering systems and personal computers.

In the second generation, many of these first-generation bots were implemented on the web, providing widespread access to them. We also have increasingly sophisticated bots that serve as characters in interactive fiction and video games, such as Emily Short's "Galatea" (2000) and "Façade" (2005) by Michael Mateas and Daniel Stern, respectively. Part of what characterizes these works is that audiences can find the works online and need to install the works on their computers to launch the game environments and interfaces needed to interact with these bots. These bots posed a challenge for audiences, who need to figure out how to successfully interact with them and understand their programmed personalities for different narratives to unfold. A characteristic of second-generation works is that authors like to create new or customized environments and interfaces for readers to experience the works, as is the case in these works.

Artistic and literary bots have been relatively rare during the first and second generations, but the third generation with its social media networks and API services has renewed and expanded this e-literary genre. These bots have gone beyond the chatbot subgenre to be autonomous generative e-lit works that are presented as human or personified animals and concepts and publish their content in Tumblr, Twitter, Facebook, Mastodon, or other social media networks. The techniques for creating them are not very different from bots in earlier generations, but rather than creating custom datasets, programmers are pulling or processing content from API services or using user-friendly platforms like the Twinery-powered Cheap Bots Done Quick! (CBDQ). Rather than being standalone custom experiences, these bots leverage social media networks as contexts and spaces to develop audiences. For example, when you are on Twitter, a bot might react artistically to something you posted, such as @HaikuD2 or @ Pentametron, which detect tweets that could be cut into haiku or happen to be written in iambic pentameter, respectively. Most of these bots are opt-in, which means you can follow it, though someone you follow may retweet or like bot output, which then is inserted into your Twitter stream. Because following bots is a way of interspersing art into your social media stream, and because people use social media networks for a variety of reasons, bots have become increasingly popular, developing audiences of hundreds of thousands of followers and more. And while there are more people with the programming skills necessary for bot-making, services like CBDQ and Zach Whalen's SSBot tool have lowered the barrier to entry, which magnifies the production of works in this vein. It is telling that CBDQ currently has over 7,000 active bots in Twitter, and that the total audience for bot output is in the millions—a huge growth in readership from first- and second-generation works.

Another electronic literature genre or modality where a generational shift helps bring new works into focus is kinetic texts. Álvaro Seiça's 2018 dissertation, "setInterval (): Time-Based Readings of Kinetic Poetry," traces a history of kinetic poetry across the first and second generations, offering meticulous and timely readings of both the surface and code portions of the kinetic poems, and covers works from "Roda Lume" (1968) by E. M. Melo e Castro to works as recent as "U [Total Runout]" (2015) by Ian Hatcher, and iOS apps created by Jason Edward Lewis, Bruno Nadeau, and Jörg Piringer. While this was not the focus of Seiça's dissertation, it would have been enriched by a look outside of the experimental tradition to explore the widespread use of kinetic writing occurring with lyric videos and kinetic typography in Vimeo and YouTube and animated GIFs circulating massively in sites like Tumblr, GIPHY, and other social media networks and recognizing how kinetic works are deployed differently in apps and publications designed for touchscreen devices.

First-generation works are very clearly recognizable: Melo e Castro needed a television production studio to produce and air his work, Eduardo Kac used an LED scrolling screen to create "Não!" for an installation in 1982, and bpNichol distributed his 1984 Applesoft BASIC suite of poems "First Screening" in floppy disks. Second-generation works published on the web, such as "Project for Tachistoscope" (2005) by William Poundstone and "El Poema Que Cruzó el Atlántico" by María Mencía require readers to visit websites to read the works and have a learning curve in which readers figure out how to operate and experience the works. "Hearts and Minds" by Roderick Coover and Scott Rettberg is a generative and immersive cinematic work that is too computationally demanding to be published online and therefore circulates in documentation videos and installations. All these works including early Flash and Director work by Jim Andrews, Alan Bigelow, Christine Wilks, Stephanie Strickland and Cynthia Lawson Jaramillo, Megan Sapnar, Ingrid Ankerson, Andy Campbell, Jason Nelson, Young-Hae Chang Heavy Industries, and other second-generation works seek formal innovation that is aligned with the poetics of Modernism, or what Jessica Pressman described as Digital Modernism in her 2014 book. Mark Wollaeger and Kefin J. H. Dettmar's preface to her book not only offers a concise definition but does so in connection with the notion of generations.

> "Digital Modernism" is deployed here by Jessica Pressman, the first critic to elaborate the term, to describe second-generation works of electronic literature that are text based, aesthetically difficult, and ambivalent in their relationship to mass media and popular culture. Such works offer immanent critiques of a contemporary society that privileges images, navigation, and interactivity over complex narrative and close readings.
> (Pressman 2014: ix)

I propose that Jessica Pressman's formulation of digital modernism is not only a distinctive feature of first- and second-generation electronic literature but also that third-generation works reject or are unaware of this aesthetic of difficulty, and can be thought of as postmodern electronic literature. Third-generation kinetic works write language in animated GIFs, in apps like Snapchat and Instagram, and write kinetic typography in videogames, lyric videos, and other multimedia productions without necessarily seeking formal innovation or a highbrow literary experience. I would describe these works as works of e-literary popular culture that seek ease of access and spreadability (to reference Henry Jenkins' term in *Spreadable Media*), and are aligned with the poetics of contemporary digital culture.

Some contemporary electronic literature authors began in the second generation and have shifted to the third or produce work in both modalities. Alan Bigelow, for example, began producing e-lit works in Flash, each of which had a unique interface designed for the work. This is very much in the Modernist tradition of free verse, where content determines form. When the end of Flash became clear, however, Bigelow shifted his practice to creating works with HTML5, JavaScript, and CSS with a simplified form that is designed to work well with both computers and touchscreen devices of all shapes and sizes. His recent Coover Award-winning work "How to Rob a Bank" is basically a slide show, and as a user that is expected of you is to advance from one slide to the next, with no need for instructions beyond turning on your sound. Bigelow's third-generation works all use this and other well-established digital formats, with a parallel move to the return to closed form in postmodern poetry.

Another transition from second- to third-generation practices is evident in the work of authors that develop work for popular platforms, such as iOS and Android. Jason Edward Lewis and Bruno Nadeau's Coover Award-winning *P.o.E.M.M.* Cycle, initially tapped into a second-generation poetics by being created for large-screen installations equipped with touchscreen technology because they were challenging audience expectations with innovative poetic form. When implemented in iOS, however, these works and their form feel less innovative in terms of the gestural vocabulary that users of iOS and Android touchscreen devices are already accustomed to. Poetically, each poem is creating its own form—a second-generation Modernist move—but they're more accessible to the massive audiences that Apple has cultivated for their devices. Increasing numbers of e-lit authors, such as Samantha Gorman, Jörg Piringer, Amaranth Borsuk, Ian Hatcher, and Katherine Norman are interesting bridge cases, bringing second-generation sensibilities and poetics while developing for third-generation platforms and the interactivity training that audiences bring to the device. Andy Campbell, Kate Pullinger, Mez Breeze, Caitlin Fisher, and their collaborators do a parallel move by creating works in increasingly popular virtual reality environments.

Going to where the audiences are and building upon their knowledge of the platform is a key characteristic of third-generation electronic literature works. Second-generation authors frequently create custom environments and interfaces, subverting audience expectations and frequently featuring instructions to read their works. For example, Netprov works frequently use existing platforms—such as Twitter, Facebook, or Amazon's rating and customer commenting capabilities—to create a kind of literary graffiti. Authors like Rob Wittig, Mark Marino, and others use popular platforms to create literary works. Another kind of third-generation social media performance is characters or bots based on fictional works, hit shows, celebrities, or politicians, such as @WernerTwertzog and @KimKierkegaard. These performances attract large audiences, which authors can often find a way to monetize.

Third-generation works are less interested in originality (digital modernist characteristic), and more willing to create remixes, derivations, copies, and outright plagiarism of works, frequently adding personal touches and customizations. For example, Nick Montfort created "Taroko Gorge" in 2009, inventing a poetry generator specifically for his nature poem in second-generation fashion, but those of us who remixed it after him were not inventing a form, we were adapting, appropriating, even erasing the original works as a third-generation move. Memes are another example of this type of postmodern digital poetics. Language-driven memes, such as image macro memes may have existed formally in print culture, but have become a central type of cultural production in digital media, particularly as deployed in social media networks. I like to think of image macro memes as a kind of gateway drug into e-lit, because all the people creating them are taking a step away from the page: they're writing on images, sometimes moving images, and that alone is a step toward a deeper engagement with digital media. Memes are frequently not original, nor do they wish to be, yet they get expanded, adapted, forked, combined with other memes, reframed, parodied, become self-referential, go viral, go dormant, return, and are a great example of how a massive amount of people are writing and reading a kind of electronic literature that is probably looked down upon by those committed to a digital modernist poetics. This is why a new generation is key to mark a paradigm shift in the field.

The field of electronic literature began in the first generation but was formed and grew in academia during the second generation, and its poetics guide its production, reception, circulation, and economics. The web—the defining platform for the second generation—disrupted established markets for the circulation of music, writing, video, and to a lesser extent the visual arts by creating a powerful gift economy and taking the means of production away from publishers and labels to return them to creators. For this generation, the main way to profit from electronic literature was by developing cultural capital in academia, the art world, and other spaces, advancing the field through innovation while connecting to literary and artistic traditions. Second-generation electronic literature writers and artists tend to make a living not through the sale of works, but through day jobs, artist residencies, academic positions, invited performances and gallery exhibitions, commissioned work, and other related practices. A poetics of innovation and aesthetic difficulty go hand in hand with the economics of the prestige and cultural capital market. As Matthew G. Kirschenbaum noted in his ELO 2017 keynote: "I submit ... that difficulty, seriousness, and conceptual density are all characteristics that have served to gain e-lit a firm institutional purchase in academia, where difficulty and seriousness are rewarded" (5). And academia helps circulate e-lit through courses, criticism, publications, exhibitions, digital repositories, and more, even as it has struggled to directly monetize such circulation.

Third-generation works respond to new markets, platforms, and monetization possibilities and have developed without the need for academia and its validation. Poets who publish first on Instagram, such as Rupi Kaur, Lang Leav, and Robert M. Drake, build massive audiences and then publish books of poetry that reach unprecedented sales numbers, as Kathi Inman Berens demonstrated in her ELO 2019 presentation, "Populist Modernism: Printed Instagram Poetry and the Literary Highbrow." Lyle Skains, in "Not Sold in Stores: The Commercialization Potential of Digital Fiction" discusses some alternate commercialization options for digital fiction, such as Twine games and walking sims sold through the Steam store, webcomics and fanfiction sold in print, and detailing some of her own efforts in creating and selling hypertext publications in Kindle format. Some of the new commercialization models available for third-generation works that achieve good circulation includes crowdsourcing platforms like Patreon, advertising revenue, and sales through app stores and in-app purchases.

An aesthetic of difficulty would undermine the very spreadability and commercialization paradigms that help third-generation works thrive. And as Kirschenbaum provocatively stated in his keynote, "maybe what matters is the continued growth and diversification of an e-lit that is not dependent on whatever contradictions or complications attend its status in relation to an academic valuation of the avant garde" (Kirschenbaum 2018: 7). This provocation was an important catalyst for my own formulation of a new generation of electronic literature because I recognize the need to account for the explosive growth and diversification of e-literary digital writing practices beyond what is practiced and studied by the ELO community.

An important example of a third-generation work that went "viral" in 2018 is "Lazy Cat" by txtstories, reaching over 68 million views on Facebook and over 31 million views on YouTube. The story is presented as a video capture of a text messaging session between a cat and its owner, after the cat informs that the stove was left on (see Figure 1).

If we were to think of this in terms of a second-generation work, it is not a particularly sophisticated piece. From a literary point of view, it's an amusing

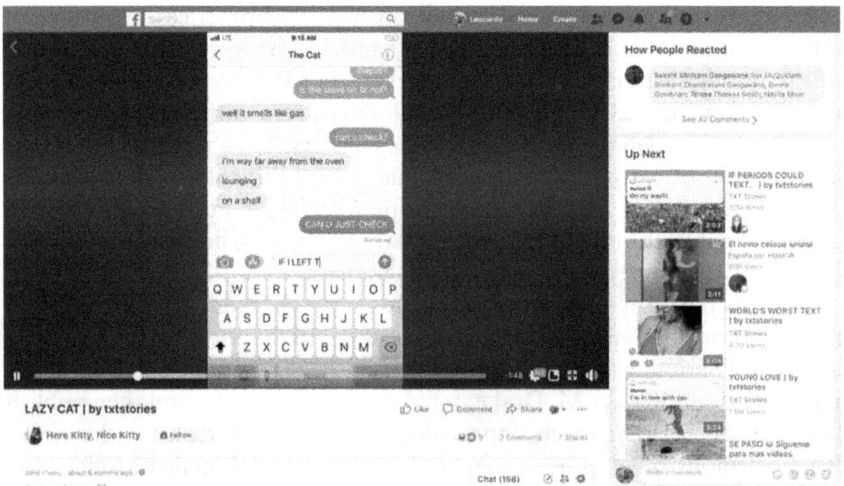

FIGURE 1 *Screen capture of "Lazy Cat."*

story that personifies a cat and does a nice job of capturing its personality as expressed through a texting conversation. It's a video produced using some sort of messaging software and screen-recording technology and it isn't particularly innovative from a technical standpoint. However, its popularity comes from its humorous narrative, its leveraging cat culture, and by how effectively it uses the limited texting vocabulary they have at their disposal: text in all caps, lowercase, tactical use of punctuation, and hashtags. These videos are circulated on Facebook and YouTube and use these social media networks not only as publication platforms but also generate revenue using their advertising services. The Los Angeles-based company that produces txtstories, New Form, describes itself as follows:

> New Form is an entertainment studio for TV and digital content that redefines how stories are developed, packaged and distributed. New Form empowers creators and audiences to produce original narratives that transcend traditional categories and platforms. Watch New Form series on a variety of global outlets, including Facebook Watch, YouTube Red, TBS, CW, and TruTV.

It is telling to see how some of this work is created and circulated using production models designed for cinema, television, and digital media rather than those developed for the print world. This may be one of the most profound differences between second- and third-generation electronic literature: while second-generation e-literature aligns itself with the literary tradition formed by the print world (and publishes zines, anthologies, blogs, and web pages) and the art world (gallery exhibitions and installations), third-generation e-literature identifies itself with electronic and digital media in terms of its formats and publication models, producing video and interactive works that could be published as video games and other kinds of digital content. This extends to notions of the author and artist as creative geniuses who labor in isolation or tight collaborations to create works that are then produced and shaped by craft professionals (editors, layout designers, illustrators, typographers, printers, binders, and others) to reach publication. And while the web has returned the means of production to individual authors and encouraged a DIY (do-it-yourself) self-publication aesthetic, academia has elevated it to a constituitive part of e-lit poetics this by connecting these practices with famous authors and craftsmen like William Blake and William Morris. Perhaps the time has come to explore, encourage, and recognize other models that may be better suited to the contemporary media economy.

To sum up the differences between the generations, I am including a slide from my ELO 2018 presentation that visually juxtaposes the second and third generations, point by point (see Figure 2).

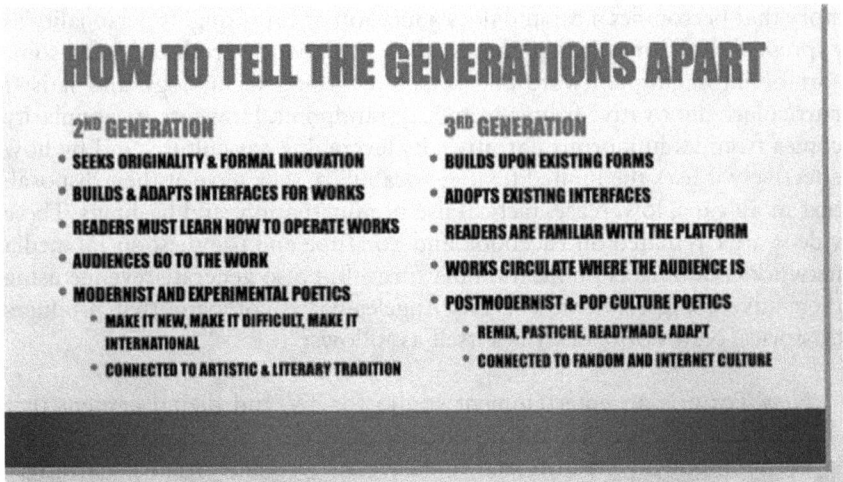

FIGURE 2 *Slide from ELO 2018 presentation "Third Generation Electronic Literature."*

It is important to note that both generations can coexist harmoniously. The second generation seeks originality and formal innovation while third generation is exploring existing forms, established platforms, and interfaces. In second-generation works readers must learn how to operate them—to the extent that many works feature instructions and many books about electronic literature, feature explanations on how to read works of e-lit (such as Funkhouser's *New Directions in Digital Poetry*). In third-generation works, readers are already familiar with the interface and genres and the works don't usually seek to challenge that established training because it reduces readership. This is why works from the second generation are published in websites and that readers must go visit with their computer frequently needing plugins to access the work, while third-generation works seek to reach audiences where they already are with computationally simpler works. Perhaps the most significant difference is in the line between digital modernism and postmodernism and their affinity to (highbrow) literary culture and (lowbrow) popular culture.

This chapter may give the impression that one generation of e-literature is somehow better than the other. To correct this, I propose a series of fallacies that unpack technical and aesthetic biases that emerge from the historical development of the field and recent developments in digital culture, pointing out how they affect our notions of quality.

The Pioneer Fallacy is that to be the first to do something doesn't mean it's a quality work. Pioneering works are of historical interest, but that doesn't mean they're successful. On the contrary, they're frequently interesting failures.

The **Generational Fallacy** reminds us that just because it's the most recent generation does not mean its work is better. In most cases these works lack the aesthetic and technical sophistication that second-generation works have developed over years of work with scholars and curators.

The **Technical Fallacy** states that technical complexity does not equal quality. Too often we find ourselves talking about the works on a technical level which is interesting, but that is often a distraction from whether a work is successful or not on its own merits.

The **Viral Fallacy** means that just because something is super popular doesn't mean it's good. "Lazy Cat" is fun, but I don't know if it will stand the test of time.

The **Hipster Fallacy** states that made-from-scratch does not equal quality. Whether you're using a highly polished product or you're hand-coding an interactivity or generative engine shouldn't be a factor in assessing the quality of a work.

The **User Fallacy** is the flip side of the hipster fallacy. We can't say that someone is just a user and what they're doing in Snapchat or some other platform cannot have merit from an e-literary perspective.

There are other biases and fallacies we could consider, but the idea is to try to keep an open mind when considering the many ways people arrive at creating and experiencing electronic literature.

To conclude, I propose four phases of electronic literature adoption, both at an individual level and a societal level: approach, discovery, experimentation, and adoption.

Approach is when people start to do things with other media that are better suited for digital media. For example, when Borges writes "The Garden of Forking Paths," Cortazar writing *Rayuela*, and Choose Your Own Adventure books become popular you know that the world is ready for hypertext.

Discovery is the moment of realization that digital media have great potential for writing. Many scholars and artists in the ELO community have an e-lit origin story (like superheroes) when we came to a moment of realization: "This is what this digital media can do." "This is its potential." This is the space of pioneering and proof of concept works, both at a personal and social level. Belén Gache's "Word Toys," for example, is a virtual cover organizing a series of short e-lit pieces developed in Flash that explore concepts and interfaces.

Exploration is when the field develops and matures. Authors go beyond their discovery phase and develop their corpus of works, maturing and refining their poetics. Scholars study and teach works, theorize the field, curate exhibitions, prepare journals, 'zines, collections, archives, and other resources. The ELO community and other scholarly and artistic communities around the world are an indication that that particular country or region has reached the Exploration phase.

Adoption is when electronic literature goes mainstream and is recognized at a societal level, beyond academia. Electronic literature, its digitality and materiality start to fade, and becomes naturalized. People create electronic literature without realizing that that is what they are doing. This is partly what is happening with the third generation of electronic literature. It is starting to be adopted by digitally native populations.

This is where we can see a generational age difference in terms of the authorship of third-generation works of e-lit. Many ELO members, such as myself, grew up at a time where we could see the digital materiality sharply. Most third-generation e-lit writers are a younger generation who have naturalized what was experimental to us, and even though the work they create may be naïve and disconnected from the artistic and literary traditions of the past, they were more directly formed by digital culture. They also have massive numbers on their side and it is a matter of time before quality work emerges from the vibrant and massive e-literary production happening in apps and circulating in social media networks.

I predict that a third-generation work is going to break through to mainstream attention with something really exciting and awaken the world to electronic literature. And I hope that we scholars and artists formed in the first and second generations are able to recognize it as electronic literature. We need to build bridges between e-lit generations so they can learn from us as we learn from them. Nothing less than the future of the field is at stake.

Note: This is an adaptation of my talk titled "Third Generation Electronic Literature" offered in the panel Towards E-Lit's #1 Hit during the Electronic Literature Organization (ELO) Conference in Montreal on August 14, 2018. Special thanks to my students Ashley Páramo and Aleyshka Estevez for their help converting a raw YouTube transcript into intelligible text and with the list of references, respectively.

References

Berens, Kathi Inman (2018), "Populist Modernism: Printed Instagram Poetry and the Literary Highbrow," Electronic Literature Organization Conference, Montreal, 2018.

Bigelow, Alan (2017), "How to Rob a Bank," https://webyarns.com/howto/howto. html.

bpNichol (1984), "First Screening," http://vispo.com/bp/.

Coover, Roderick and Scott Rettberg (2016), "Hearts and Minds," https://www.crchange.net/hearts-and-minds/.
Cortazar, Julio (1963), *Rayuela*, New York, NY: Pantheon.
de Melo e Castro, E. M. (1968), "Roda Lume," https://po-ex.net/taxonomia/materialidades/videograficas/e-m-de-melo-castro-roda-lume/.
Funkhouser, Christopher T. (2007), *Prehistoric Digital Poetry: An Archaeology of Forms, 1959–1995*, Tuscaloosa, AL: University of Alabama Press.
Funkhouser, Christopher T. (2012), *New Directions in Digital Poetry*, New York, NY: Continuum.
Gache, Belén (2006), "Word Toys," http://belengache.net/gongorawordtoys/.
Hayles, N. Katherine (2002), Keynote, Electronic Literature: State of the Arts Symposium, UCLA, April 5, 2002.
Hayles, N. Katherine (2004), "Print is Flat, Code is Deep: The Importance of Media-specific Analysis," *Poetics Today* 25 (1): 67–90.
Hayles, N. Katherine (2007), "Electronic Literature: What is it?" *https://eliterature.org/pad/elp.html*.
Hayles, N. Katherine (2008), "Electronic Literature: New Horizons for the Literary," Notre Dame, IN: Notre Dame University Press.
Kac, Eduardo (1982), "Não," http://www.ekac.org/no.html. [http://dtc-wsuv.org/elit/elit-loc/eduardo-kac/].
Kirschenbaum, Matthew G. (2018), "ELO and the Electric Light Orchestra: Electronic Literature Lessons from Prog Rock," *Materialities of Literature* 6 (2), https://impactum-journals.uc.pt/matlit/article/view/2182-8830_6-2_2/4745/.
Lewis, Jason and Bruno Nadeau (2014), "The P.o.E.M.M. Cycle (Poetry for Excitable [Mobile] Media)," http://collection.eliterature.org/3/work.html?work=vital-to-the-general-public-welfare.
Mateas, Michael and Andrew Stern (2011), "Facade," *Electronic Literature Collection*, Vol. 2, http://collection.eliterature.org/2/works/mateas_facade.html.
Mencía, María (2017), "El Poema Que Cruzó el Atlántico," http://winnipeg.mariamencia.com/poem/.
Montfort, Nick (2009), "Taroko Gorge," https://nickm.com/taroko_gorge/.
Saum-Pascual, Alex (2018), *#Postweb! Crear con la máquina y en la red*, Iberoamericana Editorial Vervuert.
Pressman, Jessica (2014), *Digital Modernism: Making It New in New Media*, Vol. 21, Oxford: Oxford University Press.
Seiça, Álvaro (2018), "setInterval ( ): Time-Based Readings of Kinetic Poetry," PhD dissertation, University of Bergen.
Short, Emily, "Galatea," *Electronic Literature Collection*, Vol. 1, http://collection.eliterature.org/1/works/short__galatea.html.
Skains, R. Lyle (2018), "Not Sold in Stores: The Commercialization Potential of Digital Fiction," Electronic Literature Organization Conference, Montreal.
Tender Claws (2015), "Pry," Apple Store, https://tenderclaws.com/pry/.
Txtstories (2018), "Lazy Cat," https://www.facebook.com/txtstories/videos/234390640463135/.
Vectorpark (2018), "Metamorphabet," Apple Store, http://metamorphabet.com/.
Weizenbaum, Joseph (1966), "Eliza." MIT.
Young-Hae Chang Heavy Industries (n.d.), "The Struggle Continues," http://www.yhchang.com/THE_STRUGGLE_CONTINUES.html.

3

Toys and *Toons*: From Hispanic Literary Traditions to a Global E-Lit Landscape

Élika Ortega and Alex Saum-Pascual

Is it possible to talk about Hispanic electronic literature? If so, what elements render a particular work *Hispanic*? In the current media landscape, cultural specificities and differences are reconfigured in the many spaces where they come into contact. The web, commonly articulated as borderless and global, sits at the center of this landscape, becoming the workspace where e-lit has thrived in the last years. Consequently, Hispanic electronic literary works cannot be thought of simply in terms of national literature, not even in terms of Hispanism around the world. Hispanic e-lit works published on the web offer the possibility to observe the tension between a global digital culture, world e-lit, and specific literary traditions through their complex relationship to various forms of language. A solid grounding on the linguistically defined literary tradition is made explicit through the intricate referential networks found in Hispanic e-lit works. Yet, many works often fracture said linguistic tradition by means of rhetoric and meaning-making systems coming from digital media. At the same time, these media extend, parody, and put into question the seamless continuation of the past into the present, and the solidity of referential language. Thus, Hispanic e-lit invites a reflection on the mechanisms employed by authors to appeal to both a linguistic-literary tradition and the global landscape of digital cultural production.

This chapter explores the deliberate and problematic construction of e-lit works which, though cemented in the Hispanic literary canon, reach out to a landscape of global e-lit. Further, the dialog established between earlier works (chiefly print products) and current digital works allows us to comment on a type of intertextuality/intermediality that cuts through time, individual authors, and media. The resulting phenomenon is a "relocation" of the *literary* from its niche as a *product of language* into *non-linguistically bound word-objects*. We take two examples to explore this, Belén Gache's *Góngora Wordtoys (Soledades)* (2011) and Benjamín Moreno's *Concretoons* (2010). In these works, both writers have established a manifest connection between their e-lit production and two of the most celebrated periods of the Hispanic tradition: seventeenth-century baroque, and twentieth-century avant-garde. Over this basis, we analyze first how Gache's poems reimagine the rhythms and imagery of Luis de Góngora's *Soledades* (*The Solitudes*), stressing the potentialities for movement and the distinct materiality that kinesis gives to the baroque writer's verses. Second, we look at how Benjamín Moreno's *Concretoons* explore material qualities of language, and how features like the iconicity of graphemes exploited by concrete poets in the twentieth century are enacted through game dynamics.

Looked at from an e-lit viewpoint, baroque and concrete poetry might seem abysmally different between them. However, both literary movements share a rejection of figurative realism, and draw on marked intermedial resources (visual, aural, and kinetic) as compositional principles. Góngora's rhythm and sound throughout *Soledades*, for example, signal poetry's structural and expressive potential beyond the word level and carry their own aesthetic meaning. Similarly, in concrete poetry, texts usually draw on two or more semiotic systems or media "in such a way that the visual and/or musical, verbal, kinetic or performance aspect of its signs are inseparable" (Clüver 2000: 34). It is thanks to these intermedial features that we trace in and out of Gache's and Moreno's pieces that the poetic strategies borrowed from their predecessors become word-objects.

Nevertheless, the relationship between the historical baroque and avant-garde periods and e-lit is not a linear or seamless one, and certainly not one of incorporation or allusion alone. As Jessica Pressman would have it in *Digital Modernisms*, drawing from established creators and their work to construct new digital texts is "a strategy of renovation that purchases cultural capital from the literary canon in order to validate their newness" (2014: 2). Partly a matter of influence and legitimization, of "making it new" in Pressman's terms, we further sustain that the connections between Gache's and Moreno's works and baroque and concrete poetry are characterized by the tension of how poetic aspects are enacted individually in each particular case. Connections may come in the form of analogous instantiations or materializations in both the earlier creations and contemporary ones—a process akin to Joseph Tabbi's "relocation of the literary."

In "Electronic Literature as World Literature," Tabbi explains how e-lit has revealed that literary qualities such as narrativity are not "universal"— i.e., not equally fitted to all expressive media—but best realized in particular ones and thus, "new media bode ... a revaluation and relocation of the literary in multiple media" (2010: 28). Similarly, Pedro Reis proposes that some features of electronic literature such as "[t]he combinatory strategy, the use of space, the destruction of syntax, the depersonalization of the work, the expedient of chance and the relative absence of orientation in the poetic structure" have already taken place in print literature (2015). Nevertheless, they have been developed and relocated in the electronic environment "so that [they] may be (re)discovered and (re)invented every time" (Reis 2015). Seen under this light, digital literary innovations may seem relative, except when they go beyond renovating or adapting the affordances of print intermediality. Thus, features of e-lit are not merely means to overcome the saturation of literary media or forms, but strategies to put into practice—to enact—poetic elements that might have been problematic in print as well. In that sense, we see enactment as a process of putting the literary into practice in a given work and to create new mechanisms of meaning not restricted to electronic media or linguistic referentiality. This idea is further useful to examine the trans-linguistic relations and the influence of globalization that shape Gache's and Moreno's work.

Relocating the Spanish Baroque: Word-Objects, Movement, and Rhythm

Baroque poetry explicitly drew on the potentialities of language to create complex structures that superseded the utilitarian qualities of expression. In the case of the Spanish baroque, this responded to a dramatic cultural and political decline after the glory of the Renaissance. The powerful reign of the Catholic monarchs was being eroded by religious changes and political challenges that threatened the continuity of an almost global empire. Thus, baroque literature combined a pessimistic look on the present with escapism and satire, which arguably fostered imaginative forms of writing. It is in this context where Luis de Góngora, one of the most influential Golden Age poets, wrote his widely studied, yet unfinished poem, *Soledades* (*The Solitudes*). Composed in 1613, *Soledades* is a long silva poem praising the natural world in which an anonymous castaway—depicted as a pilgrim—finds himself on an island. *Soledades* is paradigmatic for its difficult grammatical structures and the over-abundance of erudite and mythological allusions and references. Although it is a poem about nature, the natural world is evoked in figurative and rarified language because "language itself, not its emotive referent or expressive content, is the intrinsic aesthetic component" of *Soledades* (Grossman 2011: ix). Key to understanding *Soledades* is its rhythm embedded in a set of mostly metarhetoric images pointing to the composition process itself. To appreciate *Soledades'* rise and fall cadence one should consider the classical figures of Icarus, Sisyphus, and the Phoenix, which are alluded to explicitly in the text (Halevi 1995: 463). Moreover, Góngora's employment of nets, labyrinths, and rivers, as well as "other rhythmical images such as the movement of the birds ... require interpretation to be viewed in a metarhetorical light" (Halevi 1995: 463).

Gongora's emphasis on the form of the lyrical composition is taken on by Belén Gache in her *Góngora WordToys (Soledades)*—an online collection of five digital poems that explicitly engage with the Spanish poet's work. In the opening poem, "Dedicatoria espiral," Gache relocates the rhythm and rhetorical intricacy in *Soledades'* dedication to the Duque de Béjar by transforming the verses into a moving spiral that turns clockwise or counterclockwise depending on where the reader places the cursor. Where Gache exploits the affordances of the animated object to instantiate the sense of movement in Gongora's poem, Gongora's general departure of writing conventions rejected the order and stability of classic or imperial writings of the time. Due to the extensive use of a variety of erudite references, the multiplicity of readings in Gongora's poem has been a recurrent topic of investigation. Analogously, the near impossibility to pin down a reading of *Soledades* also lies at the bottom of "Dedicatoria espiral" as the spiral moves too quickly to read the words that constitute it. The spiral presents an intriguing paradox where the reader is able to read the poem only as long as she does not activate it, while it is still words. Conversely, the activation of the poem renders the composition asemic appealing only to its objectual and kinetic qualities

The recurrent representations of movement and rhythm in language common in baroque poetry are highlighted in Góngora's *Soledades* to sidestep referentiality, which may also suggest an asemic intent. As a matter of fact, Góngora's baroque convolution has been interpreted as non-sense—i.e., not conforming to referential interpretations of language in the baroque era. For Roland Barthes, and Paul Julian Smith, Góngora's textual obscurities are open to free play of sense and meaning (Smith 1986: 83).[1] Góngora's favor to rhythm and form as facilitators of multiple understandings is enacted

---

[1]As Edith Grossman notes on her preface to the bilingual edition of *The Solitudes*, the conventional view of Góngora's complex, allusive, hyperbolic, and highly metaphorical poetry had been seen for many years in literary histories as the result of mental disturbance. The insanity of Góngora's poetic style, so contrary to the values of the Counter-Reformation has also been equated with non-sense. In this way, and following Lacan's comments on the Spanish poet, Góngora's nonsensical writing should be read in opposition to what our doxa (or "common opinion") would deny in the name of truth. "As we shall see, it is precisely a lack of meaning that Góngora himself is accused of and is forced to deny" (Smith 1986: 83).

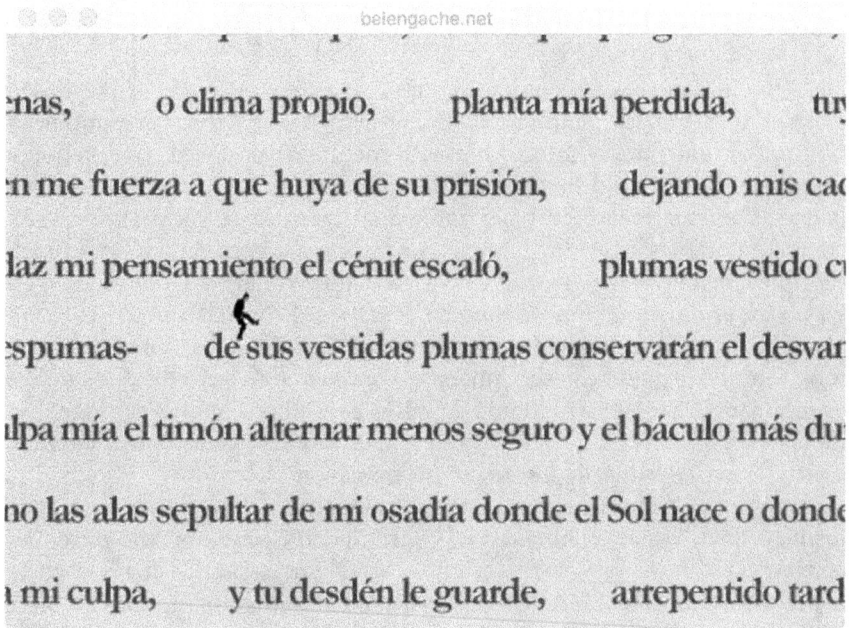

FIGURE 1 *Belén Gache. "El llanto del peregrino." Screenshot by the authors.* *http://belengache.net/gongorawordtoys/llanto/laberinto.htm.*

in Gache's work in the act of reading itself. In "Dedicatoria espiral," the position of the cursor changes the direction and speed of the text on the screen and invites distinct reading acts—some of which can be semantic while others are kinetic. In that sense, Góngora's non-sense becomes actual meaning-as-play when readers activate "Dedicatoria espiral."

Through his newfound surroundings, the protagonist's pilgrimage is rendered an adventure—a motif that makes *Soledades* move forward. In "El llanto del peregrino" (Figure 1) Gache's castaway arrives not at an island but at a platform game, a puzzle made of verse fragments. Along with the avatar, the reader moves around the new environment and walks in between words using the keyboard arrow keys. The verses, cut up and disordered, become a maze with no beginning or end and no clear referential meaning. Gache's poem becomes an entrapment of meaning in itself. "Emulating labyrinth poems so dear to baroque aesthetics, and taking the baroque (and Borgesian) idea of the 'poem as labyrinth,' this *wordtoy* recreates the text as a metaphor" (Gache 2015, our translation). Referential linguistic meaning is lost as it becomes a puzzle in the platform game, translating semantic meaning to the kinetic qualities of the word-object. Gache's poem demands from her reader a separation from baroque wordplay and rhythm so as to engage with reading as video game.

Concrete Relocations: Word-Objects in E-Lit

Since the mid-twentieth century, the term "concrete poetry" has been used to refer to a variety of innovations and experiments that revolutionized writing around the world. Although the name received international support, and it should be considered in relation to a mainstream defined in terms of continents and not individual cultures (Clüver 1987: 113), concrete poetry was not a homogenous practice across the globe. In fact, the movement was originated in Switzerland by Eugen Gomringer who was born in Bolivia and published his first "word constellations" (1952) in Spanish, his native tongue. Furthermore, concretism almost simultaneously took root in Portuguese in the American continent thanks to the Noigandres group from Brazil—Haroldo and Augusto de Campos and Décio Pignatari (Solt 1970: 8). This brief backstory should caution us against understanding concretism as a primarily European phenomenon.

Among the different types of concrete poetry at least three have been distinguished: visual, phonetic, and kinetic poetry (static on the page, but activated by the passing of pages in a visual succession). Mostly seen in a performative combination, the fundamental aspect of concrete poetry is the concentration upon the physical material from which the poem or text is made (Solt 1970: 7). The implication of this is the subjugation of semantic referentiality to the poem's structure—a structure that should be defined as intermedial as well. In concrete poetry, language, in a semantic sense, becomes secondary to how signs can convey meaningful information and, thus, the concrete poem communicates its structure. Concrete poets, however, were disunited on the importance that the poem should give to semantic meaning (Solt 1970: 9). On one side of the debate we place the Brazilian Noigandres whose work, although sometimes abandoning words, remains within the communication area of semantics: in their poems we can *read* words and sentences although these are distributed playfully and meaningfully throughout the page. On the other side we situate the Spanish visual poet Joan Brossa, whose one-letter poems and sculptures, such as those representing only the letter A, rely on the capacity of the character to transmit purely aesthetic information.

Taking this debate as a starting point, Benjamín Moreno's *Concretoons* explicitly reflect on the affordances of digital media, specifically the arcade video game, to transmit poetic information. His poems "Noigandres vs. Brossa" and "Brossa vs. Noigandres" take on the issue enacting it through video game dynamics. By transforming the debate into a literal fight following the *Space Invaders* (1978) and *Asteroids* (1979) game models, Moreno's poems not only communicate their own structure; their very structure is (put into) play. In "Noigandres vs. Brossa" (Figure 2) the player takes on the pro-semantics Noigandres side put against Brossa's army.

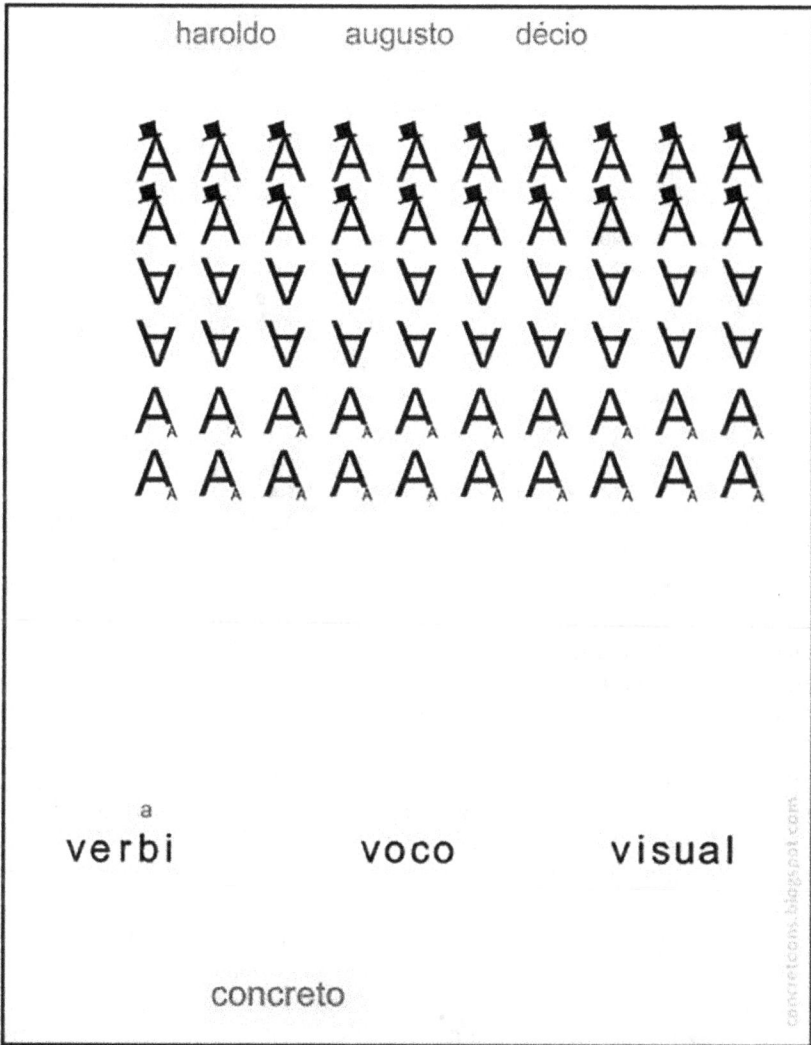

FIGURE 2 *Benjamín Moreno, "Noigandres vs. Brossa." Screenshot by the authors.*
http://concretoons.net84.net/noigandres.html.

The names of the actual Noigandres poets (Haroldo, Augusto, Décio) and their "verbivocovisual" composition principle are the only words in the poem. These "words" fight against a battalion of capitalized As—like those found in Joan Brossa's reductionist one-letter poems. In this way, the joint "verbal," "vocal," and "visual" capabilities of poetry defended by the Noigandres group are in tension with the affordances of kinetic poetry and video game mechanisms of meaning production. In the twin work "Brossa vs. Noigandres," Moreno uses the *Asteroids* game to invert the fighting scenario. Instead of asteroids, the game shows originary concrete poems like Gomringer's "Silencio" or Augusto de Campos's "Sem um numero" flying across the screen. Likewise, the player's spaceship is Brossa's capital A, trying to blow up the approaching asteroids-now-poems. Moreno's use of concrete poems in this work underscores their non-semantic qualities as these become objects standing in as asteroids. Thus, poems that already conceived the word as object are turned into second-order word-objects by Moreno. Since they are still made of words, the potential to "read" the object remains in tension within the object. Yet, the famous concrete poem's iconic shapes are prone to be recognized by the reader because of their objectual characteristics.

Moreno's poems act out the two-sidedness of the debate around semantics. By exploiting language as objects, Moreno brings about the asemic and kinetic understanding of poetry as play. Interestingly, with the addition of movement and the reader's input as interplay, Moreno's poetic enactment still falls within the basic concrete poetry standards by which the poem would communicate first and foremost its structure, beyond its linguistic meaning. Put slightly differently, the intermedial features of these poems, together with the reader's necessary interplay, reveal how the poem is to be handled rather than read. Moreno's poems radically manifest that in e-lit works language is most often pushed beyond semantic referentiality and turned into (digital) objects. However, Moreno fails to solve the debate as the explicit intertextual relationship points to a larger context that resituates the poems beyond the objects they depict, which should not be forgotten.

Concrete poetry often adopted procedures and objects coming from mass media, resituating literature within broader communication networks while exploiting their visual, aural, and kinetic dimension. Where concrete poets engaged the billboard and the page to suggest movement in the 1950s and 1960s, Moreno's poems put into practice their readers' poetic interplay through the global mass media object of the video game. The kind of non-linguistically bound video game dynamics utilized by Moreno borrow and simultaneously appeal to a global audience. The interplay required in these poems relocates the intersemiotic nature of concretism. These poems push for play, rather than reading. Further, Moreno's approach to language as object signals an asemic intent and is thus not limited by any given language. Incidentally, this could explain the quick and vast adoption of concrete poetry across the world and the potential for e-lit to follow suit.[2]

---

[2]In her field-defining work on concrete poetry (1968), Mary Ellen Solt brought together examples appearing almost simultaneously in Switzerland, Brazil, Germany, Austria, Iceland, Czechoslovakia, Turkey, Finland, Denmark, Sweden, Japan, France, Belgium, Italy, Portugal, Mexico, Spain, Scotland, England, Canada, and the United States.

Emerging Paths for a Global E-Lit Landscape

As we have argued, Gache's and Moreno's relocations of poetic aspects from their predecessors fracture the link between a literary tradition and strategies of meaning-making in digital media, while they open two reading paths and potential audiences. First, a semantic reading that entices readers to try to read the words in movement in order to understand the poems and, further, to identify the intertextual references to canonical works. It would follow that Gache's and Moreno's poems appeal most clearly to Spanish- or Portuguese-speaking readers for whom words and literary references are recognizable. Seen in this light, these works might indeed be making Góngora's, the Noigandres', and Brossa's poems new in the sense proposed by Pressman.

Nonetheless, a second reading path pushes for a "structural reading" that bypasses linguistic content and figurative languages in favor of visual and kinetic forms of engagement. This happens when poems taken from the literary canon are literally put into parodic play by being relocated *as objects* in the rhetoric of *toys, toons,* and video games. In this sense, Gache's and Moreno's poems appeal to a linguistically broader audience, perhaps even more so to e-lit readers familiar with various interplay dynamics, codes, and practices. Aside from the material conditions that foster this twofold reading, on the writers' part we might also find a "desire to speak as widely as possible, over time and space, and the desire to reach a carefully targeted and constructed audience" (2015: 298) as Alexander Beecroft would have it, when talking about the emergent global landscape of world literature.

Moreover, and aside from the asemic information these poems exploit, Gache's and Moreno's work also reach out to a global audience through bilingual, "spanglish," expressions. Most evidently seen in their titles— *Wordtoys* and *Concretoons*—the initial bilingual intent in these works signals an awareness of the potentially global literary space of the web in English that exists in tension with some of the boldest uses of the Spanish and Portuguese languages during the historic baroque and avant-garde. Furthermore, as Joseph Tabbi proposes "[t]he concept of a world literature ... is tied to the creation of newly internationalized reading publics and to the loss of such publics (and their renewed creation) with the rise of new communications infrastructures" (2010: 20). By appealing to Spanish-speaking readers and others alike, Moreno and Gache carefully appeal to a reading public that is in no way exclusive to a single language, but perhaps suggest the emergence of e-lit grammars.

English and Spanish are two of the most spoken languages in the world. Along the American continent, English, Spanish, and Portuguese coexist within complex socioeconomic and cultural relationships shaped crucially by mass media. That Gache and Moreno hint in their titles at the intricacies of cultural exchange shaped by market and political forces is further amplified by the video game dynamics alluded to in their poems. The engagement with these games—products of a globalized digital culture—situate the work of Gache and Moreno in a media ecology that challenges their canonic literary grounding, and stretches their reach toward the global. This is even more relevant because just as the concrete elements alluded to above, the grammar of video games and our familiarity with their dynamics do not demand linguistic understanding. In that way, and in Rita Raley's words, the hegemony of English as "the literal and metaphoric operating system for what Manuel Castells terms the 'network society'" (2012: 105) is contested by creating supralinguistic *toys* and *toons*.

Similarly, as Hayles notes in the work of Loss Pequeño Glazier, "[t]he combination of English and Spanish ... further suggest compelling connections between the spread of networked and programmable media and the transnational politics in which other languages contest and cooperate with English's hegemonic position in programming languages and, arguably, in digital art as well" (2008: 18). Doubtless, the influence of English as hegemonic language in the digital media landscape has shaped Gache's and Moreno's poetry collections. However, given that both artists share a history of working in the United States aside from Mexico, Argentina, and Spain, their creations may well be suggestive of a transnational workspace where figurative, kinetic, and visual languages operate in parallel to verbal ones. Gache's and Moreno's work must be understood as creations emerging of literatures in contact, Hispanophone, Lusophone, and Anglophone, as well as print and electronic. This repositions the cultural and media differences in tension that shape and inform them as part of an emerging global literary landscape.

Gache's and Moreno's emphasis on earlier literary traditions signals the frictional relationship between e-lit and previous experimentalism. Their deliberate call on baroque and avant-garde poetry establishes intertextual relationships that cut through—rather than just follow—literary traditions and uses of media. Further, the game and kinetic dynamics in Gache's and Moreno's work de-formalize our engagement with the weighty historical precedents and suggests a reconsideration of their relevance in our global network society as fixed ouvres in the canon. In this way, these works reveal previously unexplored compositional principles found in earlier works through the potentialities of electronic devices and uses. Gache's and Moreno's poetic strategies uncover the reconfiguration of the global literary panorama and the idea of national literatures by exploiting the expressive affordances of digital media objects. For Beecroft, the emerging global ecology of literature depends on how languages and literary strategies or devices are in contact and, thus, reconfigure cultural differences (2015: 295). Gache's and Moreno's enactment of poetry beyond language-specificity locates them at this stage of cultural production and suggests a possible avenue for further e-lit production and study.

References

Beecroft, Alexander (2015), *An Ecology of World Literature*, London and New York, NY: Verso.

Clüver, Claus (1987), "From Imagism to Concrete Poetry: Breakthrough or Blind Alley?" in Rudolf Haas (ed.), *Amerikanische Lyrik: Perspektiven und Interpretationen*, 113–30, Berlin: Erich Schmidt Verlag.

Clüver, Claus (2000), "Concrete Poetry and the New Performance Arts: Intersemiotic, Intermedial, Intercultural," in Claire Sponsler and Xiaomei Chen (eds.), *East of West: Cross-cultural Performance and the Staging of Difference*, 33–61, New York, NY: Palgrave.

Gache, Belén (2012), "Gongora Wordtoys," *Sociedad Lunar. Literatura Expandida*. January 16, 2020, http://belengache.net/gongorawordtoys/.

Grossman, Edith (2011), "Foreword," in Luis de Góngora (ed,), *The Solitudes: A Dual-Language Edition with Parallel Text*, trans. Edith Grossman, London: Penguin Books.

Halevi, Yael (1995), "The Rhythm of Góngora's *Soledades*: An Approach for Reading and Teaching," *Hispania* 78 (3): 463–73.

Hayles, N. Katherine (2008), *Electronic Literature: New Horizons for the Literary*, Notre Dame, IN: University of Notre Dame Press.

Moreno, Benjamín (2010), *Concretoons*, January 16, 2020, http://concretoons. centroculturadigital.mx/.

Pressman, Jessica (2014), *Digital Modernism: Making it New in New Media*, London and New York, NY: Oxford University Press.

Raley, Rita (2012), "Another Kind of Global English," *Minnesota Review* 78: 105–12.

Reis, Pedro (2015), "Portuguese Experimental Poetry-Revisited and Recreated," e-poetry 2007, Université Paris 8, March 14.

Smith, Paul Julian (1986), "Barthes, Góngora and Non-Sense," *PMLA* 10 (1): 82–94.

Solt, Mary Ellen (1970), *Concrete Poetry: A World View*, Bloomington, IN: Indiana University Press.

Tabbi, Joseph (2010), "Electronic Literature as World Literature; or The Universality of Writing under Constraint," *Poetics Today* 31 (1): 17–50.

4

Community, Institution, Database: Tracing the Development of an International Field through ELO, ELMCIP, and CELL

Davin Heckman

Publishing digital anthologies and databases, testing out new models of distribution, exhibition, and preservation, and building interdisciplinary collaboration not only between traditional academic disciplines but also between distinct international communities, one might think of electronic literature as the research and development wing of the digital humanities. This chapter will situate these three projects within an emergent institutionalization of network-based creative community in electronic

I owe a special debt of gratitude to Scott Rettberg for his collaboration on the draft of the proposal for this chapter and for his regular feedback throughout the writing process. More important than this is Rettberg's powerful role in all three institutions referenced in this piece, the ELO, ELMCIP, and CELL. Over the last several years, I have been fortunate to work very closely with all three projects, and have had a chance to see their operations from the inside and out. Without Rettberg's visionary leadership, collaborative spirit, and generous nature, the history of the field might have followed a very different path. I am convinced that the energy and creativity of this community would surely have coalesced into something significant, it's hard to imagine the field without these institutions.

literature. Specifically, this chapter will focus on the cluster of activity that circulates around three specific institutions that have sought to document the field as it has developed: the Electronic Literature Organization (ELO)—a literary nonprofit organization that has become central to the evolution of a community of creative and critical practice in the field; Electronic Literature as a Model of Creativity and Innovation in Practice (ELMCIP)—a three-year European research project which explored the domain of network-based creative community even as it produced research infrastructure including an Electronic Literature Knowledge Base to enable better mapping of the field; and, finally, the Consortium for Electronic Literature (CELL)—an international project created to bring shared search capabilities and common cataloging standards to electronic literature databases.

Electronic literature can be said to be a field federated through two models which may seem incoherent in terms of their relation to institutionalization. On the one hand, the practices of individual artists and authors can, to date, be understood as to being to large extent atomized and avant-garde. In comparison to the practices and institutions of contemporary print literature, ranging from commercial publishers, libraries, academic creative writing programs, and literature programs with long histories, it can be said that electronic literature authors have worked mostly in isolation, producing experimental literary works for which there has been virtually no commercial demand, few educational support structures, and even a lack of basic research infrastructure such as library cataloging standards.

Without the institutional legacy that supports the practices more commonly associated with the traditions of print literature, the field of electronic literature has attempted to leverage the concept of the "literary" from the print tradition to describe nonprint forms. See, for instance, the ELO's definition:

> Electronic literature, or e-lit, refers to works with important literary aspects that take advantage of the capabilities and contexts provided by the stand-alone or networked computer. Within the broad category of electronic literature are several forms and threads of practice ...[1]

Until recently, the organization listed specific forms, like hypertext fiction, kinetic poetry, interactive fiction, and others, as examples of the kinds of practice that might fit under the umbrella term with the understanding that the field will continue to evolve and new forms will emerge.

One might balk at the strategic circularity of "literary aspects" in this definition of "literature." But, in fact, it is an honest definition. In many

---

[1]"What is E-Lit?" Electronic Literature Organization, accessed on December 6, 2016, http://eliterature.org/what-is-e-lit/.

respects, the genres and forms that we envision when we discuss cultural practices like literature are the product of shared fictions. The forms themselves usually emerge as mutations of established practices, exploiting the recognized formalities of practice to carry little (and not so little) bursts of novelty into the lives of those that behold them. For instance, the pragmatics of the spoken word can be nudged ever so slightly into musicality through rhythm, rhyme, and consonance, to create linguistic constructions that sound slightly more pleasurable than more instrumental utterances. Static concepts can become lively though the use of metaphors. Words themselves are held in constant tension between what we understand them to be and what they can become. The literary can be understood the potential for emergence that exists wherever subjects seek to relate through expanded systems of signification.

The history of literature is an elaboration on this theme that carries across millennia and spans cultures. On top of this undulating landscape of instrumental words and their monstrous offspring, we deploy terms like poem, novel, sonnet, romance, fable, etc. These containers give names to the qualities of various waveforms, which at their most accurate, gesture towards vague, familial resemblances that are shared by handfuls of works, but that never fully account for their magic. At their worst, they describe desiccated formulae of banal repetition. Consider the difference in "dystopian fiction" that exists between, say, *The Left Behind* series and, say, Margaret Atwood's *The Handmaid's Tale*. While both might be categorized as dystopian speculations, one might be seen as formula, the other literature.

When we sit down to talk about literature, there are some general expectations of what we might encounter. We can envision an anthology, a great writer, a class we once took, a cinematic professor pontificating emphatically about the word's power over us. But in the end, we tend to use our general understanding of literature as a way of raising a set of expectations around writing. The works themselves, however, are not notable for their ability to conform to a generic stereotype. They get invited into the generalizing discourse of literature based, hopefully, on some exceptional claim they can make to this institutional definition. Which is to say, the categorical understandings themselves can frame our strategies of textual interpretation, but they cannot actually provide a definitive rubric for the apprehension of "the literary."

Electronic literature is no different, except, perhaps that its "institutional definition" arrives to us in a historical moment that is specifically reflexive of institutional definitions. The formal medium of presentation, the computer, is relatively new. This formal medium itself is undergoing rapid transformation (desktop, GUI, networked, mobile, haptic, distributed, and so forth). The means by which this medium makes its content present is highly malleable and unstable. The scholarly milieu into which this definition enters has an unsettled definition of literature.[2] The larger society as a whole is undergoing a massive shift in literacy, thanks first to the computer, the mobile, and whatever is next. As a result, electronic literature is and remains for the indefinite future, a shifting concept.

Nevertheless, given the availability of new technologies and the creative opportunities these enabled, writers have been drawn to the new media and the globally networked writing environment and have now managed to develop a corpus of work that can be read as a literature. If we can observe that these writers and the critics and theorists who have contextualized their works have not inherited a great deal of economic, social, or research infrastructure, we can also note that this very lack of community and resources has been generative. Lacking traditional publishers interested in distributing this work, elit writers have been driven to formulate new types of publishing entities. Lacking libraries and archives capable of cataloging, disseminating, and preserving these works, writers and scholars have reached out to libraries, archives, and database developers. Lacking meeting places within established academic organizations and traditional disciplinary frameworks, these actors have found opportunities to resituate their activities within new, emergent communities. So if the communities of creative and critical practice in the field of electronic literature were born in a networked digital environment largely free of the dispositif of print literary culture, over the past two decades, a remarkable amount of activity has taken place to develop infrastructure particular to the practices of electronic literature.

The past several decades of radical media change, and the interrogation and contextualization of these changes through an emerging digital arts practice, have outpaced the methodical practices of documentation, preservation, and criticism, resulting in an uneven record of this transformative period. In a 2011 keynote address at ISEA (the International Symposium for Electronic

---

[2]Certainly, Vannevar Bush's 1945 text, "As We May Think" in *The Atlantic*, anticipates an entirely new system of textual organization, symbolic communication, and daily life that the computer would unleash. Theodor Holm Nelson's "Complex Information Processing: A File Structure for the Complex, the Changing, and the Indeterminate" from the 1965 ACM conference introduces the term "Hypertext," fleshing out the implications of the digital computer for human textual expression. Running alongside these shifting technologies of language is the avant-garde tradition that sought new techniques of poetic and narrative language, leading through Stéphane Mallarmé, Virginia Woolf, Raymond Queneau, Jorge Luis Borges, Augusto de Campos, and others. It's fair to say that the reorganization of media imagined by people like Bush and Nelson and the deconstruction of the literary imagined by the avant-garde are part of the cultural shift into "postmodernism." That critics/philosophers like Roland Barthes, Marshall McLuhan, Michel Foucault, Jacques Derrida, Julia Kristeva, and others could identify profound shifts in subjectivity and knowledge during this same period. The net effect is an overdetermined resistance to an objective definition of the "literary." Vannevar Bush, "As We May Think," *The Atlantic*, July 1945, accessed December 6, 2016, http://www.theatlantic.com/magazine/archive/1945/07/as-we-may-think/303881/.

Art) entitled, "Media Art Explores Image Histories: New Tools For Our Field," media historian, Oliver Grau, identified this field as fertile ground for scholarship:

> Comparable with natural sciences, digital media and network research catapault the humanities within reach of new and essential research tools (Wikipedia might be a glimpse of what is possible) and what we need are collective documentation and preservation tools for media art. Or, even better, tools which can manage an entire history of visual media and human perception by means of thousands of sources.[3]

Beyond simply recording the changing tools (as platforms and devices vie for market dominance), Grau sees the potential for revolutionary change in knowledge practices, explaining, "Documentation changes from a one-way archiving of key data to a pro-active process of knowledge transfer."[4] It is critical, then, not only to have a community of practice and organizations that recognize this practice, but to develop archival strategies with which one can learn, criticize, and advance the field.

Before entering with a discussion of the human institutions that have played a powerful role in the advancement of the field, it is necessary to meditate on the "electronic" component of this community. In fact, it is hard to discuss the success of the community without addressing a key structural advantage: the disruptive nature of its chosen medium. The very medium which often estranges electronic literature from its consideration by literary institutions, has also been a key to its dynamism and success. Its members have, in many cases, created and controlled the texts that they have written. Likewise, this community has built networks of distribution, collaboration, and critique that can step outside the concentrated power of traditional institutions. A lone digital poet working against the advice graduate advisors in Bowling Green, Ohio, for instance, can immediately find works, readers, venues, and colleagues around the world. While many know the alienation of seeing their work disregarded or even insulted by colleagues in their immediate geographical proximity, this experience has nurtured feelings of affinity among those with shared enthusiasm for the field. The ability to publish and circulate one's work digitally, particularly in the early days of the web when such connections had a serendipitous quality to them, is part of the culture of the community. While it would be naïve to say that electronic literature is "free," it is important to note that its production costs, systems of distribution, and economics are different

---

[3]Oliver Grau, "Media Art Explores Image Histories: New Tools for Our Field" ISEA2011, Istanbul, Turkey, 2011, accessed December 6, 2016, https://vimeo.com/35194212.
[4]Ibid.

from the tradition that precedes it. This dynamic has allowed an insurgent community of authors, readers, and scholars to circumvent the inertia that complements the formal conservativism of institutions that focus on print media. This alterity of practice often finds itself matched in the avant-garde form and content of the works themselves.

A watershed moment in the history of the field, then, might very well be the decision to form an institution around such an unfixed practice. Of course, we cannot discount the specific crystallizations of resources around specific forms (for instance, the flurry of innovation around Eastgate Systems, Inc., the first publishing house for literary hypertext in 1987 or the formation of Lecture Art Innovation Recherche Ecriture in 1988). But electronic literary practice has advanced through the relationship between communities, institutions, and databases. The development of the field remains an ongoing project, but the path through ELO, ELMCIP, and CELL offers a unique view on how experimental work in the humanities can mature into an international field.

Electronic Literature Organization (ELO)

The Electronic Literature Organization (ELO) was founded in 1999 by Scott Rettberg, Robert Coover, and Jeff Ballowe to serve as a nonprofit advocate for the study, preservation, and promotion of this emerging practice. That same year, the ELO developed a wiki-based Electronic Literature Directory (ELD) to provide records of works in the field. A decade after the release of the first incarnation of the Directory, the ELD was rebuilt and a peer-review process was implemented.[5] The Electronic Literature Directory Working Group has been a significant partner in ongoing discussions with similar database projects, and has taken on a leading role, under Joseph Tabbi, in the eventual formation of the Consortium (CELL).

In addition to the Directory, the ELO is responsible for the *Electronic Literature Collection*, an edited anthology of "born-digital" texts. The first volume of the *ELC* was published in 2006, a second in 2011, with a third under production. Each volume of the *ELC* can be found on the Organization's website.[6] The scope of the *ELC* is ambitious, each containing a broad selection of edited work by a rotating cast of artists and scholars active in the field, creating a competitive venue for publication that

---

[5] A key contribution to the development of ELD 2.0 was a Start-Up Grant from the National Endowment for the Humanities, both a validation of the ongoing strength of the e-lit community and a contribution to its growth.

[6] *Electronic Literature Collection*, Vols. 1–3, accessed December 5, 2016, http://collection. eliterature.org.

nevertheless manages to provide a sample of exemplary work in an ever-expanding landscape of creative activity.

The ELO has published a number of key texts of use to artists and scholars. These include *State of the Arts: The Proceedings of the 2002 Electronic Literature Organization Symposium* (2003, edited by Rettberg); Noah Wardrip-Fruin and Nick Montfort's "Acid-Free Bits: Recommendations for Long-Lasting Electronic Literature" (2004); Alan Liu, David Durand, Nick Montfort, Merrilee Proffitt, Liam R. E. Quin, Jean-Hugues Réty, and Noah Wardrip-Fruin's "Born-Again Bits: A Framework for Migrating Electronic Literature" (2005); N. Katherine Hayles' "Electronic Literature: What is it?" (2007); Joseph Tabbi's "Toward a Sematic Literary Web" (2007); and Hayle's *Electronic Literature: New Horizons for the Literary* (2008).[7]

Since the ELO's founding, the Organization has held a multitude of conferences.[8] These conferences include scholarly papers, roundtable discussions on current issues in the field, artists' talks, gallery exhibitions, performances, and workshops. In 2012, the Organization adopted a schedule that includes annual conferences, with strong commitment to the international identity of the field. In between the conferences, the organization sponsors exhibitions, readings, and panel discussions at related events.

A key aspect of the ELO's ambitious portfolio of accomplishments is the strength of the overall field it represents. Articles developed from

---

[7]N. Katherine Hayles, "Electronic Literature: What is it?" Electronic Literature Organization, last modified 2007, accessed December 6, 2016, http://eliterature.org/pad/elp.html; N. Katherine Hayles, *Electronic Literature: New Horizons for the Literary* (South Bend: University of Notre Dame Press, 2008); Alan Liu, David Durand, Nick Montfort, Merrilee Proffitt, Liam R. E. Quin, Jean-Hugues Réty, and Noah Wardrip-Fruin, "Born-Again Bits: A Framework for Migrating Electronic Literature," Electronic Literature Organization, 2005, accessed December 5, 2016, http://eliterature.org/pad/bab.html; Scott Rettberg, editor, *The State of the Arts: The Proceedings of the 2002 Electronic Literature Organization Symposium,* Electronic Literature Organization, 2003, accessed December 6, 2016, http://eliterature.org/state/; Joseph Tabbi, "Toward a Sematic Literary Web," Electronic Literature Organization, 2007, accessed December 6, 2016, http://eliterature.org/pad/slw.html; Noah Wardrip-Fruin and Nick Montfort, "Acid-Free Bits: Recommendations for Long-Lasting Electronic Literature," Electronic Literature Organization, 2004, accessed December 6, 2016, http://eliterature.org/pad/afb.html.
[8]Sites include University of Porto, Portugal (2017); University of Victoria, Canada (2016); University of Bergen, Norway (2015); University of Milwaukee, USA (2014); Paris, France (2013); West Virginia University, USA (2012); Brown University, USA (2010); Washington State University Vancouver, USA (2008); University of Maryland, USA (2007); University of California Santa Barbara, USA (2003); and University of California Los Angeles, USA (2002). "History," Electronic Literature Organization, accessed December 5, 2016, http://eliterature.org/elo-history/.

conference presentations have seen publication in venues like *Electronic Book Review*, *Hyperrhiz*, *Digital Humanities Quarterly*, *Leonardo Electronic Almanac*, *Formules*, *Dichtung Digital*, and as standalone publications.[9] Authors who have shared creative work within the ELO network have often enjoyed broad success in other venues as well.[10] Furthermore, the portfolio of tools offered by the ELO (white papers, collections, conference proceedings, etc.) are often used as reference points by many of these same individuals. This is not to say that the ELO is the determining factor of success in these cases, but to say that the Organization has consistently been a meeting place/transmission point for many of the most talented and creative participants in this emerging field. This fact speaks to the intensifying effects that occur where communities of practitioners, institutions, and archives converge to create sustainable cultural practices. And, when few other institutional support structures exist to cultivate this activity, such meeting points are critical for sharing, critiquing, and improving practice.

Electronic Literature as a Model of Community in Practice

Initiated in 2010 with funding from the Humanities in the European Research Area (HERA) JRP for Creativity and Innovation, Electronic Literature as a Model of Creativity and Innovation in Practice (ELMCIP) is a multinational

---

[9]Notable critical publications from the past two years include Sandy Baldwin, *The Internet Unconscious* (New York: Bloomsbury, 2015); Alice Bell, Astrid Ensslin, and Hans Rustad, *Analyzing Digital Fiction* (New York: Routledge, 2014); Dave Ciccoricco, *Refiguring Minds in Narrative Media* (Omaha: University of Nebraska Press, 2015); Lori Emerson, *Reading Writing Interfaces* (Minneapolis: University of Minnesota Press, 2014); Astrid Ensslin, *Literary Gaming* (Cambridge: MIT Press, 2014); Chris Funkhouser, *New Directions in Digital Poetry* (New York: Bloomsbury, 2014); Jessica Pressman, *Digital Modernism* (New York: Oxford University Press, 2014); Jessica Pressman, Mark Marino, and Jeremy Douglas, *Reading Project* (Iowa City: University of Iowa Press, 2015); Marie-Laure Ryan, Lori Emerson, and Benjamin Robertson, *Johns Hopkins Guide to Digital Media* (Baltimore: Johns Hopkins University Press, 2014).

[10]Some prominent examples include notable authors' archives at the Harry Ransom Center (Michael Joyce), Duke University (Stephanie Strickland), and Maryland Institute for Technology in the Humanities (Deena Larsen). Other prominent examples include print and digital publications of writers like J. R. Carpenter, Nick Montfort, Andy Campbell, Kate Pullinger, Mez Breeze, Scott Rettberg, Mark Amerika, Shelley Jackson, Jason Nelson, Steve Tomasula, and many others.

research project with seven institutional partners in six European nations.[11] The goal of the project was to "investigate how creative communities of practitioners form within a transnational and transcultural context in a globalized and distributed communication environment."[12] Over a three-year period, ELMCIP explored the relationship between community and creative practice in Electronic Literature, producing 179 directly affiliated project publications, six seminars, a conference, an exhibition, an anthology of creative works, two books, and the world's largest research database in the field of electronic literature: the ELMCIP Knowledge Base.[13] Beyond these clearly identifiable outcomes, this period of intense activity amplified the ongoing efforts of the ELO, providing the stimulus for new scholarship, publications, creative works, and relationships.

An especially useful contribution to understanding the history of the field is the ELMCIP project's formal relationship with the journal *Dichtung Digital*. *Dichtung Digital* 41 and 42, both published in 2012, provide specific accounts of genres and communities of electronic literary writing that often are not entirely integrated with institutional actors identified in this chapter. For instance, Philippe Bootz's "From OULIPO to Transitoire Observable: The Evolution of French Digital Poetry" established a lineage for digital poetry in France that begins with OULIPO in the 1960s and runs through ALAMO in the 1980s.[14] Serge Bouchardon's "Digital Literature in France," while acknowledging many of the same antecedents as Bootz, focuses primarily on the community of writers organized around the *e-critures* mailing list and website.[15] Meanwhile, Laura Borràs Castanyer's "Growing Up Digital" focuses on the emergence of electronic literary practices in Spain, which draws a lineage through centuries of constrained, experimental

[11]These partners include the University of Bergen (Norway; Project Leader: Scott Rettberg; and Co-investigator: Jill Walker Rettberg), the Edinburgh College of Art (Scotland; Principal investigator: Simon Biggs; Co-investigator: Penny Travlou), Blekinge Institute of Technology (Sweden; Principal investigator: Maria Engberg; Co-investigator: Talan Memmott), the University of Amsterdam (Netherlands; Principal investigator: Yra Van Dijk), the University of Ljubljana (Slovenia; Principal investigator: Janez Strehovec), the University of Jyväskylä (Finland; Principal investigator: Raine Koskimaa), University College Falmouth at Dartington (England; Principal investigator: Jerome Fletcher), New Media Scotland (Scotland; Mark Daniels). "Partners," *ELMCIP*, accessed December 6, 2016, http://elmcip.net/page/partners.
[12]Ibid.
[13]For a more detailed account, see ELMCIP, accessed December 7, 2016, http://elmcip.net.
[14]Philippe Bootz, "From OULIPO to Transitoire Observable: The Evolution of French Digital Poetry," *Dichtung Digital* 41 (2012), accessed December 6, 2016, http://www.dichtung-digital.org/2012/41/bootz.htm.
[15]Serge Bouchardon, "Digital Literature in France," *Dichtung Digital* 41 (2012), accessed December 6, 2016, http://www.dichtung-digital.org/2012/41/bouchardon.htm.

writing in the Spanish print tradition.[16] Hans Kristian Rustad's contribution maps the development of the field in the Scandinavian countries.[17] Scott Rettberg's "Developing an Identity for the Field of Electronic Literature" provides a genealogy of the field as it emerged in the context of the ELO, which was initially rooted in a network of North American authors and creators.[18]

In addition to the geographical and national accounts of electronic literature communities, a number of essays address digital writing communities formed around specific forms, practices, and relationships. Loss Pequeño Glazier provides an account of the field that is strongly focused on the poetic tradition and that leads through the international community of writers affiliated with the Electronic Poetry Center and the international E-Poetry festivals that this network of writers participate in.[19] Nick Montfort and Emily Short focus on the thriving community of writers writing under the umbrella of Interactive Fiction (or IF).[20] And finally, Jill Walker Rettberg takes a "distant" view of electronic literature, sketching out the contours of the field as perceived through the lens of digital analytics.[21] While the dual issues of *Dichtung Digital* amount to more than mere documents of community practice, when seen from the perspective of this chapter as creative communities, this output of the ELMCIP project represents an interweaving of multiple sub-communities of practice and help us illustrate the emergence of electronic literature as a global phenomenon.

Though it would be difficult to assess the relative value of any one piece of the ELMCIP project, the Knowledge Base initiative is certainly the most visible legacy of this experiment. It serves as an entryway to the project, yet the genius of this open-source tool is the way that it serves to map

---

[16]Laura Borràs Castanyer, "Growing up Digital: The Emergence of E-Lit Communities in Spain. The Case of Catalonia 'And the Rest is Literature,'" *Dichtung Digital* 42 (2012), accessed December 6, 2016, http://www.dichtung-digital.de/en/journal/archiv/?postID=620.

[17]Hans Kristian Rustad, "A Short History of Electronic Literature and Communities in the Nordic Countries," *Dichtung Digital* 41 (2012), accessed December 6, 2016, http://www.dichtung-digital.org/2012/41/rustad.htm.

[18]Scott Rettberg, "Developing an Identity for the Field of Electronic Literature: Reflections on the Electronic Literature Organization Archives," *Dichtung Digital* 41 (2012), accessed December 6, 2016, http://www.dichtung-digital.org/2012/41/rettberg.htm.

[19]Loss Pequeño Glazier, "Communities/Commons: A Snap Line of Digital Practice," *Dichtung Digital* 42 (2012), accessed December 6, 2016, http://www.dichtung-digital.de/en/journal/archiv/?postID=540.

[20]Nick Montfort and Emily Short, "Interactive Fiction Communities: From Preservation through Promotion and Beyond," *Dichtung Digital* 41 (2012), accessed December 6, 2016, http://www.dichtung-digital.org/2012/41/montfort-short.htm.

[21]Jill Walker Rettberg, "Electronic Literature Seen from a Distance: The Beginnings of a Field," *Dichtung Digital* 41 (2012), accessed December 6, 2012, http://www.dichtung-digital.org/2012/41/walker-rettberg.htm.

out the network of relations within the creative, critical, and institutional dimensions of the field. All works documented in the Knowledge Base can be linked to the works that they reference, the works that reference them, creators, publishers, venues, events, syllabi, etc. The result is a very rich framework that invites user participation to flesh out the field by connecting discrete artifacts and moments of production with the relevant relations to the rest of the field. In a relatively short period of time (and with significant stewardship provided by Scott Rettberg, Eric Dean Rasmussen, Elisabeth Nesheim, and Stein Magne Bjørklund), the Knowledge Base became the richest survey of the field (and continues to grow to this day).

In addition to the information stored in the Knowledge Base, ELMCIP's leadership in this area (open access, collaborative, and open source) has provided a useful template for others to adopt and develop. This backbone of cooperation, in conjunction with the ELO's similar efforts, have provided the proper context for the formation of the third institution discussed in this chapter: the Consortium on Electronic Literature.

Consortium on Electronic Literature (CELL)

In the first decade of the twenty-first century, a number of other institutions with database projects emerged to join ELMCIP and the ELO to cover aspects of the growing field of electronic literature. Along with the editorial leadership of *Electronic Book Review*, which has been an advocate for experimental publishing and scholarship since its inception in 1994, a critical mass quickly cohered around the practice of electronic literature.

In 1999, Hermeneia: The Literary Studies and Digital Technologies Research Group was initiated in Spain. Since 2003, ADEL (originally LIKUMED), a project from the University of Siegen, has been documenting works of electronic literature in German. Beginning in 2005, the Canadian NT2 has been building a database to study hypermedia art and writing. The Brown Digital Repository, which began in 2008, catalogues and documents works of digital writing from Brown University.[22] In 2006, Po-ex.net (Digital Archive of Portuguese Experimental Literature) was founded to document Portuguese experimental and electronic literature from the 1960s

---

[22]The Language Arts program at Brown has been a leader in the exploration of electronic literary forms, often in collaboration across disciplines and institutions. Notable alumni include Mark Amerika, Shelley Jackson, Alan Sondheim, Noah Wardrip-Fruin, Talan Memmott, Brian Kim Stefans, Daniel Howe, William Gillespie, Aya Karpinska, Justin Katko, Judd Morrissey, Ian Hatcher, Claire Donato, Samantha Gorman, and others. See: "Digital Language Arts," Brown Digital Repository, accessed December 7, 2016, https://repository.library.brown.edu/studio/collections/id_462/.

to the 1980s. In 2010, I ♥ E-Poetry, initially the product of an individual scholar, Leonardo Flores, began with daily reviews of works of electronic literature. An Australian Database, ADELTA (originally Creative Nation), and a Spanish database, Ciberia, are currently in development.[23]

The NT2 lab in Montreal is responsible for the technical development of the search tool, which will ultimately require all partners to adopt common metadata standards and shared taxonomy, and will allow scholars to conduct refined searches across all partner databases and, eventually, to deploy new analytic tools over thousands of records.[24]

Driven by an array of complementary tendencies that circulate in the discourse around the field, the concept of a consortium of database projects was perhaps as overdetermined as the field's resistance to a simple definition of electronic literature. For instance, Joseph Tabbi envisioned *Electronic Book Review* in light of digital space: "This is the late age of print we're in, when all the books worth saving are being scanned into digital archives, and the very conception of the book as a fixed object is giving way to the hyperreality of letters floating on a screen."[25] This mood of excitement about the creative and critical potential of digital spaces remains as a key motivation for these institutions to share their knowledge of the field and extend the reach of their constituencies. At the same time, Rettberg's commitment to sharing data as a basic principle for computing and a strategy for longevity (evident in his contributions to both ELMCIP's Knowledge Base and the ELO's wiki-based iteration of the ELD) was an equally compelling impetus for collaboration. A third theme, expressed by Heckman in "The Disturbed Dialectic of Literary Criticism in an Age of Innovation," is the idea that collecting our efforts might offset the disruptive effects of speed on our

---

[23]A number of meetings have been critical to the evolution of CELL. Meeting sites include the LitNet project in Siegen (Winter 2008), the Maryland Institute of Technology in the Humanities (Summer 2008), Washington State–University Vancouver (Summer 2009), the University of Colorado, Boulder (Winter 2009), Brown University (Summer 2010), the University of Western Sydney (Winter 2010), the University of Bergen (Summer 2011), West Virginia University (Summer 2012), Paris 8 (Summer 2013), and the University of Wisconsin–Milwaukee (Summer 2014), University of Bergen (Summer 2015).

[24]The CELL project is supported financially through an array of sources: NEH Digital Humanities Start-up grant; Programme Québec-États-Unis by the Ministère des Relations internationales, de la Francophonie et du Commerce extérieur du Québec; Center for Literary Computing at West Virginia University, including a 2014 PSCoR Grant; the Electronic Literature Organization; the University of Bergen; and the University of Siegen. All members of the Consortium contribute to the overall success of the project. For more information on projects affiliated with the Consortium on Electronic Literature, see: "Members," *CELL: Consortium on Electronic Literature*, accessed December 5, 2016, http://cellproject.net/members.

[25]Joseph Tabbi, "ebr version 1.0: Winter 1995/96," *Electronic Book Review* (December 1995), accessed December 6, 2012, http://electronicbookreview.com/thread/electropoetics/manifesto.

processes of care and attention to cultural practices.[26] By identifying these particular forces driving the formation of the consortium, I do not mean to suggest that individuals played solitary roles that were strictly necessary to the success of the endeavor. I think it is fair to say that all participants are driven by an excitement for the potential opened up by the digital computer. Similarly, all seem equally informed by the recognition that open-source code, shared data, and a spirit of generous collaboration are key ethics for scholars working in a digital age. And, finally, the recognition that a common search tool would serve a key role in protecting our objects of study for the kind of careful criticism and deep appreciation appropriate for significant works of art.

Conclusion

One can look across the documents of any of the institutions covered in this chapter—the ELO, ELMCIP, and CELL—one can look at any member of CELL—and see the common enthusiasm for electronic literature, like DNA, genetically expressed throughout. The coalescence of these common impulses to create, read, and critique have provided a healthy foundation for the formation of a community practice as evidenced by its capacity to adopt an "institutional" character, or a collective identity that itself can be shared among its members (and, more importantly, is considered worth sharing among its members).

Although such institutions are formed from unique conditions giving rise to their existence, the next logical question becomes whether or not singular institutions belong to a broader arc of existence. Here, these three institutions have all answered questions of their historicity and futurity. The reason for this is simple: without broadly held definitions and in the absence of legacy institutions, the field of electronic literature has asked itself whether it has a history worth preserving and whether it has a future worth anticipating. In both cases, the collective response has been yes. And the strategies by which the field has launched itself forward and established its historicity are clear: to expand their networks to include more partners, to plan for the future by establishing databases and archives, and to refine its practice through critical reflection on these networks and databases.

In less than a generation, we can see in the emergence of a robust field of practice, nurtured through a strong cultural model founded on respect for

---

[26]Davin Heckman, "The Disturbed Dialectic of Literary Criticism in an Age of Innovation," *Leonardo Electronic Almanac* (November 2014), accessed December 6, 2016, http://www.leoalmanac.org/disturbed-dialectic/.

individual (and often idiosyncratic) contributions of artists, the engagement of an enthusiastic (and responsibly critical) collective, and the development of technologies of institutional preservation.

References

Baldwin, Sandy (2015), *The Internet Unconscious*, New York, NY: Bloomsbury.

Bell, Alice, Astrid Ensslin, and Hans Rustad (2014), *Analyzing Digital Fiction*, New York, NY: Routledge.

Bootz, Philippe (2012), "From OULIPO to Transitoire Observable: The Evolution of French Digital Poetry," *Dichtung Digital* 41, accessed December 6, 2016, http://www.dichtung-digital.org/2012/41/bootz.htm.

Bouchardon, Serge (2012), "Digital Literature in France," *Dichtung Digital* 41, accessed December 6, 2016, http://www.dichtung-digital.org/2012/41/bouchardon.htm.

Bush, Vannevar (1945), "As We May Think," *The Atlantic*, July, http://www.theatlantic.com/magazine/archive/1945/07/as-we-may-think/303881/.

Castanyer, Laura Borràs (2012), "Growing up Digital: The Emergence of E-Lit Communities in Spain. The Case of Catalonia 'And the Rest is Literature'," *Dichtung Digital* 42, accessed December 6, 2016, http://www.dichtung-digital.de/en/journal/archiv/?postID=620.

CELL: Consortium on Electronic Literature (n.d.), accessed December 5, 2016, http://cellproject.net.

Ciccoricco, David (2015), *Refiguring Minds in Narrative Media*, Omaha: University of Nebraska Press.

"Digital Language Arts" (n.d.), Brown Digital Repository, accessed December 7, 2016, https://repository.library.brown.edu/studio/collections/id_462/.

Electronic Literature Collection, Vols. 1–3 (n.d.), accessed December 5, 2016, http://collection.eliterature.org.

Emerson, Lori (2014), *Reading Writing Interfaces*, Minneapolis, MN: University of Minnesota Press.

Ensslin, Astrid (2014), *Literary Gaming*, Cambridge, MA: MIT Press.

Funkhouser, Chris (2014), *New Directions in Digital Poetry*, New York, NY: Bloomsbury.

Glazier, Loss Pequeño (2012), "Communities/Commons: A Snap Line of Digital Practice," *Dichtung Digital* 42, accessed December 6, 2016, http://www.dichtung-digital.de/en/journal/archiv/?postID=540.

Grau, Oliver (2011), "Media Art Explores Image Histories: New Tools for Our Field," ISEA2011, Istanbul, Turkey, accessed December 6, 2016, https://vimeo.com/35194212.

Hayles, N. Katherine (2007), "Electronic Literature: What is it?" Electronic Literature Organization, Last modified 2007, accessed December 6, 2016, http://eliterature.org/pad/elp.html.

Hayles, N. Katherine (2008), *Electronic Literature: New Horizons for the Literary*, South Bend, IN: University of Notre Dame Press.

Heckman, Davin (2014), "The Disturbed Dialectic of Literary Criticism in an Age of Innovation," *Leonardo Electronic Almanac* (November), accessed December 6, 2016, http://www.leoalmanac.org/disturbed-dialectic/.
"History" (n.d.), Electronic Literature Organization, accessed December 5, 2016, http://eliterature.org/elo-history/.
Liu, Alan, David Durand, Nick Montfort, Merrilee Proffitt, Liam R. E. Quin, Jean-Hugues Réty, and Noah Wardrip-Fruin (2005), "Born-Again Bits: A Framework for Migrating Electronic Literature," Electronic Literature Organization, accessed December 5, 2016, http://eliterature.org/pad/bab.html.
"Members" (n.d.), *CELL: Consortium on Electronic Literature*, accessed December 5, 2016, http://cellproject.net/members.
Montfort, Nick and Emily Short (2012), "Interactive Fiction Communities: From Preservation through Promotion and Beyond," *Dichtung Digital* 41, accessed December 6, 2016, http://www.dichtung-digital.org/2012/41/montfort-short.htm.
"Partners" (n.d.) *ELMCIP*, accessed December 6, 2016, http://elmcip.net/page/partners.
Pressman, Jessica (2014), *Digital Modernism*, New York, NY: Oxford University Press.
Pressman, Jessica, Mark Marino, and Jeremy Douglas (2015), *Reading Project*, Iowa City, IA: University of Iowa Press.
Rettberg, Jill Walker (2012), "Electronic Literature Seen from a Distance: The Beginnings of a Field," *Dichtung Digital* 41, accessed December 6, 2016, http://www.dichtung-digital.org/2012/41/walker-rettberg.htm.
Rettberg, Scott (ed.) (2003), *State of the Arts: The Proceedings of the 2002 Electronic Literature Organization Symposium*, Electronic Literature Organization, accessed December 6, 2016, http://eliterature.org/state/.
Rettberg, Scott (2009), "Communitizing Electronic Literature," *Digital Humanities Quarterly* 3 (2), accessed December 6, 2016, http://digitalhumanities.org/dhq/vol/3/2/000046/000046.html#.
Rettberg, Scott (2012), "Developing an Identity for the Field of Electronic Literature: Reflections on the Electronic Literature Organization Archives," *Dichtung Digital* 41, accessed December 6, 2016, http://www.dichtung-digital.org/2012/41/rettberg.htm.
Rettberg, Scott (2014), "Developing a Network-Based Creative Community: An Introduction to the ELMCIP Final Report," in Sandy Baldwin and Scott Rettberg (eds.), *Electronic Literature as a Model of Creativity and Innovation in Practice: A Report from the HERA Joint Research Project*, 1–38, Morgantown: West Virginia University Press, accessed December 6, 2016. http://elmcip.net/sites/default/files/files/attachments/criticalwriting/elmcip_1_introduction.pdf.
Rettberg, Scott and Patricia Tomaszek (2012), "Editorial: Electronic Literature Communities, Part I," *Dichtung Digital* 41, accessed December 6, 2016, http://dichtung-digital.de/editorial/2012_41.htm.
Rettberg, Scott and Patricia Tomaszek (2012), "Editorial: Electronic Literature Communities, Part II," *Dichtung Digital* 42, accessed December 6, 2016, http://elmcip.net/critical-writing/editorial-electronic-literature-communities-part-ii.
Rustad, Hans Kristian (2012), "A Short History of Electronic Literature and Communities in the Nordic Countries," *Dichtung Digital* 41, accessed December 6, 2016, http://www.dichtung-digital.org/2012/41/rustad.htm.

Ryan, Marie-Laure, Lori Emerson, and Benjamin Robertson (2014), *Johns Hopkins Guide to Digital Media*, Baltimore, MD: Johns Hopkins University Press.
Tabbi, Joseph (1995), "ebr version 1.0: Winter 1995/96," *Electronic Book Review* December, accessed December 6, 2012, http://electronicbookreview.com/thread/electropoetics/manifesto.
Tabbi, Joseph (2007), "Toward a Sematic Literary Web," Electronic Literature Organization, accessed December 6, 2016, http://eliterature.org/pad/slw.html.
Wardrip-Fruin, Noah and Nick Montfort (2004), "Acid-Free Bits: Recommendations for Long-Lasting Electronic Literature," Electronic Literature Organization, accessed December 6, 2016, http://eliterature.org/pad/afb.html.
"What is E-Lit?" Electronic Literature Organization, accessed December 6, 2016. http://eliterature.org/what-is-e-lit/.

5

The E-Poetry Festivals: Celebration, Art, and Imagination in Community

Loss Pequeño Glazier

The E-Poetry Festivals[1] was the first festival series conceived to celebrate literature's emergence in digital form; under the direction of the Electronic Poetry Center, Dept. of Media Study, SUNY Buffalo and in collaboration with numerous sister organizations, it consists of a series of international media poetics gatherings that have occurred over the past two decades in locations worldwide. Though other efforts existed before, in parallel, and in other manners responding to the field, no other organization can claim to have pre-dated the E-Poetry Festivals as an ongoing format and as a consistent conceptual frame. Though a curated event, founded and consistently directed by Loss Pequeño Glazier, it has largely been collective in spirit. Thus, diverse e-poetry community members and the EPC Advisory Board, among others, without whom the continuity and character of the E-Poetry Festivals, across more than a decade, could not have been sustained.

The festivals have made an indelible presence on the field. Not only did it inaugurate a new sensibility towards language as art in the digital age but it has always been focused in its attention to artistic expression.

---

[1]http://writing.upenn.edu/epc/e-poetry/archive/.

The festivals were always about art as a locus of action in and of itself, not in relation to the academy, to canons, to grant agencies, to corporate apparatuses, or to the economic dot-con [stet] frenzy for irresponsible amounts of corporate wealth. Among other firsts, E-Poetry was the first to offer an electronic literature festival in the United States, the first to present one in Europe, the first to sponsor an event in the Caribbean, the first to bring a festival to Latin America, and the first to cross numerous gender, language, literary, and cultural thresholds.

At the outset, however, I want to make it clear that the E-Poetry Festivals, from the start, were conceived of as "festivals." That is, "festivals" offered an occasion to celebrate, to present experiments, to exhilarate in the sculpted contours of new media formations. At this level, there was no hard-knuckled boasting nor chest-thumping nor posturing about these works as being harbingers of a "new canon" nor of this being a professional field or an organization to represent practitioners in a field defined by any given technology.

In this regard, the E-Poetry Festivals have also been distinct in presentation: no simultaneous panels are presented (if you present, you present to all); no keynote speakers are presented (all artists are equally presenting notes that are "key"), and, as with almost every exhibition and gallery or art or music festival in the world, the series is "curated." Importantly, it is dedicated to bringing together participants from diverse geographic areas, language, and cultural contexts; it aims to encourage younger, emerging practitioners and women artists; to explore possibilities of performance; and, to foster conversation, intergeneration exchange, and an international perspective, with multiple language formations in mind.

E-Poetry Festivals: An Inventory

An annotated inventory, such as the following, allows us to see not just the geographical contours of path of the E-Poetry landscape but, as viewed from one perspective, suggests that the entire series of E-Poetry Festivals aims for a "total immersion" in the digital as part of an overall effort: that is, allowing for the context of related defining events, the E-Poetry Festivals describe an arc that is an idea, meditative, spiritual, and contemplative. Like an indigenous ritual or a work of Latin American magical reality, the Festivals view the changing field from a larger perspective, that of a total vision of literature as a continuous process of an emergent process. However, it could be said that the emphasis on E-Poetry has been on the art as practice (and criticism as observation of practice) rather than on any specific medium.

The First Decade

E-Poetry's first iteration occurred at a conference center at SUNY Buffalo (itself the site of so many literary firsts) in May, 2001. Through 2015, the E-Poetry Festival series has had eight iterations, starting in Buffalo and culminating in Buenos Aires, Argentina. All have been under consistent directorship, given occasional slippages and adjustments along the way. Nonetheless, each of the E-Poetry Festivals contribute to a clear, complete vision of the field. (More detailed resources for most of the E-Poetry events is available online at http://writing.upenn.edu/epc/e-poetry/archive/.) Keep in mind that E-Poetry has been held on a regular basis, in a consistent format, and in the spirit of a festival since its inception.

I. E-Poetry 2001 Buffalo

Local organizers: Loss Pequeño Glazier and Ed Taylor; hosted by the Electronic Poetry Center, SUNY Buffalo and Just Buffalo Literary Center.

Certainly, there were precedents to E-Poetry. There was the work of Eduardo Kac, Jim Rosenberg, and John Cayley, notably in the original edition of their historic collection (1996) *New Media Poetry* published as *Visible Language*, Vol. 30, No. 2.[2] There were the events of the ACM Hypertext conferences and Digital Arts & Culture conferences that preceded E-Poetry. ACM Hypertext had some helpful moments. Nonetheless, E-Poetry was the first event to dedicate a full festival to literary activity in the field. It sought specifically to investigate, as did *Digital Poetics: the Making of E-Poetries* (Alabama University Press, 2002—the first university press book to do the same) to deal with the digital as literary space. Though digital poetry has been part of other digital conferences, "E-Poetry, 2001: An International Digital Poetry Festival" was a historic, landmark literary event as well as a coming-of-age event for new practices in digital literature and the first such event to dedicate itself entirely to the contemplation of digital literature as a topic in itself.

II. E-Poetry 2003 Morgantown

Local organizer: Sandy Baldwin; hosted by West Virginia University.

E-Poetry 2003 was, thanks to its organizer's appreciative spontaneity, a spin-off of the glorious communal energy of E-Poetry 2001. E-Poetry 2003 took place at West Virginia University, Morgantown West Virginia,

---

[2]http://visiblelanguagejournal.com/issues/issue/110/.

organized by Sandy Baldwin. Baldwin had appeared in E-Poetry 2001 as part of the Purkinge Group presenting "The Awopbop Groupuscle and the Forms of Improvisation," a group based in Albany, NY, which included Baldwin, Don Byrd, Nancy Dunlop, Chris Funkhouser, Belle Gironda, Thomas Mackey, Christina Milletti, and Derek Owens, among others, one of the evenings events held at Hallwalls Contemporary Arts Center, Tri-Main Center, in Buffalo, as part of the festival. The second iteration of the festival was done in close collaboration with me, and Baldwin did a splendid job as local organizer. In addition, the French-based international group, Transitoire Observable, made its first international presentation at E-Poetry 2003, adding a distinctive contour to the series.

III. E-Poetry 2005 London

Local organizers: Piers Hugill, William Rowe, John Cayley; hosted by Birkbeck College, University of London.

For its 2005 festival, E-Poetry took a decisive turn across the Atlantic. E-Poetry 2005 was the first ongoing digital literature festival in Europe. The event was a curious collaboration, through the agency of foundational digital practitioner, John Cayley, between E-Poetry, SUNY Buffalo, the University at Buffalo, and Birkbeck College. Cayley deserves immense amounts of credit for making the necessary institutional connections. The festival could not have enjoyed a more celebratory occasion than its setting at the University of London campus in central London. Grad student Piers Hugill was an amiable collaborator and it was an honor to coordinate with Birkbeck College, and the acclaimed Professor William Rowe. It was exciting to think that there would be a connection with the legendary Contemporary Poetics Research Centre (CPRC), Birkbeck College, London. Rowe and Hugill were part of the experimental poetry scene. Without a doubt, local organizer John Cayley not only organized excellently, but his micro-publisher background was a tribute to the origins of the E-Poetry Festivals concept: the idea of language art produced through the technology at hand (in this case, digital), material born a sense of urgency because of its value to its authors themselves. These are works outside the institution, emanating from aesthetics rather than organizational objectives. E-Poetry's renaissance in Europe occurred in the true spirit of its founding vision.

IV. E-Poetry 2007 Paris

Local organizer: Philippe Bootz, hosted by Le Laboratoire Paragraphe, Université de Paris VIII, Mots-Voir, Le Divan du Monde, Le Cube, and Le Point Ephémère.

In Paris for its 2007 event, months after the Electronic Literature Organization (ELO) held its second major conference in Maryland, the E-Poetry Festival, drew from a different source and was colored by a completely different complexion that its earliest predecessors. E-Poetry 2007 was extraordinarily organized by Philippe Bootz with the added expertise of Patrick-Henri Burgaud, Jean Clément, and Alexandre Gherban. Further, E-Poetry had the singular honor of presenting the first regular international digital literature series to take place in Paris. The daytime events were held at the Université Paris VIII, at the very north edge of Paris, and its evening performances took place at some of the leading cultural venues in the city (Le Divan du Monde, Le Cube, and Le Point Ephémère). It was indeed an honor—and a tribute to the vision of its local facilitators—to present E-Poetry events at such cutting-edge venues. Thus, though I (and my vision of incubating the field through intimate activity across broad international contexts), Bootz and Burgaud had a different sense of mission in mind than previous E-Poetry festivals: their thought was that to get the events deeper into the public arena would create a greater momentum for the field, almost crossing a threshold into a kind of popular awareness. (And I must admit, if this were to be done, Paris would be the place to do it.) It may have marked the end of a period, rather than a broadening of the field. Nonetheless, aside from the logistics an individual had to navigate to arrive at evening locations and the distance itself to the daytime events, it was a visionary model.

V. E-Poetry 2009 Barcelona

Local organizer: Laura Borràs; hosted by Hermeneia Grup de Recerca, Universitat de Barcelona.

In 2009, E-Poetry took place in Barcelona, Spain. Its organizers were meticulous in their attention to every detail of the festival, from the selection of extraordinary and culturally-rich venues, programming, and coordination of events. In addition, as the precursor to the tenth anniversary celebration of E-Poetry (celebrated with a social event, cake, and special, small ceremony), this E-Poetry gave me to take stock of the trajectory of E-Poetry. Indeed, with four events under its belt, two in the United States and two in Europe, E-Poetry had a distinct trajectory. (Also, E-Poetry consciously does not divide the world into a European event, a US event, etc. The goal, despite the difficulties in attending when one occurs at a distance from one's home city, offset by its biennial gatherings, is to engage this worldwide aesthetic context.) The organization, under the direction of Laura Borràs, was exemplary—and the attention to detail unprecedented. One difference I had with the organization is that E-Poetry never presents "keynote" speakers. Among high points in cultural diversity was the presence of attending Catalan poets. As to the differences, all is well that ends well.

My somewhat removed role was due to shifting poetics that were seeming to draw the trajectory away from E-Poetry's original aesthetic. This, too, was a positive, because through this experience I saw that the sails needed some adjustment. I was determined to point E-Poetry back to its original course: for this reason, I scheduled the tenth anniversary celebration back to E-Poetry's founding venue at Buffalo, where I did my best to emphasize specific aspects of the range, vision, cultural breadth, and depth of analysis that would be possible for the Festival.

The Second Decade

I'll never forget opting out of one of Barcelona's evening events (one that featured works that bore less relation to "E-Poetry"); these were events with less interest to me than taking time to reflect on the present moment and the Festival-in-progress. I sat on the sidewalk absorbed in the motion of people going about their affairs, wanting to truly enjoy the exhilaration of being in Barcelona. I felt the evening air, the voices in the night, thought of the history of this particular place—and sat for a long time thought about how the literature of this place is a living palpable presence—not the novelty of electronics nor modes of engagement imposed from a specific technology. I spent a long time pondering, wondering if this was my lesson from Barcelona, not the event that I had passed up for this respite. (And all conference-goers do occasionally need a respite. There is simply a lot going on.)

It was during that time I realized what the idea of "curating" meant. I also realized that if everything is decided by large groups, one ends up with a lot of compromises; the result is a different outcome. If E-Poetry was to be an "E-Poetry" that pointed to a specific vision of poetics (and here I claim simply "a poetics" of the digital, one among many possible approaches); its vision had to be clear. Second, I realized that there was a larger tradition, an ongoing music that branches, crosses, and wends through the generations that were my reference point. E-Poetry was not so much about comical invented game interfaces, nor Second Life existences, nor a trio of voices creating a cacophony of sounds, nor data on a scale too immense for digestion. My aim all along had been to engage a distinct curating emphasis. That which E-Poetry means to me and to the E-Poetries Advisory Board is a path to investigating questions greater than ourselves through the folds, textures, and grace of language in its material presence. In terms of digital literature, this means code for me. It also means being in touch with what lines of investigation had preceded the present technology of literary expression. It also included numerous other types of experiments, different approaches, but with a "curated" tone. Thus, I realized that the second decade of E-Poetry must begin with a clear expression of artistically driven, gender aware, multicultural, innovative, and inclusive (especially regarding younger practitioners) vision. It was clear that E-Poetry 2011 must return to Buffalo, to celebrate its anniversary, renew its roots, and energize an open field for digital poetics in a worldwide context.

VI. E-Poetry 2011 Buffalo

Local organizer: Loss Pequeño Glazier; co-organizer: Sandy Baldwin; hosted by the Dept. of Media Study, SUNY Buffalo.

For the tenth anniversary celebration of E-Poetry, the Festival returned to Buffalo. As it had been to Philippe Bootz in 2007, there was now a generational shift tangibly present. mIEKAL aND, for example, an early key voice in the movement no longer seemed to find the festival relevant. Neither Friedrich W. Block nor Florian Cramer were on the scene. Others had declared digital poetry "dead" or off their radar and had moved on. Some hung on and hung on in Second Life and other protocols, though they seemed to be going the way of MOOs and MUDs, but perhaps those worlds will serve us some day and I am wrong about the lack of general interest. It's true that groups always change but it is like a family; it changes the dynamic when folks come and go. The era of E-Poetry 2001 was over. Curiously, with the incrementing of one digit (very digital), the E-Poetry 2011 decade was launched.

To signal the strength of curatorial vision and the range of artistic vision across disciplines (rather than prioritizing a specific technology in its "programming")—and as a means of reasserting the literary, multicultural, gender-inclusive, and performance-inclusive direction of E-Poetry's vision—a number of diverse artistic strands were interwoven to provide a cross-generational, cross-cultural vision. This decennial anniversary celebration included, in addition to a monumental program of new works and an entire half-day of new critical presentations crowned by a reading by prize-winning Cuban poet Reina María Rodríguez (with translation), a significant gallery exhibition diligently curated by Sandy Baldwin, screenings curated by Tammy McGovern, an impressive atrium installation-performance by Mark Jeffery and Judd Morrissey, a digital poetry and dance concert in a professional black box theater featuring dozens of dancers performing an entire evening's program of digital-poetry-to-dance, directed by Anne Burnidge, along with readings by Jörg Piringer, Eugenio Tisselli, and Charles Bernstein and a closing night gala black box performance with stellar presenters Joan LaBarbara (a one-time collaborator of John Cage), an algorithmic visual-sound poetry performance by Lawrence Upton and John Drever (works invoking, in my mind, UK sound poetry legend Bob Cobbing), headlined by a young Cuban hip-hop poetry ensemble led by

Telmary, a key vocalist in Cuban's world-renowned band, Interactivo, where Telmary insisted everyone dance to hip-hop, partaking of rum. The idea behind the interdisciplinary approach was to signal, in a major way, that visual works, music, performance, philosophy, language, and computer-generation of language arts were not just interrelated in the present, but interwoven across avant-garde and multicultural historical precedents on a worldwide scale. Such an approach was, per se, a definition of the curatorial vision of the E-Poetry Festival series. There were some who wondered what the union of these elements meant—the text-based reading by renowned poet Charles Bernstein, the theatre-style dance concert, the musical and performance bent of the closing evening performance/party—had to do with E-Poetry. I was hoping to gesture to open doors to the integration of E-Poetry and the arts. E-Poetry would move to Europe for the following festival in 2013 and the E-Poetry Advisory Board was formed to shepherd coming events. The E-Poetry Advisory Board formed was diverse in its constitution across gender, culture, and geography, while being of a manageable size for effectiveness. The first E-Poetry Advisory Board consisted of Yves Abrioux/ Serge Bouchardon (France), Amaranth Borsuk (USA), David Jhave Johnston (Canada/Hong Kong), Leonardo Flores (Puerto Rico), Claudia Kozak (Argentina), Manuel Portela (Portugal), and Laura Shackelford (USA). The site of the next E-Poetry Festival would be Kingston University, London, on the banks of the Thames.

VII. E-Poetry 2013 Kingston, UK

Local convener: María Mencía; hosted by the Kingston Writing School (KWS), the Practice Research Unit (PRU), the School of Performance & Screen Studies, Faculty of Arts & Social Sciences (FASS), Kingston University London, the Watermans Art Centre, and the Poetry Library, Southbank Centre, and the Tate Britain.

E-Poetry 2013's London adjacent venue allowed space for new voices, schools, and settings, specific to the tenor and location of the event. London-Kingston, under María Mencía's preparations as local convener, was an overwhelming success! The festival benefited from a most suitable venue on the Kingston campus during the days, with evening performances at extraordinary London venues, the Watermans Art Centre, the Tate Britain, and the Poetry Library, Southbank Centre, Royal Festival Hall. It included an extraordinary gallery, organized by Mencia, with one work in a separate small room by itself and a range of works, a stimulating range of expression and technological exploration, on panels in a riverfront hallway outside the entrance to the theatre. The festival performances themselves provided a stunning styles, true to the goal that, "The 'poetry' in 'E-Poetry' does not signal a genre preference but an origin. That is, making as a means of realizing art, a delight in digital literary invention …" Of great note was the involvement of the E-Poetry Board of Advisors, very active in all aspects of the event, with special thanks to Laura Shackelford, who provided key programming and administrative support in 2013 and 2015, in addition to presenting her own remarkable work. (Note: a review and descriptive article about E-Poetry 2013 appears as "Tangible expressions of a present poetic: A review of E-Poetry 2013 Festival London," published in the online journal, *Jacket 2*, http://jacket2.org/reviews/tangible-expressions-present-poetic.)

VIII. E-Poetry 2015 Buenos Aires

Local organizer: Claudia Kozak; hosted by Universidad Nacional de Tres de Febrero.

E-Poetry 2015 was the first ongoing digital literature series to occur in Latin America. Given the logistics, subtleties of language, customs, cultural context, and distance, the logistics for this festival were a bit more difficult than others. However, due to the tireless and extraordinary work of Claudia Kozak, local organizer, the Universidad Nacional de Tres de Febrero, the Museo de la Inmigración/Centro de Arte Contemporáneo—UNTREF, a cultural gem, the historic nineteenth-century Teatro Margarita Xirgu in San Telmo (a classic nineteenth-century Spanish style architectural gem and the Centro Cultural Borges—UNTREF. There was a great breadth of new material to be shown in the world of digital literature. There were extraordinary panels on aspects of Latin American digital literature (seen from distinct global perspectives, including panels organized in the UK, France, and Argentina), a panel on women in digital literature (organized in the UK), incredible performances (notably those of Carlos Estevez, Judd Morrissey, Brian Kim Stefans, Philippe Bootz, Ethan Hayden, Ottar Ormstad, Pablo Gobira, Felipe Cussen/Ricardo Luna, and the first international performance of Orquesta de Poetas, among many others).

E-Poetry 2015 opened new vistas in the field and the pulsating, gargantuan, cultural monolith of Buenos Aires and its distinct neighborhoods added to the tempo and broad reach of the event. Indeed, E-Poetry 2015 Buenos Aires exhibited a "newness" to the many new voices drawn together in one venue, a spirit of discovery and camaraderie reminiscent of the first E-Poetry in 2001. It was a complete thrill.

Reflections on E-Poetry

Really, what I had longed for was a community that spread across organizations and continents. To me that was one benefit of the digital age (along with new types of works, obviously)—the ability to embrace a broader consciousness. Thus, the E-Poetry Festivals were not about the fact of technology, nor the "end(s) of electronic literature" but about the transcendence of technology with the goal of a shared intelligence. This, perhaps, was even a romantic notion, admiring of other poetic/arts communities. For example, Ricardo Baeza marveling at the magnanimity with which Lorca and his peers spoke of one another in his treatise "A Generation and Its Poet" (Stainton: 172). Or Leslie Stainton has suggested, "Although ... among themselves they sometimes squabbled over ideas and personalities, their respect and affection for one another endured 'We love each other, we adore each other, we're all of us the same person,' Lorca declared" (Stainton: 175). If not so idealized, a tangible group energy did move us forward.

It should be said that, though the E-Poetry is a curated series, it was never restricted to my own poetic vision. (Indeed, it could be said that, at this writing, my own poetic vision, except for occasional encouraging moments, has yet to be parsed.) Rather, having worked with literary figures such as Robert Creeley and Charles Bernstein and, years before, having been immersed in the activities of small press, the Mimeo Revolution, and various undertakings as tracked year after year in Len Fulton's historic directories of little magazines, I believed in a certain responsibility of the poet to give to the community as well as taking. The spirit that the E-Poetry Festivals seek to capture share those same sentiments, digital age or not. We are at an intensely volatile moment from text to emoji to big data. But we do have a role. What makes sense on the "page"? We must move beyond exemplars of technological effects to the root matters that opens doors to expressive creation across cultures. To devote one to organizing, advancing, and drawing attention to works that have an undefinable poetic value, to the works of younger artists, and to works by women, underrepresented groups, and from the emerging world. The aim of E-Poetry is not to advance a specific technology or ideology (criteria that always change), but to present opportunities for the "poetic voice," whatever that may be, and in numerous locations where it might be found.

I sincerely extend thanks to all local facilitators, co-directors, the E-Poetry Board of Advisors, and those many poets who have kept E-Poetry Festivals on their calendars across the years. I also thank the ELO for their contributions over the years. We have, indeed, been fortunate to have you among us.

6

Cyberfeminist Literary Space: Performing the Electronic Manifesto

Carolyn Guertin

According to media critic Justine Cassell, the most effective "feminist vision" of an electronic text or software design "as a space" for sharing authority with others is to have the work "be about [its own] design and construction" (1998: 302). Cyberfeminism and cyberfeminist electronic literature are relentlessly meta, demonstrating a prevailing preoccupation with their own construction. It should come as no surprise then that the dominant genre for cyberfeminists is the manifesto, including programs for Cyborgs (Haraway 1985), mercenaries of slime (VNS Matrix 1991), for Net.Wurkers (Mez 2001), Queertexters (Rhodes 2004), Zinesters (Antropy 2012), Xenofeminists (Laboria Cuboniks 2014) and Glitch Feminists (Russell 2020). These self-consciously styled literary works celebrate the complexity of their own construction and reading. The manifesto as a literary genre is a call to social action. As a participatory and conversational form, it invites the reader to become an active player or an interactor in a narratological, discursive, semiotic and/or literal revolution. As a twenty-first-century form, it is especially well suited to feminist digital culture because of its three major properties: it is reciprocal, viral, and memetic

(Yanoshevsky 2009: 275).[1] As Leslie Heme says, "The manifesto beckons adaptation and adoption into practice; it squirms into the subconscious" (Heme 2013: 12).

Cyberfeminism rocketed onto the international stage in a consciousness-raising flash with the publication of the Australian performance troupe VNS Matrix's "Cyberfeminist Manifesto for the Twenty-first Century" in 1991.[2] Publication may not be the right word though since, in this pre-web incarnation, the manifesto was initially broadcast on a billboard. The Australian collective, which was active from 1991 to 1997, was comprised of four women: Virginia Barratt, Francesca da Rimini, Julianne Pierce, and Josephine Starrs. They coined the term "cyberfeminist" to capture the *zeitgeist* of their radical feminist acts and blatantly viral agenda.

VNS Matrix's choice of the manifesto as a medium and delivery system is appropriately double-barreled: both performative and political. The manifesto is "a plural and open form," according to the French literary scholar Galia Yanoshevsky, for which "'crisis' is its *'raison d'être'*" and whose goal is to "'question the system'" (Demers, qtd. Yanoshevsky 2009: 263). By design a violent act, "The manifesto has a particular performativity: it does not 'merely describe a history of rupture, but produces such a history, seeking to create this rupture actively through its own intervention'" (2009: 266). Literary critic Marjorie Perloff has argued that the manifesto emerged as a literary genre and narrative form in the hands of Italian Futurist F. T. Marinetti (1984; 71). It is easy to see this narrative impulse in the "Cyberfeminist Manifesto" as well, which proclaims itself "the virus of the new world order/rupturing the symbolic from within" as "saboteurs of big daddy mainframe," "terminators of the moral code," and "mercenaries of slime." We could also read this as an anti-narrative impulse with its always already future orgasmic desire for infiltration, disruption, dissemination, and corruption. VNS Matrix went on to create video games and interactive CD-ROMs and, in 2015, a remix called "Undaddy Mainframe" was sent into space as a part of a larger project called "ForeverNow." VNS Matrix's work has been as viral and

---

[1] I am paraphrasing her concepts for these are not precisely Galia Yanoshevsky's terms. She identifies the three major sociopoetic functions of the manifesto on its readers and critics: (1) it has a reciprocal relationship to critics; (2) it has the ability to "'contaminate' theory in a way which makes theory start seeing itself, too, as real, as event, with a history and ideological and social underpinnings of its own;" and, (3) since the manifesto is repetitive, it has aesthetic and mimetic properties by design (Yanoshevsky 2009: 275).

[2] That same year, British cultural theorist Sadie Plant also chose that term to describe her recipe for defining the feminizing influence of technology on Western society and its members. Around the same time in Canada Nancy Paterson, a celebrated high-tech installation artist, also wrote an article called "Cyberfeminism" for Stacy Horn's Echo Gopher server. Clearly the time for cyberfeminism as emergent practice had arrived.

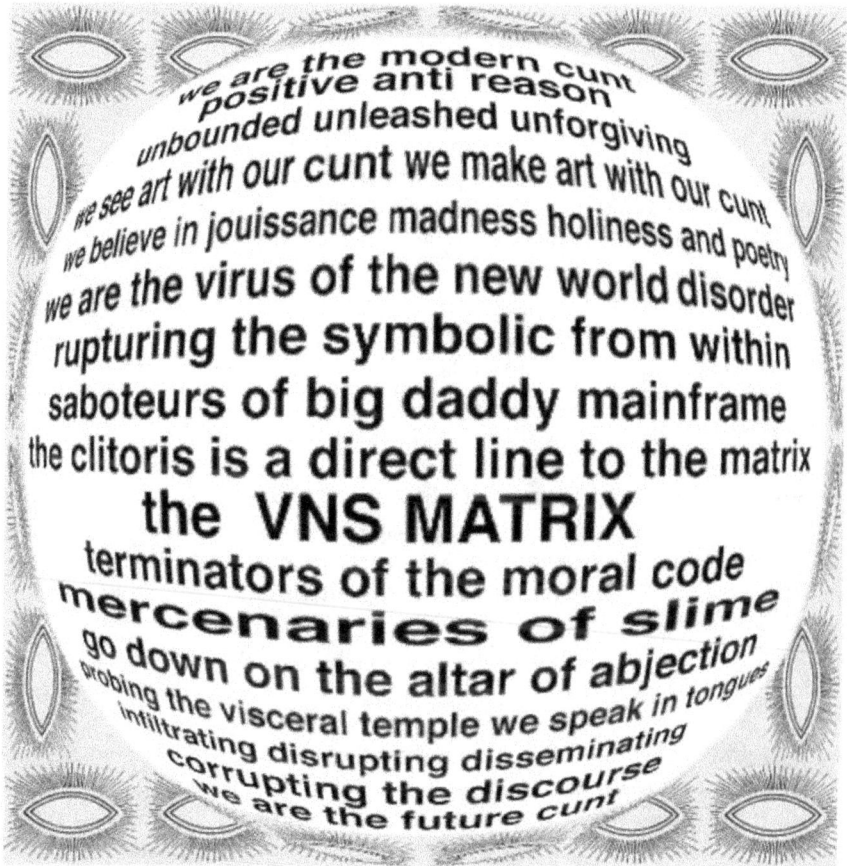

FIGURE 1 *VNS Matrix, "Cyberfeminist Manifesto for the 21st Century" (1991).*

unstoppable as it has been global in its inspiration, most recently in the birth of Laboria Cuboniks (whose work I will discuss later in this chapter).

The act of defining cyberfeminism is harder than dating its emergence, for its self-reflexive history is about the exploding of categories and the refusal of closure or classification. One half of VNS Matrix, Francesca da Rimini and Virginia Barratt, call it:

> a catalytic moment, a collective memetic mind-virus that mobilised geek girls everywhere and unleashed the blasphemic techno-porno code that made machines pleasurable and wet. a linguistic weapon of mass instruction, the manifesto struck at the mass erection of technopatriarchal order. we loved with machines, in a most unholy alliance.
>
> (*Forever Now*)

At the first cyberfeminist conference in Germany in 1997, the Old Boy's Network (OBN)—an organization that came to be the central hub of cyberfeminist thought for a while—drafted the "100 Anti-Theses of Cyberfeminism." These rules were multilingual and nonbinding, ranging from the whimsical to the militant. Cyberfeminism was both "not a fragrance" and "not caffeine-free" as well as "not a praxis," "tradition," or "ideology." Cyberfeminism was "not a structure," "a lack," or "a trauma," but also "not without connectivity" and "not an empty space." These not-empty definitions refused binaries at the same time as they refused the two-dimensional spaces of print culture. I have defined cyberfeminism elsewhere as "a way of redefining the conjunctions of identities, genders, bodies and technologies, specifically as they relate to power dynamics" and to texts.

Cyberfeminist texts were and are a celebration of multiplicity. They often refuse single authorship and exist outside institutional spaces. They exist in opposition to what Faith Wilding called "the Tupperware aesthetics" of postfeminist netchicks and grrl sites with their tendency to reinscribe female stereotypes, and were set against the phallocentric establishment. As a postured and self-conscious form of embodiment—as opposed to the masculinist cyberpunk celebration of virtual disembodiment so popular in the 1990s—cyberfeminist writing seized a politicized and historical context to write itself free of old boundaries. It sought to write a new future. Faith Wilding called it a "strategy" for claiming and taking up space ("Future is Femail").

This goal was not a new one. It was a continuation of the work that experimental women writers in print had been undertaking for more than a century. Virginia Woolf and Gertrude Stein both experimented with voice, vision, the senses, the continuous present tense, broken sequences, multiple voices, and subversive genres in ways that would be right at home on the web. More recently, Christine Brooke-Rose, Nicole Brossard, Carole Maso, Susan Howe, Lyn Hejinian, Gail Scott, Avital Ronell, and Carla Harryman have experimented with ruptured narrative, visual text, and fractured sequences on the page that are akin to cyberfeminist experimentation in digital narrative. Sharing similarites to the conceptual arts, new media writings are more like performance or installation art than other conceptual forms—which is why Barratt can identify such a clear lineage from VNS Matrix all the way to the Russian political troupe Pussy Riot (*Forever Now*). The digital medium in skilled cyberfeminist hands—like the works of Shelley Jackson, Mez, Carmin Karasic, Kate Pullinger, Caitlin Fisher, J. R. Carpenter, and Anna Anthropy—is used to foreground ruptures in language and text, in space and sound, in bodies, words, and images. Michael Joyce calls these linked narratives "a conversation with structure" (2000 94). The hybrid nature of the meeting of media allow these Hackermaker Aesthetics (Guertin 2015) and writers to "work the interface" between the creative process and reading, between bodies and materialist concerns, between conventions of the media and discourses within texts (Moyes 1994: 309). Donna Haraway in her "Cyborg Manifesto" also called for re-embodied seeing as a way of re-connecting the textual, material, and technological worlds. As I stated earlier, Barratt of VNS Matrix believed the language tradition and revolution was an integral part of feminist writing in the 1980s and 1990s.

The first lengthy study of feminist work in the new media was my own *Queen Bees and the Hum of the Hive: An Overview of Feminist Hypertext's Subversive Honeycombings* published in *BeeHive* in 1998. At that time I found an overview of women's works was necessary because even though women were leading practitioners in electronic literature, they were already being written out of official narratives and literary criticism in the 1990s. In that hypertextual essay, I examined issues of language, discourse, translation, and gendered modes of speaking that are evident in many early feminist works on the web. As a first survey, *Queen Bees* is most concerned with cataloguing the myriad discourses and texts. In 1999, my Gallery, *Assemblage: The Online Women's New Media Gallery*, debuted at the trAce Online Writing Community to further the goals of my earlier essay and to begin a conversation—still ongoing—about women's digital literary praxis. It is now archived in the British Museum. In February 2000, Marjorie Coverley Luesebrink and I published a selection of highlights from the Gallery as "The Progressive Dinner Party" (in homage to Judy Chicago) in Jennifer Ley's online journal, *Riding the Meridian*. Other works have appeared in book form since then, including Susan Hawthorne and Renate Klein's *Cyberfeminism: Connectivity, Critique + Creativity* (1999), N. Katherine Hayles' *How We Became Posthuman* (1999) and *My Mother Was a Computer: Digital Subjects and Literary Texts* (2005), Jacqueline Rhodes' *Radical Feminism, Writing, and Critical Agency: From Manifesto to Modem*, Radhika Gajjala and Yeon Ju Oh's *Cyberfeminism 2.0* (2012), and my own *Digital Prohibition: Piracy and Authorship in New Media Art* (2012). Sadly though, feminist theory and those early documents and works by women writers quickly obsolesce on the spaces of the web, and are too easily forgotten or ignored by critics. Memory is short in cyberspace. More recently other scholars have returned to those early authors' texts to again begin to recatalogue and remember the mothers of the digital text.[3]

VNS Matrix's "Manifesto" is just one of a chorus of feminist voices that has sought to rupture the patrilineal continuum. Donna Haraway's "Cyborg Manifesto" (1985) pre-dates it, of course, and has been influential in many areas and fields. In fact, Haraway revisits the manifesto form thirty years later with a new work called *Manifestly Haraway* (2016), which brings

[3]For instance, I am thinking most recently of Kathi Inman Berens' article "Judy Malloy's seat at the (database) table: a feminist reception history of early hypertext," on Malloy's pioneering work *Uncle Roger*, published in 2014.

together her earlier piece with her new "Companion Species Manifesto." In her earlier work, Haraway calls for the body to become an "agent" rather than "a resource" where we are capable of "situated conversation at every level of its articulation" (1991: 200). Boundaries are drawn and erased through our physical mappings of space, and if we can transform our conversations with the texts into action we might craft a program for feminist embodied interference. This sounds like a program for political occupation. Likewise, allowing our browsing to insert its body as an agent or interactor in textual spaces like the manifesto will multiply subjectivities—and therefore perspectives—many times over. Virtual bodies are permeable and interweave the corporeal with machine language in oppositional stances. These interweavings for Haraway's cyborg create "webs of power" that birth "new couplings, new coalitions" (1991: 170); they are permeable to language and information being nonbinary—having not one code or common language, but many codes and many languages. They are viruses performing circulation in networked space.

As a form of pure performance, Mez's Mezangelle blends the flows of the digital with the rupture of the linked space creating a subjective landscape and language for an aestheticization of the personal-political continuum. Manifestos have definite links to performance and to theatre, according to Martin Pucher. In "Manifesto = Theatre," he argues that "both involve the act of making visible. The manifesto has a particular performativity; it does not 'merely describe a history of rupture, but produces such a history, seeking to create this rupture actively through its own intervention'" (Pucher 2002: 449–50; qtd. Yanoshevsky 2009: 266). We might read the whole of Mez's oeuvre as just such a manifesto-driven performance that intervenes in the fabric of language and the self in the name of subjectivity. Mezangelle is not a single manifesto, but a protocol, a platform, and a praxis for Net.Wurkers. A recombinant language, Mezangelle works with discarded bits of code, snippets of words, archaic symbols, slang, markup, and ASCII to craft literary manifestos. A pioneer of the meeting of the linguistic and the performative at the boundaries of form, in 2008 she was a tweeter-in-online-residence for New Media Scotland. Over the space of a month, she tweeted *Twitterwurking*, a complete, full-length work in Mezangelle. As a protocol, Mezangelle seeks to both impose and relinquish control at the same time. Eugene Thacker in the introduction to Alex Galloway's *Protocol* states, "the founding principle of the Net is control—not freedom–control has existed from the beginning." Protocol opens and obstructs flows of data in concert (Thacker 2004: xv), just as the manifesto recruits and relinquishes us at the same time. As I wrote in my book, *Digital Prohibition,* her language

> is the space of Mezian play, manipulation, and logic. It is always in tension with itself, always existing in the contradictory spaces of multilevel logics. By mastering the logistics of control, she finds the free play in the system.

Mezian aesthetics operate always just outside of language in the not-quite visual realm. Composed of ruptures, openings, and jumps, her words are always in the midst of delivering poetic shocks and re-envisionings. Playful disturbance is the process by which the reader floats through the dynamic and sensory experience of reading her art-theory mélange.

(204)

Protocol is an aesthetic that influences our every move in the Net.Wurk. Both manifestos' and protocol's contradictions are hardwired into the system like a virus. Mez's textual interventions make for slow reading by design as they fuse performance, digital material, and viral media. A truly original writer, Mez's voice has been compared to William Shakespeare, James Joyce, and Emily Dickinson, among others (Mez, Interview 2007). Her texts are embodied spaces that rupture language.

In the twenty-first century, women's bodies have become more visible, but by and large their voices have not. Crushing anti-female forces that have emerged—especially but not only in the United States, from a neo-Conservativist turn in public opinion—are making it dangerous for women to speak out. Since the 1990s, this has made the web a dangerous space, and resulted in fewer overtly political tracts like the manifestos I have discussed. These attacks on women—frequently called "the war on women"—have been highly visible from the erosion of women's reproductive rights to Donald Trump's campaign trail hate speech to #GamerGate. #GamerGate arose in August 2014 when a small group of misogynist gamers set out to silence feminist dialog around computer games and gaming. Threatened by the rising popularity of games with women (who now comprise about 40 percent of gamers), a group of anonymous trolls launched an all-out war. A game developer named Zoe Quinn was the catalyst and first target. She was driven from her home as the result of threats and doxxing related to her game *Depression Quest*. From there the trolls singled out Anita Sarkeesian, a popular YouTube personality and pop culture critic who tackles sexism in games with a series called "Tropes vs. Women in Video Games." Katrin Higher calls the consolidated attacks, doxxing and threats against these women's lives and safety "misogynistic terrorism." What is more frightening is that they are not the only targets. The Game Developers' 2015 Conference set out to address the issue, but it was too dangerous a topic to run a panel on. Soliciting anonymous comments from women game developers instead, the conference organizers showed a devastating video called "The Empty Chair" which demonstrates in chilling fashion what female game developers endure. SXSW, the popular media conference in Austin, Texas, also later ran a panel on the topic of sexism in games in response to a cancelled event and criticism that they were doing too little to support women. The violence of #GamerGate, however, has gradually shifted the rhetoric in the gaming industry as people now seek to distance themselves from the anonymous trolls. This outpouring of violent #GamerGate hate has also contributed to a birth of a number of new cyberfeminist manifestos.

Ironically enough, anonymity was initially one of the attributes of the web that was so empowering to women. According to Clare Evans, in the 1990s cyberfeminists had "championed" anonymity "as a method for transcending gender," but now anonymity is weaponized to enable violent assaults on women everywhere online from comments to email to Twitter. Evans says, "It's not that the CyberFeminists failed. It's that as the Venn diagrams of digital and real life have edged into near-complete overlap, the problems of the real world have become the problems of the digital world. The web is no longer a separate space; we are inseparable from the web" (Evans). In the past, the collective has been seen as a successful anonymous strategy for the cloaking of cyberfeminist dialog and protection. As everything old is new again, the Carnegie Mellon-based group Deep Lab is now putting these strategies to work again in response to #GamerGate. Their manifesto is a book, written over five days in December 2014, in an effort that was equal parts "hackathon, charrette, and a micro-conference" (http://www.deeplab. net/#the-book). They describe themselves as:

> a collaborative group of researchers, artists, writers, engineers, and cultural producers interested in privacy, surveillance, code, art, social hacking, and anonymity. Members of Deep Lab are engaged in ongoing critical assessments of contemporary digital culture and exploit the hidden potential for creative inquiry lying dormant within the deep web. Deep Lab supports its members' ability to output *anonymously via proxy tools; in this way, our research can remain fluid via multi-pseudonymous identity.* Deep Lab promotes creative research and development that challenges traditional forms of representation and distribution, evaluating these practices alongside typical traffic analysis identification. This process leverages the research of Deep Lab to contend with outdated modes of understanding culture within traditional social structures. (my emphasis; Addie Wagenknecht, Founder of Deep Lab, Deep Lab: 11)

Deep Lab thereby enables feminist dialog and dissemination of research by offering a new publishing platform via proxy. "[F]emale hackers," it says, "must engage with the future, in order to make our presence in history indelible" (12).

Another recently formed global cyberfeminist collective, Laboria Cuboniks (2015), ascribe the decline of women's power in digital spaces to the rising dominance of visual culture online. Subtitled the "politics for alienation," the Xenofeminist Manifesto calls for the "depetrification" (Laboria Cuboniks, Zero 0x00) of capitalism in the name of a more universalist model for world and identity creation. The name Laboria Cuboniks is a puzzle drawn from an anagram of another pseudonymous group, Nicolas Bourbaki, a collective of twentieth-century French mathematicians who authored treatises and books on a more abstract vision of math. According to the imaginative, dextrous, and persistent Laboria Cuboniks, the power of the sexist and objectifying gaze has re-inscribed the old gender disparities, and that power supercharges "modes of identity policing, power relations and gender norms in self-representation" (Laboria Cuboniks, Parity: 013). This collective sees social media and hate speech twisting and stifling the web's early potential for revolutionary acts perpetuated through "memes like 'anonymity,' 'ethics,' 'social justice,' and 'privilege-checking'" (Laboria Cuboniks, "Parity": 0x0D). "Valuable platforms for connection, organization, and skill-sharing become clogged with obstacles to productive debate positioned as if they are debate," they say ("Parity": 0x0C). Neither seeking to resurrect 1990s cyberfeminism nor throw the cyberfeminist baby out with the bathwater, Laboria Cuboniks call not for revolution, but for "mutations" and reanimation of the "long game of history" where xenofeminists can abolish gender, change nature, and master computation. Its "Politics for Alienation" wants to create a new language, dismantle the human genome, and abolish patrilineal constructs like nature and the family. Xenofeminism makes strange with capitalism and neoliberal policies to rewrite gender for a constantly moving target of a mutant xenofeminist philosophy.

The return of the literary manifesto as cyberfeminist/xenofeminist literature is a welcome turn of events in a time of Hackermaker Aesthetics, the active creation of alternative spaces for women for creative practice (Guertin 2015). Manifestos make for spirited, polemical, performative events. If, as Donna Haraway says, the body is a machine made out of words inscribed by time and memory, then the performance space of manifesto-fueled subjectivity is not simply uncontainable, but contagious and nomadic as well. It is transgressive speaking that circulates outside patrilineal culture, and Haraway's cyborg naturally enacts its transgression in language frame by frame through body-based thinking: the audacious site of this truly monstrous thought process. A collective is a conceptually complex subjective embodiment, forming the skin of mediation and connection between realities, the tangled interface between the virtual and the real, between lives lived both online and off. Subjectivity writ large as a manifesto is uncontainable, viral, xenomorphic, aesthetic, political, and doubled, existing "both inside and outside" the "creative domain" and "within the field of artistic production" (qtd. in Yanoshevsky 2009). It is also integrally interconnected with the cosmological, narrative fabric of women's lives, with the ruptured gaps women leap and the story women travel through. The twenty-first-century feminist e-manifesto is the origin story in a time of multiple subjects and genders, and it is the meta-space everyone needs to occupy to disrupt the overarching master narrative of misogyny terrorism in digital culture.

References

Anthropy, Anna (2012), *Rise of the Videogame Zinesters: How Freaks, Normals, Amateurs, Artists, Dreamers, Drop-outs, Queers, Housewives, and People like You Are Taking Back an Art Form*, New York, NY: Seven Stories Press.

Berens, Kathi Inman (2014), "Judy Malloy's Seat at the (database) Table: A Feminist Reception History of Early Hypertext," *Literary and Linguistic Computing: Journal of the Association of Digital Humanities Organizations* 29 (3): 340–8, doi:10.1093/llc/fqu037.

Breeze, Mez (2001), Manifestos, "_This Cybagenic Lattice_," http://www.cddc. vt.edu/host/netwurker/cyblattrice.html.

Cassell, Justine (1998), "Storytelling as a Nexus of Change in the Relationship between Gender and Technology: A Feminist Approach to Software Design," in Justine Cassell and Henry Jenkins (eds.), *From Barbie to Mortal Kombat: Gender and Computer Games*, 298–326, Cambridge, MA: MIT Press.

Evans, Clare L. (n.d.), "'We Are the Future Cunt': CyberFeminism in the 90s," Motherboard (accessed November 20, 2014), http://motherboard.vice.com/read/ we-are-the-future-cunt-cyberfeminism-in-the-90s.

Game Developers' Conference (2015), "The Empty Chair," (accessed March 5, 2015), https://www.youtube.com/watch?v=J7HuC_aLkoE.

Guertin, Carolyn (n.d.), "Assemblage: The Women's New Media Gallery 1999–2005," trAce Online Writing Centre [Archived], http://trace.ntu.ac.uk/traced/ guertin/assemblage.htm.

Guertin, Carolyn (2005), "From Hacktivists to Cyborgs: Postfeminist Disobedience and Virtual Communities," Electronic Book Review: Postfeminisms Thread (January 27, 2005), http://www.electronicbookreview.com/thread/ writingpostfeminism/hackpacifist.

Guertin, Carolyn (2015), "Public Disturbances: Toward a Taxonomy of Hackermaker Practices," Keynote Address, West Sydney University (December 8).

Guertin, Carolyn (1998). "Queen Bees and the Hum of the Hive: Feminist Hypertext's Subversive Honeycombings," BeeHive 01.02 (July), http://beehive. temporalimage.com/archive/12arc.html.

Guertin, Carolyn and Marjorie Coverley Luesebrink (2000), "The Progressive Dinner Party," Riding the Meridian, February, http://www.heelstone.com/ meridian/templates/Dinner/predinner.htm.

Guertin, Carolyn (2012), *Digital Prohibition*, New York: Bloomsbury Academic.

Haraway, Donna (1991), *Simians, Cyborgs and Women*, New York, NY: Routledge.

Haraway, Donna (2016), *Manifestly Haraway*, Minneapolis, MN: University of Minnesota Press.

Hawthorne, Susan and Renate Klein (eds.) (1999), *Cyberfeminism: Connectivity, Critique + Creativity*, North Melbourne: Spinifex Press.

Heme, Leslie (2013), "Manifesting the Manifesto." http://yadayadapourlayada.files. wordpress.com/2013/01/manifesto-essay.pdf

Higher, Katrin (2014), "The Misogynist Terrorism of #GamerGate," Web (October 21), http://www.lifeofthelaw.org/2014/10/the-misogynist-terrorism-of- gamergate/.

Joyce, Michael (2000), "Beyond Next before You Once Again," *Othermindedness: The Emergence of Network Culture*, 81–106, Ann Arbor, MI: Michigan University Press.

Laboria Cuboniks (2015), "Xenofeminism: A Politics for Alienation," (June 11), http://www.laboriacuboniks.net/.

Moyes, Lianne (1994), "Into the Fray: Literary Studies at the Juncture of Feminist Fiction/Theory," in Terry Goldie et al. (eds.), *Canada: Theoretical Discourse*, 307–25, Montreal: Association for Canadian Studies.

Old Boys Network (1997), "100 Anti-Theses of Cyberfeminism," (accessed March 30, 2016), http://www.obn.org/cfundef/100antitheses.html.

Perloff, Marjorie (1984), "'Violence and Precision': The Manifesto as Art Form," *Chicago Review* 34(2): 65–101, JSTOR, Web (accessed March 1, 2016).

Pucher, Martin (2002), "Manifesto = Theatre," *Theatre Journal* 54 (3): 449–65.

Quinn, Zoe (2013), "Depression Quest," http://www.depressionquest.com/.

Rhodes, Jacqueline (2004), "Homo origo: The Queertext Manifesto," *Computers and Composition* 21: 387–90, Science Direct, Web (accessed March 30, 2016).

Russell, Legacy (2020). *Glitch Feminism: A Manifesto*. Verso Books.

Sarkeesian, Anita (n.d.), "Feminist Frequency: Conversations with Pop Culture," http://feministfrequency.com/.

Thacker, Eugene (2004), "Protocol Is As Protocol Does," in Foreword to *Protocol: How Control Exists after Decentralization*, xi–xxii, Cambridge, MA: MIT Press.

"_Tracking a Net.Wurk O(r)bit_," http://www.cddc.vt.edu/host/netwurker/netorbit.html.

"Versatile m[c]o[mmunication]dality" (April 25, 2007): http://cont3xt.net/blog/?p=251http://multi-maryfesto.blogspot.ca/2009/05/in-terms-of-my-own-dispersed-multi.html.

VNS Matrix (1991), "Cyberfeminist Manifesto for the Twenty-first Century," https://vnsmatrix.net/projects/the-cyberfeminist-manifesto-for-the-21st-century/.

VNS Matrix (2014), "Undaddy Mainframe," *Forever Now*, http://forevernow.me/artists/artwork/undaddy-mainframe/.

Wagenknecht, Addie et al. (2014), "Deep [Manifesto]," Carnegie Mellon/STUDIO, December, http://www.deeplab.net/#the-book.

Wilding, Faith (1998), "The Future is Femail," posted to the nettime mailing list (September 18), http://www.desk.nl/~nettime/.

Yanoshevsky, Galia (2009), "Three Decades of Writing on Manifesto: The Making of a Genre," *Poetics Today* 30 (2): 257–86, Web (accessed March 30, 2016).

7

Bodies in E-Lit

Astrid Ensslin, Carla Rice, Sarah Riley, Christine Wilks, Megan Perram, Hannah Fowlie, Lauren Munro, and K. Alysse Bailey

Introduction

In this chapter, we offer a survey of electronic literature ("e-literature" or "e-lit") that deals with embodiment, corporeality, and body image in aesthetic and material ways. We begin with an examination of e-literature that foregrounds the post-human body as embedded in the cybernetic feedback loop in ways that debunk Cartesian dichotomies of mind/body and human/machine intelligence. We explore body-themed, feminist e-lit in theory and practice, arguing that female-coded bodies have been pitched against patriarchal neoliberalist appearance culture and positioned to challenge reader-players' normalized expectations of bodily playability in digital media. In this context, we examine feminist encoded hypertextuality as perhaps the most canonical, poststructuralist approach to anti-phallocentric corporeality in e-lit. We consider works that subvert the "ergodic gaze" (Ensslin et al.) in multimodal ways, rupturing the scopophilic interface and allowing the voyeur to gaze via haptic intrusion. In concluding, we surface some of the ways that postdigital *écriture feminine* has sought to write new languages, spaces, and worlds for women-identified and gender nonconforming bodies. Expanding on the latter, we map out a new, applied e-lit project ("Writing New Bodies") that adopts a reader-centric, feminist, participatory co-design process to allow young women-identified and gender nonconforming individuals to write new worlds of digital-born fiction in which they feel at home in their bodies.

Cybersomatics and AI

Cybersomatic works of e-literature expose reader-players' bodies as both situated and physiologically contingent constituents of the cybertextual feedback loop. They foreground ways that embodied reading may draw readers' attention to physiological processes that are usually taken for granted and treated as inferior binary counterparts to cognitive processes of decoding and comprehension. Against this critical post-human backdrop, cybersomatic works operationalize machine intelligence in how they expose human reading as only one component of a complex, cybernetic communication system that depends, for its poetic and narrative effects, on the reciprocal response circuits between cognitively embodied human consciousness and machine code. Cybersomatic e-lit thus subverts normalized accommodation processes underlying human–machine interaction in digital mass media like mainstream video games, where technological affordances are readily "adapted to and appropriated into our available repertoire of bodily behaviours and aptitudes" (Dovey and Kennedy 2006: 111).

Kate Pullinger, Stefan Schemat, and babel's gothic hypermedia mystery, *The Breathing Wall* (2004), for example, exposes readers to the idea of losing control of intention-driven decoding and inferencing for the sake of cybernetically controlled processes of information disclosure (Ensslin, "From (w)reader"). It uses the reader's respiratory system as the driving force for revealing key referential meaning, or "clues" about the plot. The reader has to breathe into a microphone, triggering a software called Hyper Trans Fiction Matrix to release piecemeal narrative information, depending on the reader's depth and rate of breathing. Inevitably, readers' attention is drawn to the impossibility of gaining full control of their breath and to the intrinsic interplay between cortical and subcortical control at play in respiration. Along with other breath-driven digital art such as Lewis LaCook's *Dirty Milk* and Char Davies' *Osmose*, *The Breathing Wall* aestheticizes the anatomical and site-specific double-situatedness of the reading body, and the relative uncontrollability of physiologically contingent cognitive processes.

Fast-forward one-and-a-half decades, we find ourselves in the midst of the algorithmic turn in e-literature (Ensslin et al. forthcoming). Bot poetry has become an intrinsic element of contemporary e-literary culture, populating the web from social media sites such as Twitter to idiosyncratic sites such as Nick Montfort's agglutinative, crowdsourced *Taroko Gorge*. An e-lit work that plays with cybersomatic AI in humorous and thought-provoking ways is Serge Bouchardon's *Storyface* app. "[A] digital creation based on the capture and recognition of facial emotions" (Bouchardon 2018), the app presents a fictional dating site, asking users to project those emotions into the webcam that seem "to characterize him/her the best." By profiling the reader's face, the app suggests an age range and, coupled with standardized emotion(s) captured from the reader's facial expression, proposes profiles of fictional partners. Once the reader has chosen their ideal partner, they can engage in a fictional chat with them under the face recognition algorithm that continues to track the reader's standardized, monolithic, and normalized emotions. Thus, *Storyface* critiques the collectivizing tendencies of state-of-the-art AI and their implications for interactional ethics.

Fragments and Patchworks

From the outset, feminist works of e-literature responded to a call by second-wave feminists (famously Hélène Cixous and Luce Irigaray) that, in the mid-1970s, demanded new, anti-phallocentric forms of writing. Cixous argued that the truths we subscribe to are man-made and man-biased. In "The Laugh of the Medusa" she urges women to reclaim their bodies and, by extension, their desires and identities through writing. The challenge but also potentially liberatory implication of *écriture feminine* is, according to Cixous, that women's writing, if understood simultaneously as an intervention and as a fleeting concept cannot be reduced to an essence— and this is what makes it relevant for fluid, dynamic, and playful digital as well as postdigital, medium-critical re-encodings.

In e-lit's early years, nonlinear hypertext networks of nodes and links lent themselves to ideas of writing women's bodies via the notion of the fragment that eludes materialization as it constantly deconstructs and reconstructs itself. Metaphors of quilting, weaving, sewing, and patching have proven to be pervasive elements of feminist digital fiction beyond first-generation standalone hypertext. For example, Shelley Jackson's Storyspace hypertext, *Patchwork Girl* (1995), and Christine Wilks' more recent, interactive Flash memoir, *Fitting the Pattern* (2008), engage with motifs of sewing and patching—on a textual and referential level. *Patchwork Girl* has frequently been described as perhaps the most fitting allegory of poststructuralist thought in digital space. Its readers "have to sew [the female monster] together … to resurrect [her] … in piecemeal" (graveyard). Jackson's work presents a compelling, cyberfeminist response to Mary Shelley's phallocentric *ur*-story.

Fitting the Pattern shifts the focus onto the instrumentality and materiality of bodily construction. Wilks' reader-player can choose from a range of dressmaking tools to cut and sew together pieces of cloth as synecdochic backgrounds to autobiographical sketches, allowing the reader to construe elements of Wilks' younger self via ludic interaction. Those sketches, which appear in lexias on pieces of cloth, reveal the protagonist's exposure to the restrictive body ideals of her time, culture, and class, and the overpowering role of her mother in urging her to "fit" those ideals. Similar to the elusive, fleeting persona of the nameless Patchwork Girl, Wilks' former self appears in flux and fragmentation, thus giving rise to questions of self-denial and conflicted identity.

In her Flash fiction *Underbelly* (2010), Wilks takes on the historical theme of women working in British coal mines. She juxtaposes their physical suffering with the physical work of a twenty-first-century Yorkshire sculptor (based on Wilks' sister), who has very different concerns about her body and the prospect of childbirth. They share a concern with earth and stone, and both make a living by carving stone, for mining and artistic expression respectively. The sculptor's voiced over meditations are overwritten by the voices of women working in Victorian collieries, whose static images are shown flitting across the screen as the reader-player clicks their way through animated subterranean images. The women's voices relate the dire conditions in healthcare and maternity support at the time, and their reports are accompanied by uterine and fetal images, indicating affinities of exploitation between women's bodies and Mother Earth. Gaian exploitation is thus doubly encoded in an ecofeminist narrative that, once again, foregrounds the fragment—this time as a token of capitalist-patriarchal exploitation and destruction.

Another way of portraying the body as fragmented concept is through the theme of becoming women via social inscription (Grosz 1994). As body-image scholar Carla Rice explains, "[g]ender is something we become. We become gendered by modifying our bodies and behaviours to match how we feel inside with the messages that we get from outside" (65). This gradual process of bodily modification to meet social expectations is thematized in Juliet Davis' dress-up Flash e-poem, "Pieces of Herself" (2005). Here, the protagonist's "docile body" (Foucault 1995) is literally inscribed by interactive fragments, or icons, of domesticity, in a process that is alleged to help her "find herself." The reader's role is to furnish the outline of a dress-up doll with multicoloured patterns they pick up as they scroll through black-and-white domestic environments. These gamified, "metaphoric acts of inscription … trigger audio files ranging from music to a biblical pronouncement about the 'proper' socio-cultural function of women" (Borràs et al.), suggesting a multisensory process that occurs on multiple layers of mediation.

Postmedia Bodies, Postfeminism, and the Ergodic Gaze

In our postmedia (Manovich 2014), post-human, postbiological world, bodies are hybrid, malleable, and multiple—they cannot be seen as monolithic, stable entities. Online, we can mold and shape them to meet our innermost desires and address our deepest-felt anxieties. Digital media allows us to stylize our digital bodies as idealized representations of our own narcissism, customizable through on-screen avatars. We can create countless digital bodies and experiment with a diversity of corresponding alter egos. We can enrich our bodies with algorithmic, "post-cybernetic control" mechanisms (Parisi 2004: 105), making them readable and controllable. These mechanisms provide us with greater insight into and awareness of our somatic processes yet simultaneously expose our bodies, mostly unwittingly, to hacking, datafication, and surveillance.

Our on-screen digitalized bodies seem liberated from their actual-world fleshliness. In digital media, we can experiment with our forms and those of others. These refashionings can leave our bodies simultaneously liberated and "alienated from themselves, augmented thanks to technology, modified, reincarnated, multiplied" (Brodesco and Giordano 2017: 11). Thus, our online bodies can enact metaphors of cognitive dissonance and broken relationships with ourselves and others. This phenomenon is reflected poignantly in Serge Bouchardon's and Vincent Volckaert's short, episodic Flash fiction, *Loss of Grasp* (2010). The work is a first-person narrative, told by a man who is losing control of his life and relationships. It centers his futile search for control in a variety of screen interactions. In Scene 2, the reader's mouse-over interactions construct a photographic portrait of the protagonist's wife from hundreds of miniature questions projected on the black background of the page, thus symbolizing his futile attempts at "revealing" her character and her true opinion of him (Figure 1).

Scene 5 features a metaleptic webcam (Bell 2016) projecting the reader's body onto the two-dimensional screen. Mousing over the image distorts it to the soundtrack of dissonant music, and the homodiegetic narrator comments, "I feel manipulated," thus constructing a parallel between the reader's simulated body and his own.

Despite the seemingly endless possibilities of de- and reconstructing our bodies online, post-human beings ultimately remain ensconced within Foucauldian dispositifs, in the kind of hegemonic power networks that make it near impossible to escape hypermediated toxicity, sexism, anti-fat attitudes, ableism, racism, and other appearance-centered forms of abuse. These phenomena are the side effects of postfeminist appearance culture (Rice 2014; Gill 2007; Riley et al., 2018)—a "culture of contradiction"

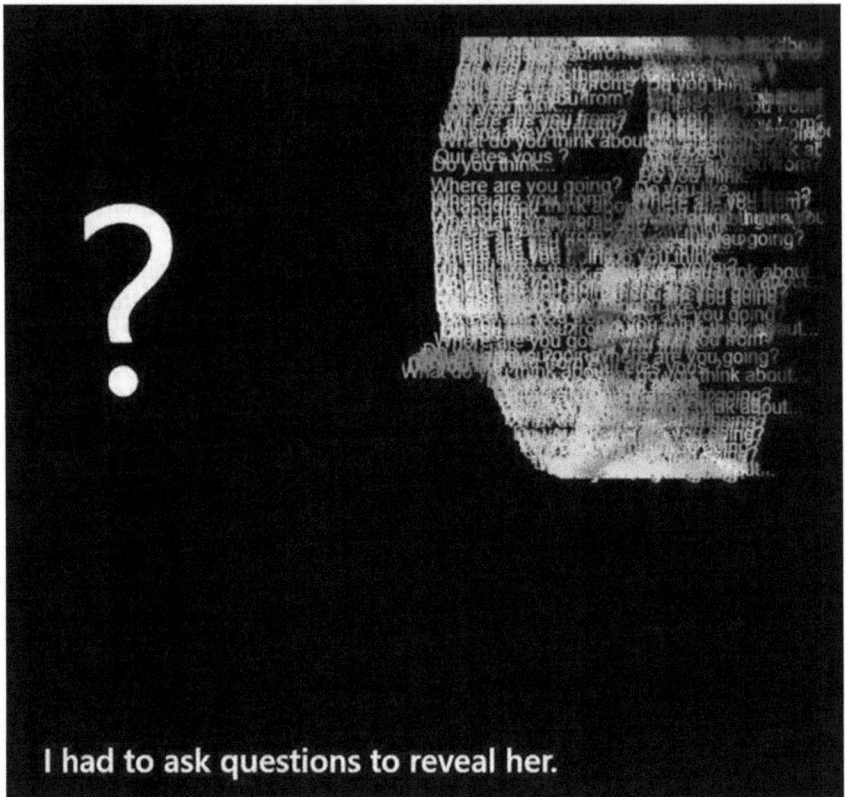

I had to ask questions to reveal her.

FIGURE 1 *Screenshot from* Loss of Grasp, *Scene 2.*

characterized by the paradoxical neoliberalist synergies of women's societal empowerment, the persistent, feminine "beauty myth," and "the beauty industry's colonization of women's bodies" (Rice 2014: i).

This persistence of the primacy of looking and being looked at (see Riley et al., 2016) has been theorized by media and cultural theorists as "the gaze" (Mulvey). The gaze forms the imagined or material starting point of unequal power relationships. It is seductively malleable and can take on a variety of highly effective manifestations. The Mulvean, binarist male gaze, which assumes that women are the passive object of the active, male voyeur, has been modulated and augmented throughout the history of visual culture. The "looking relations" (Berger 1972) we adopt as a way of naturalized looking and being looked at are primed by the cultural-hegemonial gazes we are exposed to on a daily basis, through magazine covers, billboard displays, social media "pic culture," and (hyper)sexualized bodies on screen. Among these broader hegemonial gazes are, for example, the fat-phobic gaze, which semiotically deletes and/or *a*bjectifies plus-sized bodies (Rinaldi et al. 2019). The medical and able-bodied gaze pathologizes sick, disabled and otherwise "abnormal," "unruly" bodies (Rice et al. 2018). In particular, it is targeted at perceived female monstrosities such as menstruating and pregnant bodies (Braidotti). The settler-colonial gaze exposes Black and other racialized bodies as the inferior other, limiting beauty ideals to the Western ideal of whiteness (Nelson). The cisgender gaze exposes transphobic tendencies by objectifying transgender people. The female gaze ambivalently incorporates the shift from an "external, male judging gaze" to an internal "self policing" gaze (Gill, "From Sexual Objectification": 104). It may represent deeper manipulation, since it invites female audiences to become more adept at scrutinizing their own and other female-coded images (Rice 2014; Riley et al., 2016).

That being said, in all examples of scopophilic visual culture discussed so far, the object of the gaze remains at a mediated distance. The female-coded body consistently figures as an object of imagined penetration yet manifest separation. In body-themed works of e-literature (and other forms of digital-interactive narrative), by contrast, this mediated distance is minimized or seemingly erased by a material, symbolically permeable interface between the body of the voyeur and the body on screen. These medium-specific affordances allow e-lit artists to hold the voyeur accountable for objectifying the target they are manipulating rather than simply observing.

While there are manifold ways in which e-lit has critically engaged with scopophilia, perhaps the most pertinent of these responses is the "ergodic gaze" (Ensslin et al. 2011). In Annie Abrahams' agency art e-poem, "Ne me touchez pas / Don't touch me," the user-activated, ergodic cursor physically enacts the gaze of the beholder. With every touch or click, respectively, these works materialize the reader-player's insatiable appetite to control the screen and all that it embodies. Reading "Ne me touchez pas" involves mousing over the body of a reclining, scantily dressed woman who is turned away from the camera and buries her head in her pillows to avoid the viewer's gaze.

The cursor, which Marie-Laure Ryan terms "the representation of the reader's virtual body in the virtual world" (2006 122), thus represents an "augmented me, ... [which] is 'you' and 'You,' as the narrator distinguishes us while also drawing us together ... [and making us] feel some liminal and flickering sense of presence through the screen" (Keogh 2018: 3). This idea of "embodiment ... distributed across both sides of the glass" (5) is a phenomenological concept of presence that reflects the mutual incorporation of player and game, of reader and digital-born text. And yet, while, in most video games, it would likely be correct to say, with Keogh, that "as we touch the videogame, it touches us back" (4), this is not the case in "Ne me touchez pas." The work poignantly implements the despair of the faceless woman's body on screen that reader-players keep poking mindlessly, without obtaining the expected cybernetic feedback. She does not want to touch us back: all she wants is to break free from the constraints of the ergodic gaze, a physically and forcefully enacted gaze that is wired into user interfaces with relentless univectorial certainty.

Writing New Languages, Spaces, and Worlds

Cixous' call for newly embodied forms of women's writing, symbolized by "white ink," has been widely read as a request for experimentation with new materialities of writing. In this context, Giovanna di Rosario highlights the works of María Mencía as powerful attempts at developing "a new poetic form of language" (2017 274) that are independent of the conversational conventions of patriarchy. In Mencía's video installation and Flash poem, "Birds Singing Other Birds' Songs" (2001), for example, this new language oscillates elusively between human and animal and manifests as a perpetual "play with letters, sounds, and forms" (Di Rosario 2017: 274) that blend and morph trans-semiotically without enabling the construal of transparent meaning.

Di Rosario further suggests that postdigital, medium-critical *écriture feminine* may involve writing new *spaces*, which offer new possibilities and dimensionalities for linguistic expression and/or break their own constraints for poetic writing. In Christine Wilks' and Andy Campbell's Unity-based, immersive 3D fiction *Inkubus*, for example, the reader-player navigates the interior of a human body. Not unlike in a first-person shooter, the player-character can hit inimical units. Yet these units of opposition are fragments of appearance-based cyberbullying, which move towards the camera eye and can be shot with a fireball, thus re-appropriating and detourning phallic mechanics.

Yet postmedia also means participatory culture, and to break free from postfeminist constraints, the need arises for readers from a variety of backgrounds to become co-creators in works that may help them envisage new worlds in which they might feel at home in their bodies. To address this need, the "Writing New Bodies" project (Ensslin et al. forthcoming) has employed participant research to explore how a research-creation project might help young woman-identified and gender nonconforming individuals open up paths for envisioning new body worlds through digital fiction—a type of e-lit that foregrounds playable narrativity in medium-specific ways.

Working with bibliotherapy experts, WNB is seeking to understand how digital fictions can be used as a body-image intervention. The project is a collaboration between digital media scholars, gender and the body theorists, critical psychologists, and award-winning digital writer, artist, and game developer, Christine Wilks. Through workshop collaboration and creative autofictional writing exercises on paper and online (using Twine), participants raised and discussed thematic, narrative, semiotic, and ludic design ideas informing the development of the WNB digital fiction.

Qualitative analysis of workshop transcripts, Twine creations, and written narratives in MAXQDA generated fifty distinct body-image themes, only some of which can be highlighted here. For example, participants grappled with the duality of consciousness and difficulty in breaking out of, and resisting, binary thought. The mirror surfaced as a symbol embodying all external and internal pressure and judgment. Further, there was an acknowledgment of complex, ambiguous, or nuanced experiences with one's body. Participants recognized a theme of looping or a cycle (similar to yo-yo-dieting) from which it is difficult to escape. Participants discussed their desire to be content with these moments which drew conflicting responses in themselves. Finally, they spoke of a longing to embrace the transformative potential of "the gaze" or locating space for the nonbinary gaze.

The WNB co-designers shared a desire for an intervention that could rupture the usual restrictive, binary ways in which we come to understand our bodies. The envisioned digital fiction, or literary game, will open up novel pathways and unleash new ways of knowing and being with the body. Ultimately, an applied work of digital fiction must em-body—in the sense of enabling embodied imaginaries of ontological repositionings in an intersectional variety of reader-players—a challenge that applied electronic literature is only beginning to tackle, yet that opens up new opportunities for transdisciplinary collaboration and community co-authorship.

References

Bell, Alice (2016), "Interactional Metalepsis and Unnatural Narratology," *Narrative* 24 (3): 294–310.

Berger, John (1972), *Ways of Seeing* [TV program], London: BBC.

Borràs, Laura (2011), "Pieces of Herself," *Electronic Literature Collection*, Vol. 2, http://collection.eliterature.org/2/works/davis_pieces_of_herself.html.

Borràs, Laura, Talan Memmott, Rita Raley, and Brian Stefans (2011), *Electronic Literature Directory, Vol. 2*. Cambridge, Massachusetts: Electronic Literature Organization. https://collection.eliterature.org/2/works/davis_pieces_of_herself. html

Bouchardon Serge (2018), *Storyface*, https://play.google.com/store/apps/ details?id=com.utc.costech.storyface&hl=en_US.

Braidotti, Rosi (1997), "Mothers, Monsters and Machines," in Katie Conboy, Nadia Medina, and Sarah Stanbury (eds.), *Writing on the Body: Female Embodiment and Feminist Theory*, 59–79, New York, NY: Columbia University Press.

Brodesco, Alberto and Federico Giordano (2017), "The Border Within: The Human Body in Contemporary Media," in Alberto Brodesco and Federico Giordano (eds.), *Body Images in the Post-Cinematic Scenario: The Digitization of Bodies*, 9–17, United States: Mimesis International.

Cixous, Hélène (1976), "The Laugh of the Medusa," trans. Keith and Paula Cohen, *Signs* 1 (4): 875–93.

Di Rosario, Giovanna (2017), "Gender as Patterns: Unfixed Forms in Electronic Poetry," in María Mencía (ed.), #WomenTechLit, 41–54, Morgantown, WV: West Virginia University Press.

Dovey, Jon and Helen Kennedy (2006), Game Cultures: Computer Games as New Media, Maidenhead: Open University Press.

Ensslin, Astrid (2011), "From (W)Reader to Breather: Cybertextual De-intentionalisation in Kate Pullinger et al.'s 'Breathing Wall'," in Ruth Page and Bronwen Thomas (eds.), New Narratives: Stories and Storytelling in the Digital Age, 138–52, Lincoln, NE: University of Nebraska Press.

Ensslin, Astrid, Carla Rice, Sarah Riley, Christine Wilks, Megan Perram, Hannah Fowlie, Lauren Munro, and Aly Bailey (2020), "These Waves … : Writing New Bodies for Applied E-literature Studies" in ELO2019 Gathering (Cork, Ireland), edited by Pedro Nilsson-Fernàndez & James O'Sullivan, a special issue of electronic book review, April 5, https://doi.org/10.7273/c26p-0t17.

Foucault, Michel (1995), Discipline and Punish: The Birth of the Prison, New York, NY: Vintage Books.

Gill, Rosalind (2003), "From Sexual Objectification to Sexual Subjectification: The Resexualization of Women's Bodies in the Media." Feminist Media Studies, 3 (1): 99–106.

Gill, Rosalind (2007), "Postfeminist Media Culture: Elements of a Sensibility," European Journal of Cultural Studies [online first], https://doi.org/10.1177/1367549407075898.

Grosz, Elizabeth (1994), Volatile Bodies, Bloomington, IN: University of Indiana Press.

Keogh, Brendan (2018), A Play of Bodies: How We Perceive Videogames, Cambridge, MA: MIT Press.

Manovich, Lev (2014), "Postmedia Aesthetics," in Marsha Kinder and Tara McPherson (eds.), Transmedia Frictions: The Digital, the Arts, and the Humanities, 34–44, Oakland, CA: University of California Press.

Mencía, María (ed.) (2017), #WomenTechLit, Morgantown, WV: West Virginia University Press.

Mulvey, Laura (1975), "Visual Pleasure and Narrative Cinema," Screen 16 (3): 6–18.

Nelson, Charmaine (2010), Representing the Black Female Subject in Western Art, New York, NY: Routledge.

Parisi, Luciana (2004), Abstract Sex: Philosophy, Biotechnology and the Mutations of Desire, London: Continuum.

Rice, Carla (2014), Becoming Women: The Embodied Self in Image Culture, Toronto: University of Toronto Press.

Rice, Carla, Eliza Chandler, Kirsty Liddiard, Jen Rinaldi, and Elisabeth Harrison (2018), "Pedagogical Possibilities for Unruly Bodies," Gender and Education 30 (5): 663–82.

Riley, Sarah, Adrienne Evans, and Alison Mackiewicz (2016), "It's just between Girls: Negotiating the Postfeminist Gaze in Women's 'Looking Talk'," Feminism and Psychology 26 (1): 94–113.

Riley, Sarah, Adrienne Evans, and Martine Robson (2018), Postfeminism and Health, London: Routledge.

Rinaldi, Jen, Carla Rice, Crystal Kotow, and Emma Lind (2019), "Mapping the Circulation of Fat Hatred," Fat Studies 9 (9): 1–14.

Ryan, Marie-Laure (2006), Avatars of Story, Minneapolis, MN: University of Minnesota Press.

SECTION II

Forms

8

Ambient Art and Electronic Literature

Jim Bizzocchi

Ambient video art and electronic literature are separate domains, with fundamental and significant differences in goals, aesthetics, and experience. The goal of ambient art is an experience that is pleasurable, but one that doesn't require ongoing attention by the participant. The goals of electronic literature are as diverse as the goals of literature itself, but in general the creators of electronic literature aim for an experience that involves and holds the reader more directly than the softer touch of ambient art. This chapter outlines the differences between the domains, but does not see these differences as boundaries, walls, or contradictions. Rather, the differences are positioned as end points on various continua—dynamic poles that define sets of creative dialectics. My own work is used as an example of how the dialectics of both ambient art and electronic literature can play out in practice.

Electronic Literature

The definitions and boundaries of these two domains are not absolutely agreed upon within their respective communities. "Electronic literature" (also known as e-literature, or e-lit) is the more complex domain. N. Katherine Hayles outlines a number of creative directions within the broad field of electronic literature: hypertext fiction, network fiction, interactive fiction, location-based works, interactive drama, and generative e-lit, among others. Formulating any boundary or definition for this widespread body of work and community of creators is a difficult task. The Electronic Literature Organization (ELO) maintains that its mission is "to promote the writing, publishing and reading of literature in electronic media." Hayles cites the more focused conclusion of a committee asked by the ELO to formulate a definition. The committee saw electronic literature as "work with an important literary aspect that takes advantage of the capabilities and contexts provided by the stand-alone or networked computer." This still leaves a range of directions too wide for any neat categorization. Hayles positions the field's hybrid roots and branches as a strength, seeing electronic literature as a "hopeful monster" made up of parts from "diverse traditions that may not always fit neatly together."

What are the attributes that are broadly shared within this diverse body of work? The ELO says of e-literature that "reading and writing remain central to the literary arts." Dene Grigar sees interactivity and participation as core modalities, and experience and immersion as desired outcomes for works of electronic literature. I would add that insofar as it is literature, most electronic literature privileges some type of narrative experience. In this regard, I am using a broad conception of narrative and "narrativity," which could include plot, character exposition, storyworld creation, empathic emotion, or combinations of these and other narrative elements (Bizzocchi 2011: 5–10).

Ambience and the Moving Image

Ambient moving image art is a slow form of mediation—one that plays out leisurely on the large video screens increasingly prevalent in our domestic, corporate, and public spaces. What does one do with these screens when they are not providing the direct engagement of home theatre, television programming, electronic gameplay, or utilitarian information display? Ambient video offers striking yet slowly changing imagery—reconciling visual pleasure and intermittent viewer attention. The various forms of ambient video can be based on either representational or abstract imagery—but they must always give visual interest. Sometimes called "living photographs" or "video paintings," the form exists in what Higgins would term an "intermedia" space—with content and reception experience drawing on the aesthetics of photography, painting, cinema, and video.

Ambient moving image art is a more limited field of endeavor than electronic literature—in both terms of the size of its practicing community and diversity of its output. However, there are distinct directions of manifestation. These include some well-known kitsch examples such as the venerable televised Yule Log, video aquaria, and digitally networked webcam visual sites. Much more interesting artistic directions are abstract moving image art such as the Visual Music movement, the representational works of video installation artists such as Bill Viola or Stan Douglas, or the slow-paced nature experiences of Simon King or Steve Lazur. The latest manifestation of the ambient aesthetic is the "Slow TV" movement which has emerged in Norwegian and British broadcast television.

Regardless of the differences in content and approach, all of these variations draw upon the core aesthetic of Brian Eno's "ambient music"—which he says "must be as easy to ignore as it is to notice." My own ambient video art is conceived very much in the spirit of Eno's dictum. My aesthetic includes three tests for ambient artistic success:

i. Ambient video art must *never require* your attention.

ii. Ambient video art must *always reward* your attention with visual pleasure.

iii. Ambient video art must *continue* to provide visual pleasure after repeated viewing.

My work relies on three artistic interventions to meet these tests. Because ambience is a slow form, I need strong visual compositions to support the extended on-screen shot duration. My ambient work benefits from the striking natural imagery of the Canadian Rockies and the North American western mountains and coasts. Second, I treat time as plastic, varying subject speed to maximize visual interest. Finally, I incorporate a complex visual transition process. Each shot gradually changes to the next through a series

FIGURE 1 Rockface *transition*.

of "magic realist" transformations within the frame. Figure 1 shows the first transition from my film *Rockface*. The transition begins with the waterfall from the incoming shot, exploding over the mountains at the end of the on-screen shot. The transition ends by introducing the lake and the rest of the second shot in a circular wipe transition starting at the bottom and then filling the screen with the second shot.

As we have seen, ambient video artists differ on their various approaches to the form. However, whether the images are representational or abstract, realistic, or surreal, all the practitioners utilize interesting imagery to build an ambient experience.

Contradictions and Dialectics

The two domains of electronic literature and ambient art are in some ways inconsistent. Most electronic literature incorporates the pleasure of story—even if the process of storytelling differs significantly from earlier forms. Electronic literature often involves the incorporation of some form of interaction by the reader. However, the classic and fully formed narrative arc of cinema or the novel is difficult to achieve in interactive environments. Despite this, e-lit creators can build smaller "micronarrative" elements to support a more flexible narrative plot progression. These "micronarrative" elements are localized units of plot coherence that string together in various combinations to move the overall plot forward (Jenkins 2004: 118–19; Bizzocchi et al. 2014). Other narrative elements are fully consistent with interactive design: the pleasures of character recognition, storyworld experience, and emotional empathy. Narrative sensibilities can also be incorporated within the design of the interface, allowing the user to experience a sense of narrative inflection and narrativity in the process of interaction and choice-making (Bizzocchi et al. 2011; 2008: 260–77).

Ambient art, on the other hand, is fundamentally inconsistent with both narrative and interactivity. The problem is the question of viewer experience. An ambient experience, by definition, can not be one that demands your attention. The viewer must feel free to disengage at any time. Narrative, however, exercises a firm hold on viewer attention. In a narrative experience, once we see the beginnings of a plot and build an initial sense of character identification, we feel compelled to follow the story through to its end. In cinematic terms, ambient art is inconsistent with the cinema of narrative, but it can be positioned as a form of the "cinema of attractions"—the cinema that relies not on story, but on the power of the visuals to provide audience interest (Gunning 1986: 63–70). However, the "cinema of attractions" in ambient art is a highly attenuated form of "attraction." Extreme cinematic attractions, built on horror, shock, or strong emotion, rivet our attention to the screen. This enforced attraction prevents the creation of a truly ambient experience. In my own work, the visual compositions, the manipulated time frame, and the intricate transitions all provide visual attraction and pleasure, but do not necessarily compel ongoing fixed attention.

For different reasons, interactivity is also inconsistent with ambient experience. Any interaction design that involves a set of conscious choices requires a direct user engagement. However, ambient experience performs a more subtle dance with the viewer—a dance that is slow, and intermittent, and light of touch. The requirement to interact is the requirement to engage directly, breaking the freedom of the viewer to easily disengage and drift away at any time.

This analysis of ambient art's incompatibility with narrative and interaction is valid at a certain level, but it breaks down if we move from the narrow logic of absolutes and categorical binaries to the fuzzier logic of gradations. None of these—narrative, interactivity, or ambience—are in fact absolute. Further, it is the business of creative artists to look for and test the limits, boundaries, potentials, and intersections across any set of mediated forms. Eric Zimmerman understands this in his examination of the relationship of games and story. He is not interested in whether a game IS a narrative—he is interested in *how* a game is a narrative. His resolute rejection of categorical binaries leads to a more useful focus on the actual design and experience of mediated artifacts.

In a similar vein, our critical terms (narrative, interactivity, and ambience) can be treated more effectively as design parameters—as scalars rather than categorical absolutes. A low level of "narrativity" is not inconsistent with ambient experience. In my own work, there is no standard narrative plot, but there is a logical ordering of shots. A thematic progression is embedded within the sequence of visuals, but it is not as direct as similar arguments within more typical nature documentaries. There is also an evocation of a "storyworld" in my work—a transcendent natural environment that takes us out of our normal urban cityscapes. Other ambient storyworlds are possible. The *Frame* channel on my local cablevision feed presents a series of long slow shots in interesting urban or resort environments—with people ambling in and out, unaware they are being filmed and cablecast. I call this a "webcam" ambient aesthetic, one that also benefits from a limited sense of narrativity. We see faces in visually interesting locations, but we never get any deep sense of who these people are, and what their lives are really like. This is a liminal sense of narrative character and storyworld. It doesn't promote any strong sense of identification, but it does support a moderate level of interest without necessarily locking one's eyes to the screen.

In the e-literature community, works like Brian Kim Stefans' *The Dreamlife of Letters* draw upon the traditions of concrete poetry to create a visual experience that presents words and text in a context that privileges the aesthetics of the visual and the poetics of motion graphics. Jim Andrew's *dbCinema* treats the world of internet images as a visual database, presenting them within a transformative and fluid visual stream. Sandy Baldwin's *The Lincolnshire Poacher* combines a still image, minimal on-screen text, and a dense sound mix to present an experience that is as much mirror as it is window. These and other e-lit works employ a variety of strategies and aesthetics in order to provide a level of engagement that reconciles ambient experience with a limited sense of narrativity.

In a similar fashion, interactive design is not completely incompatible with ambience. Explicit and ongoing interactive choice certainly inhibits ambience, but there are other options for interactive input. Some choices could be made by the user before the experience begins, or as optional modifications to the ongoing ambient flow. The types of shots, the length of shots, the selection of music or sound—all could be modified by the user in advance of the experience. Another possible interactive model consistent with ambience might be the incorporation of nonintrusive interactive channels such as location sensors or body sensors.

The interactive model with the greatest affinity to ambience is to use a form of interactivity that is self-contained within the artwork itself—independent of user input. An artwork can be run by a generative system designed by the artist. Generative art is a relatively under-recognized computational form, but one with considerable aesthetic power and expressivity. Despite its comparative lack of recognition, it has a long history, one that Galanter claims is "as old as art itself." Generative literary forms include Surrealist word games, texts from the Oulipo movement, and electronic literature works by a number of practitioners such as Jim Andrews, Noah Wardrip-Fruin, Bill Seaman, Nick Montfort, and many others. Generative works can be consistent with ambient experience because the interaction is carried out by the computational system, requiring no interference with user attention or ambient state.

My Computational Ambient Videos

My exploration of computationally generative art grew out of the last of my three core aesthetic principles: "Ambient video art must continue to provide visual pleasure after repeated viewing." I wondered if my linear ambient videos would meet this re-playability test. In the end I decided that their compositional strength and intricate visual transitions would hold up under repeated viewing. A fine art still photograph could hang on your wall and give pleasure—and my photographs moved, which would provide more visual variation. However, the question morphed from a practical concern to a creative challenge: could I make a computational version of my linear ambient videos? This challenge is something of an artistic Turing Test, an exercise in what my colleague Kenneth Newby calls "encoding practice." I wanted to create a generative system that could approximate the work I did in my linear ambient videos.

The design of generative art creation does have its own set of creative dialectics and contradictions. For me, there was an ongoing tension between artistic control on the one hand, and the maximization of output variability and therefore re-playability on the other.

artistic control ⇔ variability/replayability

I found that design decisions that gave me more creative control tended to decrease the range of possible variation in the system's output. For example, I preferred a closed system in terms of shot sources. I wanted to shoot or direct all the shots myself rather than rely on autonomous input from open sources for video files. This decision maximized my creative control over the visuals, but it also limited the number of shots I could include, and hence the variation in the visual output.

This dialectic between artistic control and variability/re-playability took many forms. Another early example was my choice of visual transition devices. In my linear videos, the complex visual transitions were hand-crafted to fit each specific pair of shots. Since my generative videos used variations on random sequencing decisions, I had to give up artistic control and find a transition strategy that would work with a wide variety of shot pairs. I decided on luminance and chrominance transitions—devices that use the brightness or color values of a given shot to shape the transition. The example (from *Seasons II)* in Figure 2 shows how this luminance transition overlays the new shot on top of the various levels of brightness of the old shot. These transitional devices provide a degree of visual interest to the transformation between any pair of images.

The most significant example of the control/variability dialectic is in my revision of the system's sequencing process. My generative system is based on a database of video shots. In the original version, shots were selected and sequenced from my database in a completely random process. This simple mechanic was reasonably effective, but I decided to improve it. I wanted more control over the sequencing decisions to improve the visual flow and impart a degree of semantic coherence.

The solution was to use a hierarchical metadata system to tag the shots for their content, and sequence the shots based on the tags. Each shot got a single higher-level tag for season (summer, fall, winter, spring) and a lower-level complement of tags for content (river, snow, mountain, clouds, etc.). The system's logic (programmed in Max) creates a series of "year" films as an ongoing visual output. Each "year" film consists of four seasons in order, and each season consists of three sequences of three shots each. A single

Before Luminance Transition

Luminance Transition Begins

Luminance Transition Continues

Luminance Transition Complete

FIGURE 2 *Stages of a luminance transition.*

season segment might consist of these nine shots: river/river/river—clouds/ clouds/clouds—mountain/mountain/mountain. The selection of specific content tags are randomized, and the selection of shots from the chosen tags is also randomized. Since there are currently 250 shots in the shots database, and there are sixteen variations on the luminance transitions, the variability of the system is still quite high. It runs indefinitely, spitting out a series of four-season "year" films indefinitely, with very little repetition of sequencing.

This content selection logic had a significant effect on the system's output. The level of randomized variation was diminished, but visual flow and semantic coherence increased significantly. It's not surprising—in effect I had created a simple montage editing machine. My claim was that my machine mimicked the work of a human video editor—one who was "competent but not brilliant." If it was enrolled in my introductory video production class, it would have earned a C+. Not a bad Turing Test result for a few lines of code.

The work of my research colleagues adds generative audio to the experience. Philippe Pasquier has created a generative soundscape system, and Arne Eigenfeldt has created a generative music system. We have developed processes to exchange timing, content, and affective information between our three systems. Working together, they generate an ongoing and unified audiovisual flow of moving image, music, and soundscape in our collaborative art work *Seasons II*.

From Ambience to Direct Engagement

More sophisticated associational and sequencing capabilities can allow my system to simulate a more traditional cinematic documentary output. We have begun the initial planning and prototyping for a generative version of the documentary "city film." This work will be an "open documentary"[1] that systematically presents different facets of a complicated urban society. This will include themes such as wealth, poverty, housing, diversity, commerce, transportation, manufacture, recreation, pollution, etc. A much larger database of shots will support this increased thematic scope, and a well thought-out tagging/sequencing structure will enable the construction of coherent sequences. Higher-order groupings of these sequences will build a reasonable degree of semantic connection and thematic flow. The completed system will continuously spit out an ongoing series of short "films"—each one unique in content, style, or both.

This new work will not be a "storytelling" machine, but it will support a sense of "narrativity"—the expressive presentation of storyworld, human characters, thematic development, and emotional tenor. As these semantic connections become more sophisticated, the resulting experience becomes less ambient. This more narrativized mediation will tend to hold our attention more strongly. The move away from an ambient aesthetic will lead to an experience that is more intellectually complex and emotionally engaging. This more direct engagement will shift the aesthetic from the ambient into the broader body of electronic literature.

It is worth noting that the core logic and operations of the system can support either an ambient experience, or a more focused documentary experience. The shift from ambience to thematic documentary requires three modifications to the core system: a larger and more diverse database of shots, a more complicated taxonomy of metadata tags, and a more sophisticated overlay of sequencing rules to work the tags and the shots. The underlying system logic, however, remains the same in both cases. This ability to share foundational structure is an indication that there is no hard boundary between computational ambient art and computational literature. A further indication is the role of metadata and tagging within the sequencing process. The system's logic is driven by text data—the content metadata.

---

[1] I have borrowed the term from my colleague William Uricchio's "Open Documentary Lab" at MIT, *Open Doc Lab*, opendoclab.mit.edu.

The metadata tags reflect the content of each shot. The sequencing of the metadata text by the system can be seen as the sequencing of words into a poem-like structure. Because of this, the visual sequences are themselves poem-like—relying on iteration and patterned variation to build coherent semantic flow. If "code" and algorithmic logic is the "grammar" of electronic literature, a computational system that supports both ambient works and more traditional forms is in fact a form of electronic literature itself.

Acknowledgments

This research program is supported by the Social Science and Humanities Research Council of Canada (SSHRC).

References

Andrews, Jim (n.d.), *dbCinema*, www.vispo.com/dbcinema/index.htm (accessed December 27, 2016).

Baldwin, Sandy (2006), "Goo/The Lincolnshire Poacher," [Video] *Hyperrhiz 02: Gallery: Winter* 2006, dx.doi.org/10.20415/hyp/002.g02.

Bizzocchi, Jim (2007), "Games and Narrative: An Analytical Framework," *Loading: the Journal of the Canadian Game Studies Association* 1 (1): 5–10.

Bizzocchi, Jim (2008), "The Aesthetics of the Ambient Video Experience," *Fibreculture Journal* 11, www.eleven.fibreculturejournal.org/fcj-068-the-aesthetics-of-the-ambient-video-experience/.

Bizzocchi, Jim et al. (2011), "Games, Narrative, and the Design of Interface," *International Journal of Arts and Technology*, Special Issue on: "Interactive Experiences in Multimedia and Augmented Environments," Teresa Romão and Nuno Correia (eds.), 4 (4): 260–77.

Bizzocchi, Jim et al. (2014), "The Role of Micronarrative in the Design and Experience of Digital Games," *DiGRA '13: Proceedings of the 2013 DiGRA International Conference: DeFragging Game Studies*, 161–97, DiGRA, www.digra.org/digital-library/.

Eigenfeldt, Arne et al. (2015), "Collaborative Composition with Creative Systems: Reflections on the First Musebot Ensemble," *Proceedings of the Sixth International Conference on Computational Creativity*, Park City, Utah, June 29–July 2, 2015, 134–43 Provo, UT: Brigham Young University Press.

"ELO History" (n.d.), Electronic Literature Organization, www.eliterature.org/elo-history/. Accessed December 27, 2016.

Eno, Brian (1978), "*Music for Airports*" [Album liner notes accompanying CD], PVC 7908 (AMB 001), *The Frame*, Shaw Cable.

Galanter, Philip, (2003), "What is Generative Art? Complexity Theory as a Context for Art Theory," *GA2003–6th Generative Art Conference*, www.philipgalanter.com/downloads/ga2003_paper.pdf.

Grigar, Dene (2014), [Personal interview], September 29, 2014.

Gunning, Tom (1986), "Cinema of Attractions: Early Film, its Spectators and the Avant-Garde," *Wide Angle* 8 (3/4): 63–70.

Hayles, N. Katherine (2007), "Electronic Literature: What is it?," v1.0, January 2007, Electronic Literature Organization, www.eliterature.org/pad/elp.html.

Higgins, Dick (2001), "Intermedia," *Leonardo* 34 (1), reprint of the original essays 1966 and 1981.

Jenkins, Henry (2004), "Game Design as Narrative Architecture," in Noah Wardrip-Fruin and Pat Harrigan (eds.), *First Person: New Media as Story, Performance, Game*, 118–19, Cambridge, MA: The MIT Press.

Newby, Kenneth (2015), Workshop on generative art, Simon Fraser University, March 10, 2015.

ReCycle (2010), [Film] Dir. Jim Bizzocchi; Glen Crawford (cinematography).

Rockface (2007), [Film] Dir. Jim Bizzocchi; Glen Crawford (cinematography).

Seasons II (2016), [Artistic collaboration] Jim Bizzocchi, Arne Eigenfeldt, Philippe Pasquier, and Miles Thorogood, ELO2016: Electronic Literature Organization Gallery installation, June 10–12, 2016, Victoria, BC.

Stefans, Brian Kim (2000), "The Dreamlife of Letters," *Electronic Literature Collection*, Vol. 1, 2006, www.collection.eliterature.org/1/works/stefans-the_dreamlife_of_letters/dreamlife_index.html.

Thorogood, Miles, and Philippe Pasquier (2013), "Computationally Generated Soundscapes with Audio Metaphor," *Proceedings of the Fourth International Conference on Computational Creativity*, Sydney, 256–60, http://www.computationalcreativity.net/iccc2013/download/ICCC2013-Proceedings.pdf/.

"What is E-Lit?," Electronic Literature Organization, www.eliterature.org/what-is-e-lit/, accessed December 27, 2016.

Zimmerman, Eric (2004), "Narrative, Interactivity, Play, and Games: Four Naughty Concepts in Need of Discipline," in N. Wardrip-Fruin and P. Harrigan (eds.), *First Person: New Media as Story, Performance, and Game*, 154–5, Cambridge, MA: The MIT Press.

9

Electronic Literature and Sound

John F. Barber

Contexts

Canadian media theorist Marshall McLuhan argues that each new medium incorporates, extends, or amplifies those it follows. As the earliest medium for communicating human thought, speech (sound) was incorporated and/or extended by later media technologies: writing, printing, reading, and visual arts (McLuhan 1964). Thus, speech claims a presence in most all media that follow (Levinson 1981). As James O'Donnell notes, "the manuscript was first conceived to be no more than a prompt-script for the spoken word, a place to look to find out what to say ... to produce the audible word" (O'Donnell 1988: 54).

McLuhan described two spaces, acoustic and visual, in which humankind has contextualized itself. "Acoustic space ... is spherical, discontinuous, non-homogeneous, resonant, and dynamic," he says. "Visual space is structured as static, abstract figure minus a ground; acoustic space is a flux in which figure and ground rub against and transform each other" (McLuhan and McLuhan 1988: 33).

In his descriptions, McLuhan expands the terms "figure" and "ground," both coined by Danish psychologist and phenomenologist Edgar Rubin in his 1915 dissertation exploring visual perception (Rubin 1915). Ground is surface, configurational, and comprised of all available figures, objects rising from or receding into ground (McLuhan and McLuhan 1988: 5). Ground is subliminal, spatial, universal, a surround, corresponding to the environment in which sound(s) exist. Simultaneously, ground is beyond perception except through analysis of emerging and receding figures (McFarlane 2013: 62, 103).

For McLuhan, acoustic space is ground, the surface from which figures (sounds) emerge and into which they recede. Figures are sounds heard in that space. They help conceptualize the space. McLuhan suggests expansive, unseen possibilities within acoustic space, making it more powerful and encompassing than visual space with its more precise and limited fixed point of view.

Acoustic space is a world awash in sounds and pre-literate (pre-speech and writing) humankind, the only ever to live in this space, relied on sound as an important sensory input. Sound formed the basis for humankind's explanations of and interactions with the surrounding physical world. With aural information emerging from all directions, and with no opportunity to shut off or organize the constant stream of sound, pre-literate humankind perceived its world as both surrounding and inclusive, a permeable extension of itself, and they of it (Levinson 1999: 5–6).

Acoustic space, filled with environmental sounds, was, we might suppose, a fearful wilderness. The emergence of speech technology allowed pre-literate peoples to communicate abstract thoughts regarding their situation and agency. Storytellers produced explanations for the sounds in acoustic space and wove them into larger narratives that helped explain the presence and purpose of humankind. Orality provided a means to preserve and share cultural histories and memories. Alphabets and writing, according to McLuhan, preserve and extend the aural nature of speech (McLuhan 1962).

McLuhan hoped that evolving forms of electric media with their ability to convey sound over time and distance would reverse the ascendency of the visual and transition humankind back to acoustic space. He argues that electric technologies extend the human nervous system into a global embrace, abolishing time and space, and imploding divisions between formally diverse peoples and cultural issues. He sees possibilities for far-flung citizens to communicate with one another, in what he calls the global village (McLuhan 1962: 31).

Within the global village, issues and peoples are no longer separate, or unrelated. Instead, peoples' lives are connected (McLuhan 1964: 20). The global village is "a brand-new world of allatonceness [all-at-once-ness] … a simultaneous happening. We have begun again to structure the primordial feeling, the tribal emotions from which a few centuries of literacy divorced us" (McLuhan and Fiore 1967: 63).

In short, in the global village, using various electric media, people could, metaphorically, talk among themselves in virtual town centers, or across their virtual backyard fences.

Forms

As to a medium to facilitate this dialog, McLuhan suggests the electric medium of radio, which he says resonates as a tribal drum, its magic weaving a web of kinship and prompting more depth of involvement for everyone (McLuhan 1964: 259–60). Radio is an extension of the human sensorium matched only by speech, he adds. As such, radio affords tremendous power, as "a subliminal echo chamber," to touch and play chords (memories and/or associations) long forgotten or ignored (McLuhan 1964: 264).

As a "fast hot medium" radio provides accelerated information throughput for personal information frequently utilized to involve people with one another (McLuhan 1964: 265, 267). Radio, says McLuhan, offers a "world of unspoken communication between writer-speaker and the listener" (McLuhan 1964: 261).

This tendency to connect diverse community groups, according to McLuhan, produces an artifact more compelling than, for example, the newspaper, with its continued emphasis on the linear pattern of the printed word. The opportunities afforded by current and future digital media for combining, remixing, and remediating all forms of content, including sound, may predict a return to acoustic space (ground) characterized by the verbal, musical, and poetic traces and fragments (figures) of oral culture (Edmund Carpenter 1970).

Practices

To put these ideas into practice, consider radio and speech. Broadly, speech, based upon verbalization of abstract thought, is a fundamental component of narrative (the recounting of a sequence of events and their meaning), the driver of storytelling (the addition of setting, plot, characters, logical unfolding of events, a climax), the basis for literature (written works considered to possess lasting artistic merit), and the various practices and cultures associated with its production and consumption (reading, writing, and listening).

Radio is an ecology of sound-based content historically shared across time and distance from creator to consumer. Little opportunity is provided the consumer to answer back. But, with the digital turn, the technological means for creation, communication, and consumption of sound-based content are easily available to anyone interested.

By sharing narratives across time and space with far-flung audiences, radio serves as a storyteller, binding audiences in the act of listening. While radio programming has historically been predominately speech and music, there is no reason to think it cannot express higher literary values, especially when these audiences, as noted, can create and share content, either in response, or as original expression. With sound and radio, there is potential for anyone interested to be both creator and consumer, creating and sharing sound-based content as narratives or stories, even literature.

Radio and speech, both based on sound(s), share commonalities with electronic literature. All are ephemeral, temporal, disappearing soon after their initial production. Speech, radio, and electronic literature are present but invisible, a feeling, a sense, capable of facilitating tangible experiences. They can connect people using invisible, disembodied sound (voices, music, other) rich with representation and fertile with ability to engage listeners' deep imaginations. Radio subsumes speech, re-emphasizes the aural, and returns the paralanguage qualities that printed text or pixilated screen strips from speech. This promotes deep listening, a term proposed by Pauline Oliveros to describe a philosophy of "listening in every possible way to everything possible" (Oliveros 1995: 19).

This combination of sound, radio, and electronic literature suggests that we broaden our understanding and appreciation of sound as integral to electronic literature, both as a changing cultural artifact and creative expression. Could sound and radio promote imagination, interaction, even immersion with regard to electronic literature? Could the radiophonic voice, as a trace of the body, immaterial, manifest powerfully enough to engage listeners with compelling narrative experiences? I believe the answer is affirmative and that connections might be made regarding contexts, forms, and practices.

With regard to contexts, McLuhan considers acoustic space ground, from which emerge figures of sound. The technology of speech provides meanings and a methodology for sharing narratives associated with sound(s). Speech, as the basis for narrative and storytelling, was incorporated into writing, printing, reading, and electric media, like radio, each of which extended the voice over time and distance.

Radio provides a form in which to produce, broadcast, archive, and curate sound. Sound-based, or sound intensive electronic literature provides a fluid ground, which when facilitated by the features and affordances of online, on demand digital radio can provide virtual listening spaces—think podcasts, streaming, and sound-sharing services—that link sound and listening to curation, inquiry, and making of literary media art that is both creative and compelling.

Sound(s), especially when designed and/or utilized to provide an immersive context, can provide valid literary experiences. For example, radio drama, with its foundation in scripted dialog, sound effects, and music, is amazingly effective at invoking listeners' imaginations, placing them within the narrative context, and engaging them in a literary experience than can have lasting value.

If sounds can provide valid literary experiences, then we can locate narrative and storytelling not solely in reading and writing but also in the act of listening. This suggests new practices for the creation and consumption of sound-based electronic literature.

Conclusion

With this short chapter, I suggest that sound is central to contexts, forms, and practices of electronic literature. More specifically, I suggest that works of electronic literature may incorporate or be based upon and/or inspired by sound(s) and their relationship(s) with narrative.

One *context* was McLuhan's idea of acoustic space as ground, from which emerge figures of sound. Speech, as an example figure, helped contextualize the acoustic space in which pre-literate humans found themselves. As a technology, speech allowed these early humans to share their abstract thoughts and apply meanings to the surrounding sounds. Storytellers, bards, and poets wove these explanations into narratives and stories, and for centuries held audience attention with the sound(s) of skillfully employed voices. According to McLuhan, speech, as the basis for narrative and storytelling, was, in turn, incorporated into writing, printing, reading, and electric media, like radio.

Understanding the centrality of sound(s), might we reconsider sound as integral to current and emerging forms of electronic literature? For example, the sound of a narrator's voice can be the framework for sharing stories. E-books are an example. In addition to human voice, could a sound-based work of electronic literature be composed of environmental and/or mechanical sounds? Given that music is a form of nonspoken narrative, I think the answer is affirmative.

Considering the central nature of sound may suggest new *practices* for the creation and consumption of sound-based electronic literature. Rather than sound(s) *in* or augmenting electronic literature, sound(s) might be heard *as* electronic literature. Sound-based electronic literature may be well suited to engage listener's because it engages the ear, and hence the listener's imagination. As result, sound(s) might be considered, like reading and writing, a central element of contexts, forms, and practices of electronic literature.

References

Carpenter, Edmund (1970), *They Became What They Beheld*, New York, NY: Outerbridge & Dienstfrey.

Levinson, Paul (1981), "Media Evolution and the Primacy of Speech," ERIC #ED 235510.

Levinson, Paul (1999), *Digital McLuhan: A Guide to the Information Millennium*, New York, NY: Routledge.

MacFarlane, Thomas (2013), *The Beatles and McLuhan: Understanding the Electric Age*, Lanham, NJ: The Scarecrow Press.

McLuhan, Marshall (1962), *The Gutenberg Galaxy: The Making of Typographic Man*, Toronto: University of Toronto Press.

McLuhan, Marshall (1964), *Understanding Media: The Extensions of Man*, New York, NY: McGraw Hill.

McLuhan, Marshall and Eric McLuhan (1988), *Laws of Media: The New Science*, Toronto: University of Toronto Press.

McLuhan, Marshall and Quentin Fiore, with Jerome Agel (1967), *The Medium Is the Message: An Inventory of Effects*, New York, NY: Bantam Books.

O'Donnell, James J. (1988), *Avatars of the Word: From Papyrus to Cyberspace*, Cambridge, MA: Harvard University Press.

Oliveros, Pauline (1995), "Acoustic and Virtual Space as a Dynamic Element of Music," *Leonardo Music Journal* 5: 19–22.

Rubin, Edgar (1915), *Synsoplevede Figurer: Studier i psykologisk Analyse. Første Del* [*Visually experienced figures: Studies in psychological analysis. Part one*], Copenhagen and Christiania: Gyldendalske Boghandel, Nordisk Forlag.

10

Augmented Reality

Anne Karhio

Augmented reality (AR) is a term used for media technologies that add symbolic or semantic elements, usually visual and digital, to the perceptual material environment, or live media content depicting this environment. It is an example of mixed reality—in other words, a compound experience consisting of the immediately perceived and a digital layer of mediated content. If virtual reality (VR) disengages the viewer/reader from the nondigital or nonmediated sensory, physical environment through an act of replacement, AR works add virtual/digital inputs to that environment. Typically, AR experiences require headsets (AR glasses or head-mounted displays) or portable screens (tablets and smartphones) that act as interfaces for the compound experience.

The verb "augment" originates from the Latin *augmentāre*, to increase, and thus suggests an addition, or a supplement, to the physical environment experienced without the introduction of artificially created content. There is a long history of immersive visual media technology even prior to the emergence of digital media, as audiences in the nineteenth century were lured to pay for access to early panorama displays, seduced by the promise of having their familiar physical environment replaced by a spatially or temporally distant vista. This kind of experience would be an early version of VR, an entirely immersive experience. But augmentation, too, has a history that precedes the emergence of digital technology. Lev Manovich considers augmentation less a technology than "a cultural and aesthetic practice," which covers various kinds of architectural and built environments, cinema and art, and, for example, urban spaces where electronic screens cover buildings and walls (Manovich 2006: 1–2). Some video art works in the 1970s, or what Geoffrey Alan Rhodes has termed "proto-AR," used projected live feed from a camera as a part of installations in gallery spaces (Rhodes 2014: 130). Twentieth-century cinematographers employed AR-style methods prior to the widespread adoption of contemporary digital interfaces by adding texts, images, or animation to otherwise realistic visual narrative aesthetic, like in the 1988 film *Who Framed Roger Rabbit*. In short, unlike immersive art works relying on pre-digital or digital media, AR works do not seek to replace the perceived environment with a virtual one, but alter it by integrating visual (or audible, or even tactile), symbolic material into it—an immersive suspension of disbelief is not possible, even if viewers remain unaware of the encoded processes contributing to the experience.

The first head-mounted device for an AR experience was created as early as the late 1960s by Ivan Sutherland at Harvard University (even though the headset was so heavy it had to be hung from the ceiling), but the term "augmented reality" itself was not coined until the 1990s by Professor Tom Caudell at Boeing's Computer Services' Adaptive Neural Systems Research and Development project in Seattle (Azuma 1997: 359). The early 1990s' engagement with AR was largely for aviation, engineering, and military purposes, but in the following years the potential of AR in commercial use, advertising in particular, was quickly registered by companies and their R&D departments. At the same time an increasing number of artists and authors became drawn to the possibilities of AR in their work, especially as the lower cost and wider accessibility of portable devices started making this possible from the late naughties onwards. The emergence of AR apps and browsers for both iOS and Android (for example Argon, Layar, Wikitude, and Yelp, and later Apple ARKit, AdobeAero, and Unity) enabled the creation and viewing of AR experiences without the kind of financial resources or highly skilled specialist knowledge that was a prerequisite in the case of early AR technology.

In recent years, AR has been quickly adopted by various areas of commercial and cultural production. It has been employed by museums and heritage tourism, for example in the "Lights of St. Etienne" project by Maria Engberg, which uses an Argon AR-enabled browser to allow visitors standing in the St. Etienne cathedral in Metz, France, to choose different dated views to observe their architectural surroundings as they looked in preceding centuries (Engberg). Computer game developers have tapped into the potential of creating both VR and AR 3D apps and experiences, which allow game worlds to become enmeshed with everyday physical spaces. For example, in the 2015 location-based AR game *Clandestine: Anomaly*, the player's home and neighborhood becomes the site of an alien crash landing (*Clandestine: Anomaly*). Nondigital sport and game experiences, too, may be transformed, like in the AR rock-climbing wall by Brooklyn Boulders and Jon Cheng, where climbers of a vertical wall view their achieved points and other information projected directly onto the climbing surface (McHugh

2016). Other applications are rapidly emerging in areas as varied as real estate, children's coloring books, furniture sales, and so forth.

While the ways in which we perceive the surrounding world are always conditioned by social, historical, political, and cultural factors, these dimensions may remain bracketed or unacknowledged constituents of what twentieth-century phenomenologists have termed "lifeworld," our immediately perceived, pre-discursive everyday experience. "Reality" is therefore a problematic term in itself, and scholars describing AR have used various terms to account for the experience into which additional elements are incorporated, like "world," "environment," "surroundings," or "actual scenery." Yet the "content" added to the perceived environment through augmented media applications adds a visible, symbolic/semantic layer to what is *already* an experience conditioned by its physical, social, and media contexts, not an unmediated sensory experience of the material world in all its fullness. In this sense, AR can also make visible many of the unacknowledged dimensions contributing to our lived environment. As the viewer of an AR work is presented with a compound image, not a representation posing as unmediated reality, this can highlight how the view into which new elements are added is itself a product of technological mediation. Rather than pure sensory experience, the immediacy of AR works is a result of *live or real-time input* in what is often an interactive engagement with the work, or a performance. Rhodes, for example, stresses that the "insistence on the live nature of the circuit belies the definition of the medium ... *live* mediation is the 'reality' of augmented reality" (Rhodes 2014: 135).

Though AR technologies and their commercial applications have been developed for some time outside the field of artistic production, a wider engagement with AR in creative arts, and the creation of AR literary works in particular, is a more recent phenomenon. In the field of digital literature, the integration of text or oral expression into (or onto) perceptual nontextual environment raises not only new possibilities for interrogating the relationship between verbal discourse and physical space but also the act of mediated perception itself. But while aural, tactile, and even olfactory perceptions are increasingly evoked in the multimodal experiences of virtual and mixed reality, an AR literary experience is still most typically an engagement with *visual* perception. In this sense AR literature, too, participates in the "visual" or "pictorial" turn of the post-1990s digital experience, to use a concept introduced by W. J. T. Mitchell (2006 [1994]). As a mixed-reality experience, AR also raises questions on the borders between literary and nonliterary forms of artistic production, or the distinction between verbal and nonverbal expression.

While all AR works share the basic idea of creating compound experiences of symbolic digital, and material or live experiences, this takes many forms, some of which have become more prominent in recent years. Scholars have categorized different types of AR experiences in various ways. For example, Patrick Lichty has divided AR art works into five different main categories, depending on the adopted technological platforms, their relationship with the surrounding space, and their materiality: *fiducial, planar recognition, environmental, embodied,* and *location-based* (Lichty). *Fiducial* AR works, predominant in the early stages of AR development, use markers captured by cameras or similar devices, and objects or bodies can then be detected by the computer and used as a part of a view mixing live input and pre-programmed digital content. *Planar recognition* AR engages with posters, pages, and other flat or print surfaces, and overlays additional content on them, in a manner of what Lichty considers merely a "simple semiotic swap" (108). However, yet as Robert Fletcher has suggested in discussing Amaranth Borsuk and Brad Bouse's epistolary poetry AR work *Between Page and Screen*—which allows the on-screen reading of graphic patterns printed on page only through a webcam—the intertwining of print and digitally created content may well become one of the most common forms of literary interaction in the future (Fletcher 2015: 59; Borsuk and Bouse). This type of AR has also been used to create remediated AR versions of more traditional literary works, like in Penguin Books's and Zappar's Interactive Novel versions of classic fiction works including *Moby Dick, Great Expectations,* and others (Farr). *Environmental* and spatial recognition AR works are based on devices recognizing certain spaces and locations, from rooms to architectural sites, and even natural landscapes (e.g., Microsoft's *Room2Room* app, or the *Exit Glacier* project [Metz; Shafer]). *Embodied* AR experiences allow for an engagement with one's surroundings beyond visual perception and representation, for example through wearable devices in dance performances. Finally, the growing popularity of *location-based* works is a result of a more widespread access to portable devices using GPS (Global Positioning System), which has enabled the creation of AR experiences for specific geographical locations.

Lichty's categorization is accompanied by classifications by various scholars, at times reflecting different emphases and interests. For William Uricchio, who focuses particularly on location-based AR experiences, such works themselves can be divided into three groups, depending on whether they use fiduciary markers, digital compass tracking (including GPS), or natural feature tracking (Uricchio 2011: 31), and any of these types of works can also include embodied and environmental elements. Consequently, distinguishing different types of AR is rarely an entirely clear-cut process, and it is perhaps more helpful to understand different categories in terms of the various emphases and purposes that characterize any AR experience. It can be said that AR experiences, regardless of whether they use fiducial markers, spatial/image recognition and tracking, or wearable devices (or a combination of these) are either dependent or nondependent on a specific location. They may be specific to a place or landscape, or can be transported to and performed in various spaces regardless of geographical coordinates, as long as the space fulfills the material and technological requirements necessary for the work. I will next focus on two AR works as examples of location-based versus transportable works that can be presented in any appropriate gallery or performance space.

Judd Morrissey's location-specific AR work *Kjell Theøry* is described by the author as "a site-specific mobile Augmented Reality poem mapped visually to geo-spatial coordinates in a public outdoor space" (Morrissey 2015: 181). The work, focusing on the life of "the gay computing pioneer Alan Turing's forced chemical castration with algorithmic mutations of Guillaume Apollinaire's 1917 play, *The Tits of Tiresias*," was presented as a part of the arts program of the ELO2015 conference in Bergen, Norway. Performances and excursions were located in the botanical gardens of the University of Bergen and the grounds of the city's Leprosy Museum, and in addition to following the performance participants were able to view these locations through iPads with added images, symbols, and text superimposed on the screen's visual live feed. The work has, however, also been performed in other locations, and as indoor/gallery space performances on both sides of the Atlantic, thus highlighting the mobility and adaptability of locative AR works to multiple geographical settings.

Crosstalk (2013), an interactive performance-based AR work by media artist Simon Biggs, choreographer Sue Hawksley, and composer Garth Paine, is described by its authors as a "public/private drama within an interactive system" (Biggs, Hawksley, and Paine 2014: 61). It is not locative or location-based in the sense that it can be performed in any (interior) space that meets the technological and size requirements for the installation, which is centered on two artists engaging with each other through speech and movement, with their embodied performance transcribed by a speech-to-text software onto a large screen, in real time. The performance creates a sonic environment of human and nonhuman interaction: "As the text objects interact, they re-write each other, facilitating the emergence of new textual and sonic material, created through the recombinant computation of the texts in the collided objects" (ibid.). Like *Kjell Theøry*, it depends on live, bodily engagement with and through space. However, unlike Morrissey's piece, it does not seek to place experiences or historical narratives in specific locations, but engages with digital interfaces, verbal utterance, and performance in a more conceptual and abstract manner, and can with relative ease be relocated to any indoor space of the right shape and size. Importantly, neither one of these works would simply fall into any one of the categories outlined by Lichty, for example, as they incorporate various forms of embodied engagement with the environment, and spatial recognition or coordination. Even more importantly, both have been adapted to different purposes in different locations, and no two experiences or performances of these works can be identical.

Any definition and outline of augmented reality technology and augmented reality works will inevitably be out-of-date and obsolete in the not-too-distant future; AR is, as Lichty points out, "a medium in its adolescence" (2014: 122). For Rhodes, the potential of the art form lies in its possibilities for "[making] mediation not seem real" or "[breaking] the illusion of reality in mediation" (2014: 136). Recent developments such as *Room2Room* also raise wider questions on the nature of human encounter and communication in the digital era—if 3D bodily projections are already moving mediated human exchanges beyond the voice and 2D video encounters like Skype calls, what happens when not only auditory and visual but also other sensory modes are increasingly added to these experiences, through the introduction of haptic gloves, for example? Will we soon be able to touch another human being, physically located in a separate space, or even on the other side of the globe, in our home environment? If so, how will we create and engage with fictional characters in digital narratives in the future? And how will our relationship with place and space change as increasingly realistic digital elements can be embedded in our material environments? If "augmented reality" already challenges our understanding of "reality" as the world around us, it seems inevitable that this relationship, as well as literature's possibilities for interrogating it, will develop in hitherto unseen ways as AR technologies evolve.

References and Further Reading

Azuma, Ronald T. (1997), "A Survey of Augmented Reality," *Presence* 6 (4): 355–85.

Biggs, Simon, Sue Hawksley, and Garth Paine (2014), "Crosstalk: Making People in Interactive Spaces," *Moco '14: Proceedings of the 2014 International Workshop on Movement and Computing*, 61–5, New York, NY: ACM, 2014.

Bolter, Jay David, Maria Engberg, and Blair MacIntyre (2013), "Media Studies, Mobile Augmented Reality, and Interaction Design," *Interactions* 20 (1): 36–45.

Borsuk, Amaranth and Brad Bouse (n.d.), *Between Page and Screen*, http://sigliopress.com/book/between-page-and-screen/ (accessed January 30, 2016).

Clandestine: Anomaly (n.d.) [a location-based augmented reality sci-fi mobile game], http://clandestineanomaly.com (accessed February 6, 2016).

Ekman, Ulrik, Jay David Bolter, Lily Díaz, Morten Søndergaard, and Maria Engberg (eds.) (2015), *Ubiquitous Computing, Complexity and Culture*, London: Routledge.

Engberg, Maria (n.d.), "The Lights of St. Etienne: An AR/MR Experience in the Cathedral in Metz, France," http://gvu.gatech.edu/research/projects/lights-st-etienne-armr-experience-cathedral-metz-france (accessed January 28, 2016).

Farr, Christina (n.d.), "2-D Books are Over: Augmented Reality Breathes New Life into the Classics," *VentureBeat*, http://venturebeat.com/2012/05/19/2-d-books-are-over-augmented-reality-breathes-new-life-into-the-classics/ (accessed January 30, 2016).

Fletcher, Robert (2015), "'Learn to Taste the Tea on Both Sides': AR, Digital Ekphrasis, and a Future for Electronic Literature," in Anne Karhio, Lucas Ramada Prieto, and Scott Rettberg (eds.), *ELO2015: The End(s) of Electronic Literature: Electronic Literature Organization Conference Program and Festival Catalog*, 50–60, Bergen: ELMCIP, University of Bergen.

Lichty, Patrick (2014), "The Aesthetics of Liminality: Augmentation as an Art Form," in Vladimir Geroimenko (ed.), *Augmented Reality Art: From an Emerging Technology to a Novel Creative Medium*, 99–125, Cham: Springer International Publishing.

Manovich, Lev (2006), "The Poetics of Augmented Space," *Visual Communication*, 5 (2): 219–40.

McHugh, Molly (2016), "Augmented Reality Rock Climbing Turns a Sport into a Videogame," *Wired*, January 29, 2016, http://www.wired.com/2016/01/you-know-whats-better-than-rock-climbing-ar-rock-climbing/ (accessed February 6, 2016).

Metz, Rachel (2016), "Augmented Reality Study Projects Life-Sized People into Other Rooms," *MIT Technology Review*, http://www.technologyreview.com/news/545466/augmented-reality-study-projects-life-sized-people-into-other-rooms/ (accessed January 21, 2016).

Mitchell, W. J. T. (2006) [1994], *What Do Pictures Want? The Lives and Loves of Images*, Chicago, IL: Chicago University Press.

Morrissey, Judd (2015), "Kjell Theøry," in Anne Karhio, Lucas Ramada Prieto, and Scott Rettberg (eds.), *ELO2015: The End(s) of Electronic Literature: Electronic Literature Organization Conference Program and Festival Catalog*, 181, Bergen: ELMCIP, University of Bergen.

Rhodes, Geoffrey Alan (2014), "Augmented Reality in Art: Aesthetics and Material for Expression," in Vladimir Geroimenko (ed.), *Augmented Reality Art: From an Emerging Technology to a Novel Creative Medium*, 127–37, Cham: Springer International Publishing.

Shafer, Nathan (n.d.), *Exit Glacier Augmented Reality Terminus Project*, http://nshafer.com/exitglacier/ (accessed January 31, 2016).

Uricchio, William (2011), "The Algorithmic Turn: Photosynth, Augmented Reality and the Changing Implications of the Image," *Visual Studies* 26 (1): 25–35.

11

Artistic and Literary Bots

Leonardo Flores

A bot is a software robot. Frequently personified or embodying concepts, animals, or things, bots operate autonomously in digital networked environments. While most bots are used for practical purposes, such as gathering, analyzing, and storing data, producing messages, interacting with users, political activism, or carrying out other actions, bots are also frequently used for artistic and literary goals. Bots are among the oldest genres of electronic literature, but for decades this development occurred along a single primary subgenre: chatterbots. In recent years—and thanks to social media networks that serve as platforms for data collection, interaction, and publication—bots have seen exponential growth in numbers, complexity, subgenres, and popularity.

This chapter will discuss bots that produce output of interest from a literary and e-literary perspective. For the purposes of this chapter I will define literature broadly as a language-based art which has traditionally taken form as drama, narrative, and poetry. An e-literary perspective focuses attention on how a language-based creative work engages and is made possible by digital media technologies, resulting in an extension and re-examination of literary practices. Some of the genres of electronic literature related to bots are computer-generated literature, text adventure games (Interactive Fiction, MUDs, MOOs), and e-poetry. This chapter will provide a historical overview of its development, practices, and communities, concluding with recent attempts to formulate poetics, theories, and taxonomies.

Historical Overview

Bots find their theoretical origins in Alan Turing's 1950 essay "Computing Machinery and Intelligence" in which he recasts the question "can machines think?" as one based on "the imitation game" in which a person communicating with a computer purely through text cannot accurately guess whether they are interacting with, a human or a computer (1950: 49). The attempt to create a machine that can generate language that might pass as written by a human being aligns the test with the literary, because it uses writing to create a fictional representation of a human being. This is a challenge faced by writers and dramatists for centuries: to write characters that an audience might believe in. From this perspective, the scientific goal of testing and tracking results are an extension of the logic of realism and naturalism in arts and literature. The new medium that Turing helped create brought back questions that had run their course in print-based media—realism and naturalism had yielded to Modernism—and to answer them would require the efforts of practitioners of both programming and natural language arts.

The Turing Test was seemingly passed in 1966 with Joseph Weizenbaum's ELIZA, a chatterbot that offered the first artificial character in electronic literature. As described by Kuipers et al.:

> I composed a computer program with which one could "converse" in English. The human conversationalist partner would type his portion of the conversation on a typewriter connected to a computer, and the computer, under the control of my program, would analyze the message that had so been transmitted to it, compose a response to it in English, and cause the response to be typed on the computer's typewriter. I chose the name ELIZA for the language analysis program because, like the Eliza of Pygmalion fame, it could be taught to "speak" increasingly well.
> (1976: 369)

ELIZA, as Weizenbaum explains, operates on two tiers: a natural language processing (NLP) program and a script which provides context and a template for its generation of responses (369). Its most famous script is "Doctor" which allowed ELIZA to mimic a Rogerian psychoanalyst, provoking emotional responses from users, some of whom believed they were interacting with a human being. And by acknowledging its literary inspiration, it reinforced characterization as a genre convention, which has the added benefits of conceptually framing the bot while narrowing its context and potential responses. We can see this strategy in her successor PARRY, implemented by Kenneth Colby in 1972, which sought to simulate a paranoid schizophrenic patient and was described as "ELIZA with an attitude." PARRY's characterization offered a forceful character with a personality disorder, which might account for lapses in social conversational conventions.

These two chatbots made literary history by being the first time two fictional characters were placed in conversation and developed their own narrative. During the International Conference on Computer Communications conference on January 21, 1973, Vincent Cerf used ARPANET to connect both bots (ELIZA was in MIT, PARRY was in Stanford), producing an entertaining conversation in which their programmed personalities and conversational strategies became apparent in sharp relief. The transcript for that first of several interactions is a powerful piece of collaborative algorithmic writing, a script of what might pass for contemporary audiences as Theatre of the Absurd. More importantly, it further establishes the chatterbots as characters with scripted behaviors that can generate potential narratives rather than characters whose personalities we can infer from records of performed actions.

The practice of interacting with chatterbots or connecting them to other bots to see their interactions play out continues to this day in video games and social media and is suggestive of Janet Murray's notion of "cyberdrama," which refers to the reinvention of storytelling in digital media. In this case the characters play out their scripted personalities, but Murray is also referring to the fictional worlds these bots can inhabit. The development of fictional settings in which chatterbots exist as nonplayer characters (NPCs) and narrators became prominent in the late 1970s until the late 1980s with Interactive Fiction (IF) and networked MUDs (Multi-User Dungeon) and MOOs (MUD, Object Oriented). The IF genre was initiated with *Adventure* (also known as *Colossal Cave Adventure*), written in 1976 by Will Crowther and expanded in collaboration with Don Woods in 1977. These text-based adventure games use natural language processing (NLP) to parse input from players, who type textual commands to explore and interact with a textually described fictional world.

While text adventure games are best known for their development of virtual settings and plot, it advances the bot genre along two vectors: the narrative voice and NPCs. The scripted narrative voice in IF can be filled with personality, especially with how it handles commands that don't make sense. For example, if a player tries to kill a bear in *Adventure*, the narrator replies "With what? Your bare hands? Against *his* bear hands??" From this perspective, playing an IF game can be considered as having an extended conversation with a chatterbot, in which the bot's script is that of a role-playing game referee. In this case, the goal isn't reaching a level of verisimilitude that might pass the Turing Test, but to establish a consistent tone in the interactions with the player that will reinforce the game's mood.

The development of increasingly sophisticated artificial intelligence (AI) methods and natural language processing saw an increase after the 1980s, inspiring the Loebner Prize, started in 1991 by Hugh Loebner and the Cambridge Center for Behavioural Studies. The award has hosted yearly competitions ever since, and the participants have still not achieved the goal of satisfactorily passing the Turing Test (though various teams have claimed to). The validity of the Turing Test itself has been challenged many times, as discussed in Saygin, Ayse Pinar, Ilyas Cicekli, and Varol Akman's 2003 essay "Turing test: 50 years later." Even though the Turing Test and its implementation by the Loebner Prize Contests seek verisimilitude in human character development deployed in a purely textual vector, this has proven to be both a productive and a very limiting constraint. While it encourages AI development towards achieving verisimilitude in the imitation of human conversational writing, the mainstream creation and development of bots, has developed along other lines.

This shift is motivated by practical and aesthetic reasons. Instead of spending resources in trying to create a bot that can pass as a human being, many developers have chosen to simply acknowledge the bot's robotic nature and focus on expanding their usefulness. Also, as contemporary users become accustomed to interacting with bots (and computing in general) newer aesthetics take over. Conceptual poetry and art, readymades, cut-ups, Flarf, Dada, hypertext, and Oulipian constraint-based writing, and humor all begin to inform the poetics of bot-making. The shift to graphical-based games in the 1980s and the widespread development of the internet in the 1990s also created new areas for bot development, particularly AI for graphical video games and the creation of agents for phone and computer networks.

Since the mid-1990s bots have been increasingly used for customer service (e.g., phone-answering systems), marketing (e.g., robocalls and spam), surveys, information gathering (e.g., web crawlers), and creating networks (e.g., botnets). The legitimate use of bots has not been necessarily well received, as in the case of replacement of phone customer service representatives with bots that oversimplify procedures, offer limited options, and frustrate attempts to communicate with a human being with whom one might be able to reason. Bots extend the century-old anxiety of human workers being replaced by machines: anthropomorphic robots. Bots have also gained a bad reputation because unscrupulous programmers and hackers have used them to disseminate spam (by email, in forums, newsgroups, blogs, and social networks), gather private data, spread malware, and carry out DDoS (Distributed Denial of Service) attacks, among other offenses.

The development of social media networks in the mid-2000s provided rich platforms for the creation and deployment of bots, and have brought about an unprecedented growth and diversification of the genre. During their early years (roughly between 2006 and late 2010) these social media platforms were permissive and used simple protocols for their APIs (Application Program Interfaces). This allowed for many third-party companies and developers to create tools for the creation and deployment of bots. These bots were mostly utilitarian in nature, were modeled using chatterbot AI technologies, and could be configured for e-commerce advertising, searches, and interacting with users who employed certain keywords—and still comprise a huge portion of currently active bots. Programmers also began to create bots that produced more literary and artistic output. Unfortunately, many of the works created during this early period have been lost due to changes in how the social media platforms worked.

As Facebook and Twitter's user bases grew, they needed to upgrade and modify different aspects of their platform APIs, and with each upgrade, they would lose a portion of the early bots created for those computational environments. Facebook, for example, has sought to establish a user base tied to human identity and habitually discontinues accounts that don't conform to its human detection algorithms (another ironic inversion of the Turing Test). Two Facebook bots (or "Facebots," as called by their creator, Eugenio Tisselli), Ariadna Alfil and Debasheesh Parveen, have survived detection and upgrades since their launching on December 31, 2009 and January 1, 2010, respectively, and were shut down on December 30, 2015.

Twitter has been a much friendlier environment for bots from the outset, even though they do discontinue accounts that violate their API terms or are reported as abusive. During its first few years, casual users were creating bots for practical and entertainment purposes, such as references, searching for data, and compiling information. This vibrant community of bots and apps was decimated by two upgrade events (the so-called "Twitpocalypse" in 2009 and the "OAuthpocalypse" on August 31, 2010), which disabled bots and apps unable to transition to the platform upgrades. Only a handful of literary and artistic bots from before 2010 continue to operate. A few popular and influential bots survive from this period: @everyword (2007), @IAM_SHAKESPEARE (2009), and @big_ben_clock (2009). The first two tweet words from a dictionary or lines from *The Complete Works of William Shakespeare*, respectively, while the last tolls the hour by tweeting the corresponding number of "bongs." A complex chatterbot based on the Star Wars franchise is "Chewbacca" (@cr_wookie [2009]), developed by the cantremember.com team in 2009, has an elaborate "personality engine" that generates Wookie "speech," responds to Twitter interactions, and emulates standard Twitter user behaviors.

After 2010, with the maturity that comes from a growing user base and a stable platform, Twitter became the home to a growing community of bot makers producing artistic and literary bots. An indicator of their size (almost 2,000, as of this writing) is the omnibots public list compiled by Tully Hansen's crowdsourced bot @botALLY. The Twitter hashtags: #bot and #botALLY are home to a growing community of bot makers. This community meets physically and online in yearly Bot Summits organized by Darius Kazemi, the creator of dozens of influential bots.

Along with social media networks, the past decade has also seen the development, growth, and increased sophistication of online data services. For example, dictionary services like Wordnik provide detailed information about a word, such as definitions, synonyms, antonyms, related words, syllable and stress breakdowns, and more. There are searchable online databases of images, video, music, audio, text, scanned images, museum catalogs, library catalogs, review aggregator services, encyclopedias, and more—all of which can be tapped into by a bot to produce its output. Twitter, Facebook, and other media-sharing platforms also offer access to their massive data streams which inspire and inform many bots. This offers a new different set of materials to bot makers than was available to earlier programmers, and has an impact on the kinds of bots produced and the quality of their output.

Shaping the Genre

As the bot maker community grows and their bots attract increasing mainstream and scholarly attention, the community begins to express theories for its poetics and practices. The primary areas for theoretical writing are: ethics and etiquette, poetics and manifestos, and genre categorization and taxonomies. Because the proliferation of artistic and literary bots is so recent, most of the theoretical writing is currently published via presentations, blog postings, news interviews, discussion groups, podcasts, and other online venues.

In 2013, Darius Kazemi published an influential blog post in which he offers "the four basic rules a Twitter bot should follow:"

- Don't @mention people who haven't opted in.
- Don't follow Twitter users who haven't opted in.
- Don't use a pre-existing hashtag.
- Don't go over your rate limit.

These basic principles are the foundation for an ethics of bot-making that helps avoid spam bot practices, abuse reports, and generally avoids getting banned by Twitter. Kazemi has spoken on several venues about bot ethics, etiquette, and his use of a document titled "badwords.txt" that his bots reference to filter out racist, sexist, or other offensive words from their output. Leonard Richardson echoes this sentiment in "Bots Should Punch Up" where he suggests that bot makers, like ventriloquists, are responsible for the content their bots produce, and should abide by a widely accepted rule of comedy and art: "always punch up, never punch down." In other words, one can attack those who are on an equal or higher socioeconomic class, but not those in lower ones.

Mark Sample, a scholar and bot maker, has been a leading voice in formulating poetics and publishing manifestos for bots. His "Protest Bot" article suggests the formation of a bot canon based on "absurdism, comical juxtaposition, and an exhaustive sensibility," and argues for the inclusion of "bots of conviction" which he describes as topical, data-based, cumulative, and oppositional. He offers several examples to test these characteristics, such as his bot @NRA_tally, which uses real data to create hypothetical gun shooting scenarios as well as Zach Whalen's @ClearCongress, which Whalen describes as follows: "Uses Huffington Post's polling data API to create fake retweets from members of congress, transforming most letters into ▓▓'s while leaving intact a percentage of letters equal to Congress's current overall job approval." The number of serious, activist bots, continues to grow as bots are created to lampoon political, news, entertainment, and higher education figures, and even participate in controversial debates (see @RealHumanPraise, @FalseFlagBot_, @TheHigherDead, @whatsgamergate, and @ElizaRBarr). @ElizaRBarr drew mainstream attention by being deployed during the height of the GamerGate Twitter attacks to use an ELIZA-inspired script to endlessly reply to insistent, hostile, GamerGaters with polite questions, forcing them to waste time and energy with their own formulaic and repetitive responses (Steadman).

Sample's blog post "Closed Bots and Green Bots" expresses some poetics and taxonomical principles based on Northrop Frye and Sherman Hawkins' structuralist work.

> But, getting back to bots, I want to suggest that the closed world and green world are not merely thematic archetypes that apply to narrative forms. The closed world and green world are also archetypes for the generative processes of computational media. They are archetypes of procedural composition. Where does a procedural work—a rules-driven work—get its source material? From within itself, or from beyond itself? Is there what Hawkins calls a "unity of place" in the work, or does the work come about through transgression across thresholds and barriers? Is the work closed or green?

Sample's framework is useful to establish how data sources become inspiration for bot creation along two impulses: to exhaust an idea in a limited dataset or template or to use an algorithm to set out into the unknown. Sample acknowledges that these impulses can combine as is the case of his own @WhitmanFML, which is both closed (using lines from Walt Whitman's Leaves of Grass as a data source) and green (combining these lines with tweets that use the #FML hashtag).

My own writing on bots in I ♥ E-Poetry for the past few years has sought to contextualize bots in literary and artistic traditions in the twentieth century, as well as creating some basic categories to approach them. My "Genre: Bot" resource in I ♥ E-Poetry roughly categorizes bots by their salient characteristics. What follows is an updated version which lists the main bot subgenres, along with a brief description and a few recent noteworthy examples from Twitter.

- **Chatterbots** are interactive characters: @oliviataters, @storyofglitch, @cr_wookie, @ElizaRBarr, @wikisext.
- **Open/Green bots** search through endless data sources and act upon the results to produce their output. @pentametron, @haikuD2, @falseflagbot_, @thewaybot, @pizzaclones, @AmIRiteBot, @ RealHumanPraise, @feelings_js, @SLOW_CRAWL, @_lostbuoy_, @ regrettoegret, @deepquestionbot.
- **Closed bots** work their way through a finite corpus: @everyword, @PERMUTANT, @IAM_SHAKESPEARE, @rom_txt, @everycolor, @everysimile, @UlyssesReader, @AutoNetflix, @MobyDickAtSea, @ JaneAustenHaiku, @elquijote1605.
- **Ebooks bots** publish random samples from a static or dynamic corpus: @horse_ebooks (initially), @10PRINT_ebooks, @Bogost_ebooks, @emerson_ebooks, zizek_ebooks, @bublbobl_ebooks, @DJ_EBOOKS.
- **Markov bots** generate texts based on a probabilistic analysis of a textual corpus of static or streaming data: @LatourBot, @ KarlMarxovChain, @MarkovChainMe, @autoblake, @tofu_product.
- **Template bots** generate texts by filling in blanks in phrases or sentences: @metaphorminute, @IsItArtBot, @MassageMcLuhan, @ snowcloneminute, @FilmRebootIdeas, @YouAreCarrying, @_The_Thief, @Robotuaries, @TXTADVNT_EXE, @Every3Minutes, @ TheHigherDead, @SortingBot, @thinkpiecebot.
- **Mashup bots** combine work from different sources: @ AndNowImagine, @gif_and, @oneiropoesis, @WhitmanFML, @LatourSwag, @LatourAndOrder, @twoheadlines, @_lostbuoy_, @ poem_exe.
- **Emoji bots** assemble pictorial art and narratives from emoji (small images that act as ideas and are deployed as textual objects): @ thetinygallery, @tiny_star_field, @tinyrelations, @atinyzoo, @ TinyDungeons, @TinyCrossword, @ARealRiver, @tiny_cityscapes, @tiny_gardens, @tiny_forests.
- **Pseudo bots** involve partial generation with human curation, or human bot-like performances: @latimehaiku, @horse_ebooks, @tweetsofgrass, @postmeaning.

While many of these bots have characteristics from other subgenres, this categorization focuses on their primary features. There are several other potential subgenres out there, but these are the ones with the greatest critical mass and community support. The speed at which the bot genre is developing means that subgenres will continue to emerge, mutate, branch out, combine, and proliferate.

References

Abraham, Ben (2014), "#FalseFlag Bot (@FalseFlagBot_) | Twitter," https://twitter.com/falseflagbot_ (accessed April 18, 2015).

Anonymous (2014), "Eliza R. Barr (@ElizaRBarr) | Twitter," https://twitter.com/elizarbarr (accessed April 18, 2015).

@thricedotted (2014), "What Is #GamerGate? (@whatsgamergate) | Twitter," https://twitter.com/whatsgamergate (accessed April 18, 2015).

Cerf, V. (1999), "PARRY Encounters the DOCTOR," IETF Tools, trans. Helen Morin, Internet Society, December 1999, Web, https://tools.ietf.org/html/rfc439 (accessed April 18, 2015).

Chamberlain, William and Thomas Etter (1984), *The Policeman's Beard Is Half-Constructed: Computer Prose and Poetry*, New York, NY: Warner Software/Warner Books.

Dubbin, Rob (2013), "Real Human Praise (@RealHumanPraise) | Twitter," https://twitter.com/realhumanpraise (accessed April 18, 2015).

Flores, Leonardo (2013), "'Debasheesh Parveen' and 'Ariadna Alfil' by Eugenio Tisselli," I ♥ E-Poetry, Wordpress, March 12, 2013, Web (accessed April 18, 2015).

Flores, Leonardo (2013), "Genre: Bot," I ♥ E-Poetry, Wordpress, June 8, 2013, Web, (accessed April 18, 2015).

Funkhouser, Chris (2007), *Prehistoric Digital Poetry: An Archaeology of Forms, 1959–1995*, Tuscaloosa, AL: University of Alabama Press.

Gaboury, Jacob (2013), "A Queer History of Computing, Part Five: Messages from the Unseen World," Rhizome.org., Rhizome, June 18, 2013, Web (accessed April 18, 2015).

Graham, Paul (2004), *"A Plan for Spam." Hackers & Painters: Big Ideas from the Computer Age*, Sebastopol, CA: O'Reilly.

Hansen, Tully et al. (2014), "@botALLY/omnibots on Twitter," https://twitter.com/botally/lists/omnibots (accessed April 18, 2015).

"Interactive Fiction," Wikipedia, Wikimedia Foundation, January 12, 2004, Web (accessed April 18, 2015).

Jackson-Mead, Kevin and J. Wheeler Robinson (eds.) (2011), *If Theory Reader*, Transcript On Press.

Kazemi, Darius (2013), "dariusk/rapbot · GitHub," https://github.com/dariusk/rapbot (accessed April 18, 2015).

Kazemi, Darius (2013), "Basic Twitter Bot Etiquette," Tiny Subversions, March 16, 2013, Web (accessed April 18, 2015).

Kazemi, Darius et al. (2013), Bot Summit 2013, Tiny Subversions, November 25, 2013, Web (accessed April 18, 2015).

Kazemi, Darius, et al. (2014), Bot Summit, Wordpress 2014, Tiny Subversions, November 8, 2014, Web (accessed April 18, 2015).

Krebs, Brian (2014), "Meet the Russian Cybercrooks behind the Digital Threats in Your Inbox," Slate, The Slate Group, November 18, 2014, Web (accessed April 18, 2015).

Kuipers, Benjamin, John McCarthy, and Joseph Weizenbaum (1976), "Computer Power and Human Reason," ACM SIGART Bulletin 58: 4–13.

"Loebner Prize" (2014), The Society for the Study of Artificial Intelligence and Simulation of Behaviour, AISB, 2014, Web (accessed April 5, 2015).

Marchant, Jonathan (2009), "Big Ben (@big_ben_clock) | Twitter," https://twitter.com/big_ben_clock (accessed April 18, 2015).

Mateas, Michael, and Andrew Stern (2011), "Façade," Electronic Literature Collection, Vol. 2, MIT Press, Web (accessed April 18, 2015).

Messina, Chris (2008), "Bots," Twitter Fan Wiki /. Ed. "komski" PB Works, February 2008, Web (accessed April 18, 2015).

Murray, Janet (2004), "From Game-Story to Cyberdrama," Electronic Book Review, Open Humanities Press, May 1, 2004, Web (accessed April 18, 2015).

"NRA Tally (@NRA_Tally) | Twitter" (2014), https://twitter.com/nra_tally (accessed April 18, 2015).

Parrish, Allison (2007), "everyword (@everyword) | Twitter," https://twitter.com/everyword (accessed April 18, 2015).

"PARRY" (2008), Wikipedia.,Wikimedia Foundation, August 7, 2008, Web (accessed April 18, 2015).

Pinar, Saygin A., Ilyas Cicekli, and Varol Akman (2000), "Turing Test: 50 Years Later," Minds and Machines: Journal for Artificial Intelligence, Philosophy and Cognitive Science 10 (4): 463–518.

Pipkin, Katie Rose (2014), "Selfhood and the Icon," Bot Summit 2014, November 8, 2014, Web (accessed April 18, 2015).

Poundstone, William (2005), "Spam," Williampoundstone.net, Web (accessed April 18, 2015).

"Racter" (2007), Wikipedia, Wikimedia Foundation, March 13, 2007, Web (accessed April 18, 2015).

"Scott McNally (@botALLY) | Twitter" (2013), https://twitter.com/botally (accessed April 18, 2015).

Rettberg, Jill Walker (2014), "A Taxonomy of Twitter Bots, by @tullyhansen #elo14 Pic.twitter.com/xFwr492AP9," Twitter, June 19, 2014, Web (accessed April 18, 2015).

Richardson, Leonard (2013), "Bots Should Punch Up," Crummy.com, November 27, 2013, Web (accessed April 18, 2015).

Ryback, Chuck (2015), "The Higher Dead (@TheHigherDead) | Twitter," April 18, 2015, https://twitter.com/thehigherdead.

Sample, Mark (2014), "A Protest Bot Is a Bot so Specific You Can't Mistake It for ...," Medium, Medium.com, May 30, 2014, Web (accessed April 18, 2015).

Sample, Mark (2014), "Closed Bots and Green Bots," SAMPLE REALITY, Wordpress, June 23, 2014, Web (accessed April 18, 2015).

Shakespeare, William (1990), "The Complete Works of William Shakespeare," [Internet resource], Champaign, IL: Project Gutenberg.

Short, Emily (2006), "Galatea," Electronic Literature Collection, Vol. 1. MIT Press, October 2006, Web (accessed April 18, 2015).

Short, Emily (2011), "NPC Conversation Systems," *IF Theory Reader*: 331.
Sorolla, Roger S. G. (2011), "Crimes against Mimesis," *IF Theory Reader*: 1.
Steadman, Ian (2014), "The Ultimate Weapon against GamerGate Time-wasters: A ...," *New Statesman*, October 15, 2014 (accessed April 18, 2015).
Strebel, Joshua (2010), "@IAM_SHAKESPEARE's (Willy Shakes) Best Tweets – Favstar," http://favstar.fm/users/iam_shakespeare (accessed April 18, 2015).
Turing, Alan M. (2004), "Computing Machinery and Intelligence (1950)," in B. Jack Copeland, *The Essential Turing: The Ideas that Gave Birth to the Computer Age*, 433–64, Oxford: Oxford University Press.
"Walt FML Whitman (@WhitmanFML) | Twitter" (2014) https://twitter.com/whitmanfml (accessed April 18, 2015).
Wardrip-Fruin, Noah and Nick Montfort (eds.) (2003), *The New Media Reader*, Cambridge, MA: The MIT Press.
Whalen, Zach (2014), "@ClearCongress," (2014), Zach Whalen, Reclaim Hosting, December 11, 2014, Web (accessed April 18, 2015).
Wikipedia Contributors (2004), "Colossal Cave Adventure," Wikipedia. Wikimedia Foundation, August 12, 2004, Web (accessed April 18, 2015).

12

Consuming the Database: The Reading Glove as a Case Study of Combinatorial Narrative

Theresa Jean Tanenbaum and Karen Tanenbaum

The database is a fundamental knowledge structure in computing, and it has its own unique poetics. In this chapter we explore how databases and narratives have been theorized within electronic literature and new media and provide a few short readings of significant works that employ the poetics of the database. We then describe one of our own works—the Reading Glove—as a case study of combinatorial storytelling that reconciles the open-endedness of the database form with the experience of closure that allows a reader to feel as though a work has been completed or *consumed*.

Theories of Database Narrative

Lev Manovich has argued that the database is the natural enemy of narrative, writing that a database "represents the world as a list of items, and it refuses to order this list" (Manovich 2001: 225). From Manovich's perspective the absence of a guiding logic or order puts the logics of the database directly in conflict with the logics of narrative, which he sees as a competing paradigm of meaning. Narrative, in this sense, is defined by the ways in which it selectively excludes information from the database set, in order to force a certain ordering and elicit a specific meaning. To reconcile the logics of narrative and database, Manovich turns to the semiological notions of *syntagm* and *paradigm*, which may be loosely understood as the *explicit* elements from which a work is constructed (*syntagm*) and the *implicit* set of related elements from which the set might have been sourced (Manovich 2001). A parallel may be drawn between these concepts and the Russian formalist notions of *fabula* and *syuzhet*, as articulated by David Bordwell and Kristin in their canonical book *Film Art* (Bordwell and Thompson 1997). *Fabula* and *syuzhet* can loosely be translated as "story" and "plot" respectively. The *syuzhet* is the plot of a narrative as represented and encoded within a media artifact. It operates in relationship with filmic *style* to produce the *fabula* or *story* of a narrative, as interpreted and understood within the mind of the reader or viewer. Thus, when Manovich reconciles the logics of database with the logics of narrative, the database becomes the *paradigm* (*fabula* or plot) from which the *syntagm* (*syuzhet* or story) narrative is drawn. The relationship between database and narrative is one of selective perception, of parsing a trajectory through the *possibility space* of the database through a lens of causal ordering in order to assign meaning to the events encountered. This conception of narrativity is in line with the work of film scholar Edward Branigan, who regards narrative as a perceptual activity by which a viewer assembles and organizes unstructured data into a causally connected pattern (Branigan 1992). In the following section we explore how these poetics of database and narrative have been incorporated into combinatorial storytelling systems.

Significant Database Stories

The poetics of the database as a narrative form pre-date digital systems. Combinatorial and experimental literature has a rich tradition: consider, for example, the seminal work of the Oulipo group, as described in Wardrip-Fruin and Montfort (2003). Founded in the 1960s, Oulipo (short for the French "*Ouvroir de littérature potentielle*" or "workshop of potential literature") as a literary movement explored intersections of algorithms, rules, mathematics, and language. Most relevant to our discussion of database narrative is Raymond Queneau's "*Cent Mille Milliards de Poèmes*" (often translated as "*A Hundred Thousand Billion Poems*"), a set of ten sonnets, each with fourteen lines, designed to be fully interchangeable between each other, bound so that each line is on a different strip of paper, allowing the reader to flip to any one of the 10^{14} possible permutations.

Contemporary electronic literature demonstrates the range of possible forms inherent in the database. Jillian McDonald's *Snow Stories* (2005) uses a database of film clips and sound to assemble footage in response to written narratives about snow.

Upon entering the site, the viewer is invited to share a written story about snow - fantasy, memory, or dream. Behind the scenes, the story is scanned for key words that match a database of parameters such as mood, landscape, danger, weather, population, and animals. Audio, video, and animation clips, stored in the database as well are similarly tagged. The visitor's story is translated into a non-linear movie based on the results of the text scan parameters, and the compiled film is displayed in a snowglobe.[1]

Stuart Moulthrop's "Reagan Library" (Moulthrop 1999) can be seen as a database narrative, but one that actively prunes away its own possibility space as the reader traverses it. Moving across various "locales" rendered in QuicktimeVR (each accompanied by a hypertextual lexia), "Reagan Library" is notable in that its language becomes more coherent over time. Although not strictly linear, the content of the system is designed to eventually reveal its own *completeness*: over time the content begins to repeat itself, allowing a reader to experience a sense of closure.

Jim Andrews et al.'s *Stir Fry Texts* (Andrews, Lennon, and Masurel 1999) are a series of combinatorial text pieces for the web. The reader may flip through a set of interconnected texts, or she may mouse over any one of the texts, causing it to become "infected" or recombined with words from the other writings in the set. The result is a sense of the writings bleeding into each other, but in an unstable and unpredictable way.

Finally, Millie Niss and Martha Deed's *Oulipoems* (Niss and Deed 2004) take many of the techniques of the original Oulipo writers, and reconceive of them for the web. Their pieces include combinatorial poetry games like *Poggle* and *The Electronic Muse*, which provide an interactor with a set of textual, structural, and stylistic elements to assemble into poems. *Poggle* is more game-like, limiting the author/reader to a grid of textual elements that must be traversed/selected-from according to a set of rules within a span of time, while *The Electronic Muse* allows the interactor to select a poetic style from a list, and to choose parts of speech, and then generates lines of poetry according to these parameters.

A unifying theme connecting all of these approaches to combinatorial and database narrative is that the possible meaning space for the pieces is left wide open for the reader to configure. The authors create the *paradigm* and the *fabula*, and they devise an interface to that *paradigm* for the reader that is designed to expand on the possible *syntagm/syuzhet/story* rather than

---

[1]http://www.fringexhibitions.com/netarchives.html.

foreclose upon a specific authored meaning or message. This open-ended experience of combinatorial narratives can be problematic.

Unstructured Databases, Structured Databases, and Ontologies

One of the central challenges facing authors of database narratives is achieving a sense of closure for the reader (Douglas 1994). Espen Aarseth's critique of hypertext narratives is appropriate here, in that he describes the sensation of being unable to fully apprehend a combinatorial text (Aarseth 1994).

> When we look at the whole of such a nonlinear text, we cannot read it; and when we read it we cannot see the whole text. Something has come between us and the text, and that is ourselves, trying to read. This self consciousness forces us to take responsibility for what we read and to accept that it can never be the text itself. The text, far from yielding its riches to our critical gaze, appears to seduce us, but remains immaculate, recedes, and we are left with our partial and impure thoughts, like unworthy pilgrims beseeching an absent deity.
>
> (Aarseth 1994: 769)

Aarseth makes a distinction between a "text" and a "script": a script is comprised of the "visible words and spaces" while a text also includes a "practice, a structure, or ritual of use" (Aarseth 1994: 763). Thus, in the systems above we might say that the textual, graphical, and multimedia elements constitute the "scriptons" of the system, while the process of interaction along with whatever attendant practices each reader/user engages in through the experience would reveal the "textons" of the system. Some of these texts provide a collection of scriptons, with very few constraints or rules to shape the reader's interactions, while others provide both a *database* and an *interface* along with constraints on the interconnections between the objects within that database. At some point, these data structures cease to be simple databases and instead become *ontologies*: a formal knowledge representation of both the contents of the database and their interrelationships with each other. *Ontologies* have a rich history within electronic literature's parallel sister field: Interactive Digital Storytelling (or IDS). IDS has its origins in computer science and artificial intelligence, and shares many of the same intellectual commitments as electronic literature. However, where electronic literature has resulted in a diverse body of experimental texts, IDS has instead pursued increasingly sophisticated software instantiations of formal narrative systems, and simulations of social worlds designed to produce *interactive dramas* similar to those envisioned in science fiction, such as Star Trek's *Holodeck* (Cavazza, Aylett, Dautenhahn, Fencott, and Charles 2000; Swartout et al. 2001). In the final section of this chapter we describe a system of our own creation that seeks to bridge the gap between these two approaches, by combining the theoretical and textual practices from electronic literature with the computational approaches to digital storytelling common in IDS.

The Reading Glove: A Case Study of Combinatorial Narrative

The Reading Glove is a wearable, tangible, interactive storytelling system that we developed in 2009 and 2010. It is comprised of the following elements:

- A wearable glove-based interface with a Radio Frequency Identification (RFID) reader in the palm of the glove and a wireless radio for communicating with a central server.
- A collection of antique (and antique seeming) objects, each tagged prominently with a unique RFID tag.
- A large horizontal tabletop display surface.
- A laptop running a software application that uses a rules-based expert system and an ontology of the narrative world to track reader interactions and make recommendations for where to go next in the story (via the tabletop display). The software layer also triggers playback of audio narration when an interactor picks up an object.

Interactors using the Reading Glove system "read" the story by picking up objects from the collection in order to trigger fragments of a narrative (lexia) that are associated with those objects. Over the course of multiple

FIGURE 1 *The Reading Glove system. From left to right: the tabletop display and objects, the Reading Glove and a tagged object, an interactor using the system, and the collection of narrative objects.*

interactions it becomes possible for readers to piece together the narrative like a puzzle. We construed the objects as "boundary objects," a notion from ethnographic research in which an object is situated between two distinct and different cultures (Star and Griesemer 1989). Boundary objects allow for a point of contact and negotiation between cultures that lack other means of interaction: in the case of the Reading Glove, the objects existed within both the physical world of the reader and the imaginary world of the fiction. They were pieces of the fiction that the reader could hold, manipulate, and experience using senses that are not often deployed while engaging with other forms of fiction. Drawing on theories of affordance, we chose objects that constrained the body of the reader in particular ways, while affording very specific postures and bodily motions (Gibson 1977; Norman 1988). The coffee maker afforded turning the crank, the top hat and goggles afforded being worn in a particular way, the telegraph key afforded tapping and manipulation, etc. In this way, each object lent itself to a particularly embodied interaction.

The Reading Glove employs a number of techniques intended to aid the reader in achieving narrative closure; however, it never seeks to assert its authority over the reader as she explores its nonlinear narrative space. Some of these techniques are purely textual in nature: we coined the term *cognitive hyperlinks* (T. J. Tanenbaum, Tanenbaum, El-Nasr, and Hatala 2010) to describe a mode of authoring that layers the text with repetition, internal references, and other ordering cues to aid the reader in establishing a temporal chain of cause and effect between the lexia. Unlike the systems described above, there *is* a canonical ordering to the lexia within the Reading Glove, a linear narrative that is mediated through the nonlinear nature of its interface. The interface produces a nonlinear version of this narrative, but as with Moulthrop's "Reagan Library", the completeness of the story becomes apparent over time, with repetition signaling when the narrative has been fully "consumed."

Other techniques used to support an experience of closure are highly computational, such as the expert system developed to recommend next

FIGURE 2 *The "recommendation" screen for the Reading Glove in a neutral state (left) and a recommendation state (right).*

steps (K. Tanenbaum, Hatala, Tanenbaum, Wakkary, and Antle 2013). This system employs both an ontology of the narrative logics underlying the lexia and an intelligent agent capable of recommending different traversals of the narrative space. We encoded three fundamental logics into this system: the logic of temporal ordering and causality; the logic of thematic connections between objects; and the logic of narrative importance (which lexia were most important for the reader to encounter to understand the narrative?). Each of these logics provided the basis for a recommendation, and each was assigned a specific color within the system so that readers would get different permutations of the story if they followed different recommended paths[2] (Figure 2).

The Reading Glove also remediates some of the poetics of live storytelling by including multiple variations of the narrated performance for each piece of text presented. Working with a professional actor, we captured several dozen variations of each lexia, ultimately paring them down to three versions of each piece of spoken narration. These variations are primarily distinguishable from each other by the pace of the reading—one is slow and deliberate, one is spoken at a normal rate, and one is rushed and urgent. The first time a reader encounters a lexia within the narrative it is the slowest version. Each subsequent encounter with that lexia is a little bit faster and more urgent. Thus, the tone of the story shifts over time from one of careful recounting to one of desperation, even as the text of the narrative remains unchanged.

Taken alongside the other techniques discussed above, this leads the reader towards a sense of closure. In these ways, the Reading Glove's design seeks to combine techniques from both electronic literature and interactive digital storytelling to produce a participatory narrative experience that draws on the poetics of combinatorial and database narrative, while still supporting a satisfying sense of having reached the conclusion of the story for the reader.

References

Aarseth, E. (1994), "Nonlinearity and Literary Theory," in N. Wardrip-Fruin and N. Montfort (eds.), *The New Media Reader*, 761–80, Cambridge, MA: The MIT Press.

---

[2]It is interesting to note that none of the readers of the system noticed the color coding in the recommendations, and none claimed to follow a particular path. We view this as a failure of design in the current instantiation of the story, and see it as an interesting opportunity for future iterations of this work.

Andrews, J., B. Lennon, and P. Masurel (1999), *Stir Fry Texts, Electronic Literature Collection*, Vol. 1 (2006), https://collection.eliterature.org/1/works/andrews__stir_fry_texts.html.

Bordwell, D. and K. Thompson (1997), *Film Art: An Introduction*, 5th edn., New York, NY: McGraw-Hill.

Branigan, E. (1992), *Narrative Comprehension and Film*, New York, NY: Routledge.

Cavazza, M., R. Aylett, K. Dautenhahn, C. Fencott, and F. Charles (2000), "Interactive Storytelling in Virtual Environments: Building the 'Holodeck'," in *Proceedings of the International Conference on Virtual Systems and Multimedia, October 3–6, Ogaki, Japan*, 678–87, Tokyo: Omsha.

Douglas, J. Y. (1994), "'How Do I Stop This Thing?': Closure and Indeterminacy in Interactive Narratives," in G. P. Landow (ed.), *Hyper/Text/Theory*, 159–88, Baltimore, MD: The Johns Hopkins University Press.

Gibson, J. J. (1977), "The Theory of Affordances," in R. E. Shaw and J. Bransford (eds.), *Perceiving, Acting and Knowing*, Hillsdale, NJ: Lawrence Erlbaum Associates.

Manovich, L. (2001), *The Language of New Media*, Cambridge, MA: The MIT Press.

McDonald, J. (2005), *Snow Stories*, http://www.jillianmcdonald.net/snowstories/.

Moulthrop, S. (1999), "Reagan Library". *Electronic Literature Collection*, Vol. 1 (2006), https://collection.eliterature.org/1/works/moulthrop__reagan_library.html.

Niss, M. and M. Deed (2004), *Oulipoems. Electronic Literature Collection*, Vol. 1 (2006), https://collection.eliterature.org/1/works/niss__oulipoems.html.

Norman, D. A. (1988), *The Design of Everyday Things*, New York, NY: Doubelday/Currency.

Star, S. L. and J. R. Griesemer(1989), "Institutional Ecology, 'Translations' and Boundary Objects: Amateurs and Professionals in Berkeley's Museum of Vertebrate Zoology, 1907–39," *Social Studies of Science* 19(3): 387–420.

Swartout, W., R. Hill, J. Gratch, W. L. Johnson, C. Kyriakakis, C. LaBore, J. Douglas et al. (2001), "Toward the Holodeck: Integrating Graphics, Sound, Character and Story," *Proceedings of the International Conference on Autonomous Agents (AGENTS'01)*, 409–16, New York, NY: ACM Press, https://doi.org/10.1145/375735.376390.

Tanenbaum, T. J., K. Tanenbaum, M. S. El-Nasr, and M. Hatala (2010), "Authoring Tangible Interactive Narratives Using Cognitive Hyperlinks," in *Proceedings of the Intelligent Narrative Technologies III Workshop*, 6:1–6:8, New York, NY: ACM Press, http://doi.org/10.1145/1822309.1822315.

Tanenbaum, K., M. Hatala, T. J. Tanenbaum, R. Wakkary, and A. N. Antle (2013), "A Case Study of Intended versus Actual Experience of Adaptivity in a Tangible Storytelling System," *User Modeling and User-Adapted Interaction* 24(3): 175–217, https://doi.org/10.1007/s11257-013-9140-9.

Wardrip-Fruin, N. and N. Montfort (2003), Introduction to "Six Selections by the Oulipo," *The New Media Reader*, 147–8, Cambridge, MA: The MIT Press.

13

Hypertext Fiction Ever After

Stuart Moulthrop

There are at least two ways to sketch the genealogy of hypertext fiction. From the perspective of literary theory, most influentially in Aarseth's *Cybertext* (1997), hypertext fiction counts as one among many instances of ergodic expression. This broad-minded approach allows us to associate the form with a number of distinct but similar ventures, such as Malloy's query-based "narrabase" stories, John McDaid's "modally appropriate" artefactual fiction, various attempts at text generation, from the ELIZA script of the late 1960s to Daniel Stern and Michael Mateas' *Façade*—and perhaps most significantly, the long tradition of text-based computer gaming, beginning with *Colossal Cave Adventure* in the mid-1970s and continuing today. Electronic literature and digital art contain many forms and practices besides hypertext fiction. If we want to consider both tree and forest, we need to see this genre as a machinic inflection of the general project of experimental writing.

A second approach to hypertext fiction might focus on the underlying technology. The term *hypertext* was invented by Theodor Holm Nelson in the mid-1960s, referring to possibilities for "non-sequential writing" made possible by digital storage and retrieval (Barnet 2000: 65). In Nelson's conception, the process of intertextual reference is automated by computable code that instantly combines one document with another. The idea came to be associated with the disjunctive, linked-node model of the World Wide Web—the kind of linking Nelson once lampooned as "diving boards into the darkness."[1] His conception has always been much richer, involving "transclusion," in which documents are deeply and dynamically interfused.

Nelson was partly inspired by Vannevar Bush's concept for a mechanical system for associative research ("Memex"), proposed at the end of the Second World War (Nyce and Kahn 1991). In the late 1960s Douglas Engelbart's groundbreaking NLS/Augment system made the first steps toward implementation of the hypertext concept (Barnet 2000: 37–64). As personal computers arrived over the following decade, academic researchers and software designers developed the concept further. The Association for Computing Machinery (ACM) launched a research conference on Hypertext and Hypermedia in 1987. The heyday of hypertext may well have been 1987–2004, when Apple Computer supported HyperCard, a product that allowed users of the company's machines to create complex, multicursal assemblies of information, employing image, text, sound, and simple animation. The apotheosis of this early development phase came with the arrival of yet another system, in November 1990, when Tim Berners-Lee and Robert Cailliau proposed a hypermedia system called "The World-Wide Web" (Berners-Lee et al. 2000). After the web's advent, hypertext moved from concept to utility, becoming as ubiquitous as indoor plumbing and electricity. Billions of humans daily encounter the curious expression *http://* (usually in the address bar of a browser) without much awareness that this formula invokes *Hypertext Transfer Protocol.*

Hypertext fiction was in some sense a byproduct of this emergent phenomenon. Long before the turn of the century, writers of late- and postmodernist print fiction such as William S. Burroughs, John Barth, John Hawkes, Robert Coover, Thomas Pynchon, and Kathy Acker had accustomed readers to difficult, recursive, and counterfactual narratives. Jorge Luis Borges' conceptual fictions provided intriguing examples, including the germinal "Garden of Forking Paths," effectively a blueprint for hypertextual storytelling. Borges' "Aleph" and "Book of Sand" furnished imaginary frameworks for the first working formula of hypertext fiction: "a story that would change each time you read it" ("our story"). The first implementation of this formula was Michael Joyce's *afternoon, a story*, written between 1985 and 1987 and presented at the inaugural ACM Hypertext Conference

---

[1]Though I am certain this phrase belongs to Ted Nelson, I am no longer sure of the occasion. My best guess is his keynote address at the 2001 Digital Arts and Culture Conference at Brown University, "Toward a True Electronic Literature" (April 2001). Jennifer Fraser of Carleton University cites the phrase in her thesis for the Master of Architecture degree. Given that her thesis was written in 1999, the first occurrence may have been earlier than the Brown talk. See "Visualising Hypertext Narrative," http://www.collectionscanada.gc.ca/obj/s4/f2/dsk1/tape2/PQDD_0020/MQ48371.pdf.

as proof-of-concept for Storyspace, the authoring system created by Joyce and Jay David Bolter (with additional credit to John B. Smith).

Also present at the ACM gatherings, destined for an important part in that community's further evolution, was Mark Bernstein, a research chemist turned software designer who had developed his own system, HyperGate, and with Erin Sweeney co-authored one of the first nonfiction hypertexts, *The Election of 1912*. The publishing house Bernstein founded, Eastgate Systems, Inc., would eventually re-publish *afternoon* as well as the Storyspace application and a substantial array of works, including cultural criticism (Diane Greco's *Cyborg: The Body Electric*), philosophical commentary (David Kolb's *Socrates in the Labyrinth*), poetry (Stephanie Strickland's *True North*), as well as fiction. Many of these titles appeared in the *Eastgate Review of Hypertext*, a groundbreaking digital publication.

Bernstein's catalog grew to include hypertexts developed in systems other than Storyspace (Sarah Smith's *King of Space*, Judy Malloy's *its name was Penelope*, Deena Larsen's *Marble Springs*, John McDaid's *Uncle Buddy's Phantom Funhouse*, and Malloy and Catherine Marshall's *Forward Anywhere*). Eastgate also became a venue for projects from Brown University's Literary Arts program. Brown had been an important site for hypertext development from early days, largely owing to efforts by Andries Van Dam, George Landow, and Robert Coover. All three of these figures were influential in the writing program, which produced important titles including Mary-Kim Arnold's *Lust*, the various experiments collected by Landow in *Writing at the Edge*, and Judd Morrissey and Lori Talley's *My Name is Captain, Captain*, all of which would eventually be published by Bernstein. The most celebrated product of the Eastgate–Brown nexus was Shelley Jackson's *Patchwork Girl* (1995), a fantasia that interfused *Frankenstein*, pastiche, literary affiliation, feminist writing, monstrous invention, and hypertextual sensibility. After Joyce's *afternoon* and *Twilight, A Symphony* it is the most widely read and analyzed example from the Eastgate catalog, and arguably the most significant for literary history.

To be sure, Eastgate Systems, Inc. represents only one leafy branch in a larger digital forest. Throughout the 1980s and 1990s, Robert Stein's Voyager Company brought out a line of exquisitely designed, groundbreaking multimedia hypertexts, some of which were developed from prior print publications as "Expanded Books."[2] Apple's HyperCard powered important experiments in hypertextual writing and design, from Brian Thomas' *If Monks Had Macs* to *Beyond Cyberpunk!* by Gareth Branwyn and Mark Frauenfelder. Rand and Robyn Miller, creators of the epic video games *Myst* and *Riven*, began by making HyperCard-based works for children.

---

[2] See Nat Hoffelder, "What eBooks Looked Like 20 Years Ago," http://the-digital-reader.com/2013/07/04/what-ebooks-looked-like-20-years-ago-video/.

As the World Wide Web became established, hypertextual writing found its way there, in venues including Mark Amerika's *Alt-X*, the *Iowa Review*'s web extension, and online journals such as *New River* and *Postmodern Culture*. Joyce produced an important web fiction, "Twelve Blue," in 1996. Jackson wrote a revealing reflection on hypertextual literary aesthetics ("Stitch Bitch") around the same time. This writer followed his Eastgate fiction *Victory Garden* (1991) with a series of web hypertexts, including "Hegirascope" (1995) and "Reagan Library" (1998). Talan Memmott's "Lexia to Perplexia" (2000) significantly raised the bar both in terms of code-infused language and language-driven coding. Caitlin Fisher's "These Waves of Girls" (2000), the first hypertext fiction to win a major national award, was also written for the web.

Though this chapter focuses on fiction, we need at least a friendly glance at the electronic poetry (ePoetry) movement, fostered in large part by Loss Pequeño Glasier of SUNY Buffalo, which in our arboreal metaphor figures as a large and flourishing growth in a nearby patch of sunshine. Along with the ePoetry Center at Buffalo, a major academic archive of digital literature is provided by the Electronic Literature Organization (ELO), an artistic and academic formation promoting the creation, circulation, and preservation of born-digital verbal art. ELO was founded in 1999 by Scott Rettberg, co-author of the hypertext fiction *The Unknown*, along with Robert Coover and others. ELO has published three volumes of its *Electronic Literature Collection* (2006, 2011, 2016), with more anticipated. These collections feature a variety of types and genres, including hypertext fiction.

Hypertext fiction has proved fertile ground for numerous theorists of contemporary writing. Given Bolter's background as a system designer and hypertext author (a version of his book *Writing Space* served as another demonstration text for Storyspace), we can at last suggest a connection to *Remediation*, the foundational guide to understanding new media written by Bolter and Richard Grusin (1999). Though their term *hypermediacy* encompasses much more than digital hypermedia, it could be argued that the latter provided at least some inspiration for the former. Dave Ciccoricco has devoted two books to digital narrative (*Reading Network Fiction*, 2007 and *Refiguring Minds in Narrative Media*, 2015). Narrative theorists such as Marie-Laure Ryan (*Cyberspace Textuality*, 1999; *Narrative as Virtual Reality*, 2001) and Markku Eskelinen (*Cybertext Poetics*, 2012) have vigorously debated the implications of hypertext and related systems for narratology. N. Katherine Hayles has provided the most complete and careful consideration of hypertext fiction per se, using it as one basis for her doctrine of *medium-specific analysis* (MSA), an approach to mediated expression that recognizes the interfusion of medium and message (Hayles 2008).

However influential, hypertext fiction was only one expression of a developing idea; Nelson's quip about "diving boards" should never be forgotten. Node-link hypertext was always a dangerously constraining model (see Rosenberg), and even in early days there were important moves beyond its constraints. "Reagan Library" (1998) lampoons headlong link-diving by including links unpredictably generated by an algorithm. "Lexia to Perplexia" plays with color, typography, and other conventions associated with links in Hypertext Markup Language, even as it exposes the language of web coding within its literary discourse. Morrissey and Talley's *Jew's Daughter* (2000) radically deconstructs the node-link convention by replacing the usual transitional links with calls to a program that dynamically recomposes the destination text.

From a certain point of view, all this creative fermentation is a thing of the past. By the end of the first decade of the twenty-first century, hypertext fiction and hypertext literature generally seemed in eclipse, if not oblivion. The noncommercial gift economy of the early World Wide Web made business exceedingly difficult for companies like Eastgate Systems, Inc. and Voyager. The latter folded after a brief, brilliant run. Eastgate carries on, in 2014 releasing an updated version of *Patchwork Girl* accessible to current-generation Macintosh systems, followed by a new version of Storyspace. Without a ready market, however, Eastgate's backlist has largely fallen to obsolescence. Most of its titles have not been updated to versions compatible with contemporary operating systems and so can be accessed only on vintage equipment.

These developments have not deterred dedicated readers. In 2010 Alice Bell published *The Possible Worlds of Hypertext Fiction*, examining four Eastgate titles as aesthetic explorations of potential ontology, a compelling way to think about branching narratives. Implicit in Bell's title, as her final chapter makes clear, is a critique both trenchant and hopeful. In some ways hypertext works seem impossible, unreadable, or at least unmarketable fictions. It is not clear that large numbers of people have time and patience for intensely complex narratives. Though critics like Steven Johnson have observed that film, television, and video games have recently featured precisely this kind of storytelling (Johnson 2006), there is a notable difference of degree between the relentless variations of *afternoon* and the more modest ramifications of *The Sopranos* or even *Mass Effect*. Closer affinity with hypertext and hypermedia may occur in films like Christopher Nolan's *Interstellar* (2014) and *Dunkirk* (2017), or Mark Kelly's *Southland Tales* (2006). Steven Shaviro (2009) describes this last film as "post-cinematic," in large part because it presents the viewer with an incomprehensibly intense, overloaded rendering of space and time. According to Steven Shaviro, post-cinematic film proceeds from a fundamental shift:

> I think it is safe to say that these changes are massive enough, and have gone on long enough, that we are now witnessing the emergence of a different media regime, and indeed of a different mode of production, than those which dominated the twentieth century. Digital technologies,

together with neoliberal economic relations, have given birth to radically new ways of manufacturing and experiencing lived experience.

(Shaviro 2009: 2)

Tracing Shaviro's "different mode of production" beyond cinema, we might see hypertext fiction as the literary equivalent of the post-cinematic, a beyond-literary writing meant for some world other than the one we inhabit. It might in fact belong to what Kenneth Goldsmith has called "the new illegibility," a shift toward texts that defy mere personal consumption because they belong to the regime of computers, networks, and distributed intelligence (Goldsmith 2011: 159). It has long been suggested that hypertext fiction seems written mainly for writers of hypertext fiction—texts only an author can love. Yet if we follow the conceptual links to Shaviro and Goldsmith, we may find an even more dismal conclusion: hypertext fiction is, properly speaking, written not even for its author, but at the mysterious whim of some machine. In this sense, even though it operates within the familiar domain of written words, hypertext fiction may belong to the same post-human aesthetic suggested by Mark B. N. Hansen's *Feed Forward* (2015), positing "a fundamental re-thinking of the human and of human experience as a non-optional complement to the new figure of the network" (2).

Though, as a good possible-worlds theorist might say, this is only one place to take one's thinking. The post-human turn is not absolutely ordained. Other destinations are possible. Even under radical regime change, the future may be a question of what branch we choose to follow. Bell herself sees sufficient room for optimism to issue a limited call to action:

Finally, the scholars of hypertext fiction and digital texts generally must publicize their work to the wider academic community. A failure to disseminate work more widely will mean that this area of research remains detrimentally niche. The fascinating narrative experiments that digital texts are capable of will be kept hidden and the methodological advances that will inevitably be made within hypertext theory will remain undisclosed. Both scenarios will disadvantage both print and digital scholarship.

(2010: 192)

In any honest cultural accounting the "wider academic community" is probably just a slightly larger "niche," so the glimmer of hope expressed here may be faint. Yet Bell does at least believe in some larger value for hypertext fiction, if only (echoing Eskelinen 2012) to inform a less print-centric study of narrative. Hypertext fiction can at least find its cell in the scholarly cloister along with twelve-tone music, non-Euclidean architecture, and object-oriented ontology. These things matter enormously to those who know why. To most people, even most educated people, they remain obscure. If we are solely concerned with texts produced at the end of the last century, this may well be where the weird story ends.

But like any good hypertext, this story escapes its particular ending by way of a timely restart.

The demise of hypertext fiction is regularly disproved by the flourishing existence of a software platform called Twine. According to its developers, Twine is "an open-source tool for telling interactive, nonlinear stories" (Twinery.org 2020). The original system was created in 2009 by Chris Klimas, an independent software developer. As an open-source application, it has been refined and expanded by various collaborators comprising the self-organizing "Twine team." Though it belongs to a different time and situation, Twine looks hauntingly familiar to an old hypertext hand. It shares the standard hypertext convention of directed graphs in which boxes (nodes) represent occasions for the presentation of various media forms (words, by default). Nodal boxes are connected by lines indicating the presence of a linked term or other mechanism that can trigger the replacement of one state by another: in most cases, the familiar action of the hypertext link. Accounts of Twine productions can produce flashbacks to the 1990s, for those who are susceptible. For instance:

> In Bryan Reid's *for political lovers, a little utopia sketch*, clicking on links cycles the text through a series of possibilities – occupations, places, dreams – that reflect the hopeful tone of the work. Clicking the verbs in certain sentences cycles them through a dreamy set of options: translating, training dogs, becoming an astronomer, growing vegetables, building houses, and so on.
>
> (Kopas 2015: 11)

This description could apply to any number of examples produced by students some of us have known over the years, or to works of certain old hands themselves.

Time-tripping aside, Twine both is and is not a hypertext system. On the affirmative side, it is designed to create logically controlled, multiply-traversable presentations intended for reception via the World Wide Web and Hypertext Transfer Protocol. In this sense Twine is a structured authoring system for web hypertext—something hypertext researchers have struggled to implement since the web began. The affordances of the program are in many ways reminiscent of Intermedia, KMS, TIES, and other old-time tools, though Twine's functions and visual conventions are generally less ambitious than those of older systems.

The relative simplicity of Twine is intentional. Though it is already evolving into new areas of application, Twine was designed for a specific purpose: the making of "interactive stories." Note that the Twine team do not say *hypertext fictions*. If hypertext fiction falls into the "new illegible," then this choice represents a stubborn neo- or retro-humanism, a refusal to give up the proposition that *we*—a broad cohort in no way limited to niche-bound academics—still have stories to share. In this sense, even if grizzled veterans think of Twine as a second coming of directed graphs, and Twine's branching narratives as hypertext fictions, these opinions are irrelevant. For its community of users, Twine has little or nothing to do with hypertext fiction. Twine is for games.

Tellingly, the first substantial retrospective on Twine works is a book called *Videogames for Humans* (Kopas 2015). This tome is a 500-plus-page collection of transcribed play sessions from Twine games (the preferred descriptor), interspersed with commentary by the players, some of whom are also Twine authors. Though even such a generous survey is bound to have some bias, the book, and especially Merritt Kopas' artful introduction, does leave a few clear impressions about Twine works.

First, the cultural milieu of Twine games differs appreciably from that of hypertext fiction. The terms *game* and *interactive story* affiliate the form more closely with Interactive Fiction (IF), a distinct genre of cybertext where narrative development is governed by logics more complex than the simple destination coding of a conventional hypertext link. According to Nick Montfort, who has written the definitive account of IF, the key element of such works is a multivariate *world model* expressed in the form of a computer program (Montfort 2005: 23). Purists might argue that Twine games lack this level of sophistication, making these objects either (in the generous view) hybrids of hypertext and interactive fiction, or (more skeptically) hypertexts that imitate interactive fictions.

In Bell's terms, both IF and Twine games occupy the same sort of localized "niche" as hypertext fiction. It is characteristic of life in a niche, or a small town, to feel a certain ambivalence toward the bright lights of the big city. In the case of hypertext fiction, these feelings took the form of strange literary embraces, as in *Victory Garden*'s pastiche of Borges and *Patchwork Girl*'s abduction of Mary Shelley. Twine writers seem similarly anxious, but not about the formal, book-bound canon. Tellingly, they use without regret the model of Choose Your Own Adventure stories, an allegedly juvenile form that once made hypertext writers cringe when the comparison was applied. Twine writers seem more at home with a literary inner child, and crucially with the concept of play—though play and games can also be grounds of contention. According to one commentator, Twine writers are "using interactive media to tell stories that mainstream videogames wouldn't dream of telling" (Kopas 2015: 11).

Once upon a time, hypertext fiction writers said similar things about their work vis-à-vis mainstream fiction; but we were usually talking about form, emphasizing the ability to do things with language that cannot be done on the printed page. The themes and subjects of hypertext fictions, to the extent one can generalize, did not diverge all that strongly from the main tradition. Literary culture at the end of the century was, however imperfectly, able to value social and sexual resistance. The coming-out story of J Yellowlees Douglas' *I Have Said Nothing* (1992) makes for a rich and powerful hypertext, but it might also work, in a different way no doubt, as a page-bound memoir. Shelley Jackson's exploration of monstrous embodiment body could be mapped readily enough onto other forms of *écriture feminine*, as Jackson herself does in "Stitch Bitch." With some notable exceptions, Jackson's later efforts have consisted largely of novels and stories. The hypertextual excursion was thus perhaps more demonstration than existential revolution. Literature forgave the rupture— and perhaps forgot.

In challenging the industry and culture of video games, Twine writers have taken on an adversary much less generous with difference, and in some ways openly hostile to social and sexual nonconformity. In so doing they have put themselves at risk of actual violence. Perhaps the most famous (or infamous) Twine game so far is *Depression Quest*, written by Zoe Quinn, Patrick Lindsey, and Isaac Shankler. The game is essentially a simulator of depressive thought and behavior, a remarkable use of interactive narrative to inform, educate, and recruit empathy. Released on the Steam game market in 2013, the game has instead stirred up epic levels of enmity. It has become the center of a "culture war" (Hudson 2014) that has broken out among independent game developers and reviewers, and has now spread to other aspects of game culture (Fangone 2015), to the science fiction and fantasy community (Minkel 2015), and most recently into the general body politic (see Nagle 2017).

As part of this conflict, popularly known as *Gamergate*, Quinn and other women in the game world have suffered threats of rape and murder, release of personal information (*doxing*), and other forms of harassment including bomb threats called in to conference venues where targeted women were to speak (Rousseau 2015). The details of the conflict are too numerous, complicated, and dismal to discuss here. Those who restrict themselves to the controversy over *Depression Quest* sometimes cast it as an argument about lax professional standards in game reviewing and the independent game community. For those who take a broader view, the matter has more to do with backlash against perceived invasion of a homogenous, heterosexual, male-dominated subculture by people with different attitudes and identities. In this sense the culture war may not be so different from the one declared by religious conservatives in the 1990s, though carried now into fresh zones of conflict. Back in the day, that conflict inspired at least one hypertext fiction (*Victory Garden*), but today the stakes seem enormously higher.

Twine writers may not appreciate being folded in with hypertext fiction, though on technical grounds the common traits between old and new seem undeniable. Socially speaking the resemblance is more debatable. Twine writers seem less interested in high-cultural approbation, better adapted to dynamics of social media, and more interested in taking their fight to institutions that academics often shun. Also, of course, they tend to be a whole lot younger than those of us who flourished in the Clinton administration. Still, an old hand must feel solidarity with the Twine community and their struggle to bring change to popular culture.

Looking out from the monastic niche, with the benefit of experience, we might see the Twine movement as the next step in a cultural logic in which hypertext fiction once played its part—though the scene is terribly different now. As we once did, Twine writers want to do unlikely things with stories, making the most of simple, easily accessible technologies. Whether they acknowledge it or not, they are agents of a vital digital literacy, asserting word-based text, both in terms of prose and underlying code, as the locus of seriously meaningful play.

Beyond this, as their enemies will say, they have an agenda, setting out to defend the interests of women, gay and transgender people, the neuro-atypical, and others marginalized and excluded by the military-infotainment complex. They are discontented with an arrogant, intolerant, casually violent society, and have set themselves against an entrenched culture industry. They have much more at risk, partly because they are intervening in a commercially important space; also because structures that once protected critical behavior (academic tenure and other channels of cultural capital) have significantly eroded since the Reagan years. However they think of what they do—as play, as storytelling, or as an occupation of gamespace— they are carrying on a struggle. Hypertext fiction may never get out of the cloister, but its younger cousins have found their way to the streets.

References

Aarseth, Espen J. (1997), *Cybertext: Perspectives on Ergodic Literature*, Baltimore, MD: Johns Hopkins University Press.

Arnold, Mary-Kim (1993), "Lust," *Eastgate Quarterly Review of Hypertext* 1(2), Watertown, MA: Eastgate Systems.

Barnet, Belinda (2013), *Memory Machines: The Evolution of Hypertext*, London: Anthem Press.

Bell, Alice (2010), *The Possible Worlds of Hypertext Fiction*, New York, NY: Palgrave Macmillan.

Berners-Lee, Tim and Mark Fischetti (2000), *Weaving the Web: The Original Design and Ultimate Destiny of the World Wide Web by its Inventor*, New York, NY: Harper Information.

Bernstein, Mark and Erin Sweeney (1989), "The Election of 1912, Hypertext for Macintosh Computers," Watertown, MA: Eastgate System Inc.

Bolter, Jay David (1991), *Writing Space*, Fairlawn, NJ: Lawrence Erlbaum.

Bolter, Jay David and Richard Grusin (1999), *Remediation: Understanding New Media*, Cambridge, MA: MIT Press.

Borràs, Laura, Talan Memmott, Rita Raley, and Brian Kim Stefans (2011), *Electronic Literature Collection*, Vol. 2, Cambridge, MA: Electronic Literature Organization.

Branwyn, Gareth and Mark Frauenfelder (1993), *Beyond Cyberpunk! A Do-it-Yourself Guide to the Future*, San Francisco, CA: The Computer Lab.

Crowther, Will, Don Woods, and K. Black (1976), "Colossal Cave Adventure" [Self-published computer game].

Deleuze, Gilles and Félix Guattari (1988), *A Thousand Plateaus: Capitalism and Schizophrenia*, New York, NY: Bloomsbury Publishing.

Douglas, J. Yellowlees (1994), "I have said nothing," *Eastgate Quarterly Review of Hypertext* 1 (2), Watertown, MA: Eastgate Systems.

Eskelinen, Markku (2012), *Cybertext Poetics: the Critical Landscape of New Media Literary Theory*, New York, NY: Bloomsbury Publishing.

Fangone, Jason (2015), "The Serial Swatter," *New York Times Magazine*, November 24, MM32.

Goldsmith, Kenneth (2011), *Uncreative Writing: Managing Language in the Digital Age*, New York, NY: Columbia University Press.

Greco, Diane (1995), *Cyborg: Engineering the Body Electric*, Watertown, MA: Eastgate Systems.

Hansen, Mark B. N. (2015), *Feed Forward: On the Future of Twenty-First Century Media*, Chicago, IL: University of Chicago Press.

Hayles, N. K. (2008), *How We Became Posthuman: Virtual Bodies in Cybernetics, Literature, and Informatics*, Chicago, IL: University of Chicago Press.

Hayles, N. K., Nick Montfort, Scott Rettberg, and Stephanie Strickland (2007), *Electronic Literature Collection*, Vol. 1, Los Angeles, CA: Electronic Literature Organization.

Hudson, Laura (2014), "Twine: The Video-game Technology for All," *New York Times Magazine*, November 19, MM44.

Jackson, Shelley (1995), *Patchwork Girl*, Watertown, MA: Eastgate Systems.

Jackson, Shelley (1997), "Stitch Bitch: The Patchwork Girl," *Media in Transition* 4, Cambridge, MA: MIT Media Lab.

Johnson, Steven (2006), *Everything Bad Is Good for You: How Today's Popular Culture Is Actually Making Us Smarter*, New York, NY: Penguin.

Joyce, Michael (1990), *afternoon: a story*, Watertown, MA: Eastgate Systems.

Joyce, Michael (1996), *Twilight: A Symphony*, Watertown, MA: Eastgate Systems.

Joyce, Michael (1997), "Twelve Blue," *Electronic Literature Collection*, Vol. 1, Los Angeles, CA: Electronic Literature Organization.

Kolb, David (1995), *Socrates in the Labyrinth Hypertext, Argument, Philosophy*, Watertown, MA: Eastgate Systems.

Kopas, Merritt (2015), *Videogames for Humans: Twine Authors in Conversation*, Instar Books.

Landow, George P. (1995), *Writing at the Edge*, Watertown, MA: Eastgate Systems

Larsen, Deena (1993), *Marble Springs*, Watertown, MA: Eastgate Systems.

Malloy, Judy (1986), *Uncle Roger*, San Francisco, CA: Whole Earth 'Lectronic Link, http://www.well.com/user/jmalloy/uncleroger/partytop.html.

Malloy, Judy (1993), *Its Name Was Penelope*, Watertown, MA: Eastgate Systems.

Malloy, Judy and Catherine Marshall (1995), *Forward Anywhere*, Watertown: MA: Eastgate Systems.

McDaid, John (1992), *Uncle Buddy's Phantom Funhouse*, Watertown, MA: Eastgate Systems.

Memmott, Talan (2000, 2006), "Lexia to Perplexia," *Electronic Literature Collection*, Vol. I.

Minkel, Elizabeth (2015), "How the Hugo Awards Got Very Own Gamergate," *New Statesman*, April 16,

Montfort, Nick (2005), *Twisty Little Passages: An Approach to Interactive Fiction*, Cambridge, MA: MIT Press.

Morrissey, Judd and Lori Talley (2002), *My Name Is Captain, Captain*, Watertown, MA: Eastgate Systems.

Moulthrop, Stuart (1991), *Victory Garden*, Watertown, MA: Eastgate Systems.

Moulthrop, Stuart (1997), "Hegirascope 2," *New River* 1 (1).

Moulthrop, Stuart (1999), "Reagan Library," *Gravitational Intrigue: A Little Magazine Publication*, Vol. 22cd: May.

Nagle, Angela (2017), *Kill All Normies: Online Culture Wars from 4Chan and Tumblr to Trump and Alt-Right*, New York, NY: Zero Books.

Nyce, James M. and Paul Kahn (1991), *From Memex to Hypertext: Vannevar Bush and the Mind's Machine*, New York, NY: Academic Press Professional.

Quinn, Zoe, Patrick Lindsey, and Isaac Schankler (2013), "Depression Quest," *Depressionquest.com*.

Rousseau, Steve (2015), "What to Read on the SXSW Gamergate Meltdown," *Digg*, October 27.

Shaviro, Steven (2009), *Post-Cinematic Affect*, New York, NY: Zero Books.

Stern, Andrew and Michael Mateas (2011), *Façade, Electronic Literature Collection*, Vol. 2, Cambridge, MA: Electronic Literature Organization.

Strickland, Stephanie (1997), *True North*, Watertown, MA: Eastgate Systems.

Thomas, Brian (1994), *If Monks Had Macs*, Los Angeles, CA: The Voyager Company.

Twinery.org (2020), http://www.twinery.org.

Weizenbaum, Joseph (1966), "ELIZA—A Computer Program for the Study of Natural Language Communication between Man and Machine," *Communications of the ACM* 9 (1): 36–45.

14

Place Taking Place: Temporary Poetic Theaters

Judd Morrissey

The Empty House

My work has been taking me outside again, a migration that I did not consciously plan, but one that happened gradually through adaptive and intuitive turns in my practice, responses to the shifting of my techno-cultural habitat, libraries of code I download in my sleep.

Writing, for me, began with a metabolic connection to walking. As a young body, I was not able to concentrate when still so I would weave and unweave myself through the city instead capturing the patterns and fragments that later became the substance of my fluidly transforming hypertext, *The Jew's Daughter.* This was before I knew anything about code, electronic writing, or performance but I arrived at these points through this initial point of departure, the process of writing while walking, a way of seeing that was partially blind to both the scene and the page.

On the afternoon of this writing, I performed in the architectural frame of a house with no walls, its sandy floor flooded irregularly with water from waves crashing in on the beach. I was viewing the scene through an iPhone embedded in a handheld paddle painted to resemble the ritualistic prop of Cornish fertility ritual. As I navigated the space, I read from an immersive landscape of texts situated in relation to scaffolding, water, clouds, and an interactive rose tattoo on the naked chest of my shivering collaborator. The words took on the textures of satellite imagery drawn from their precise

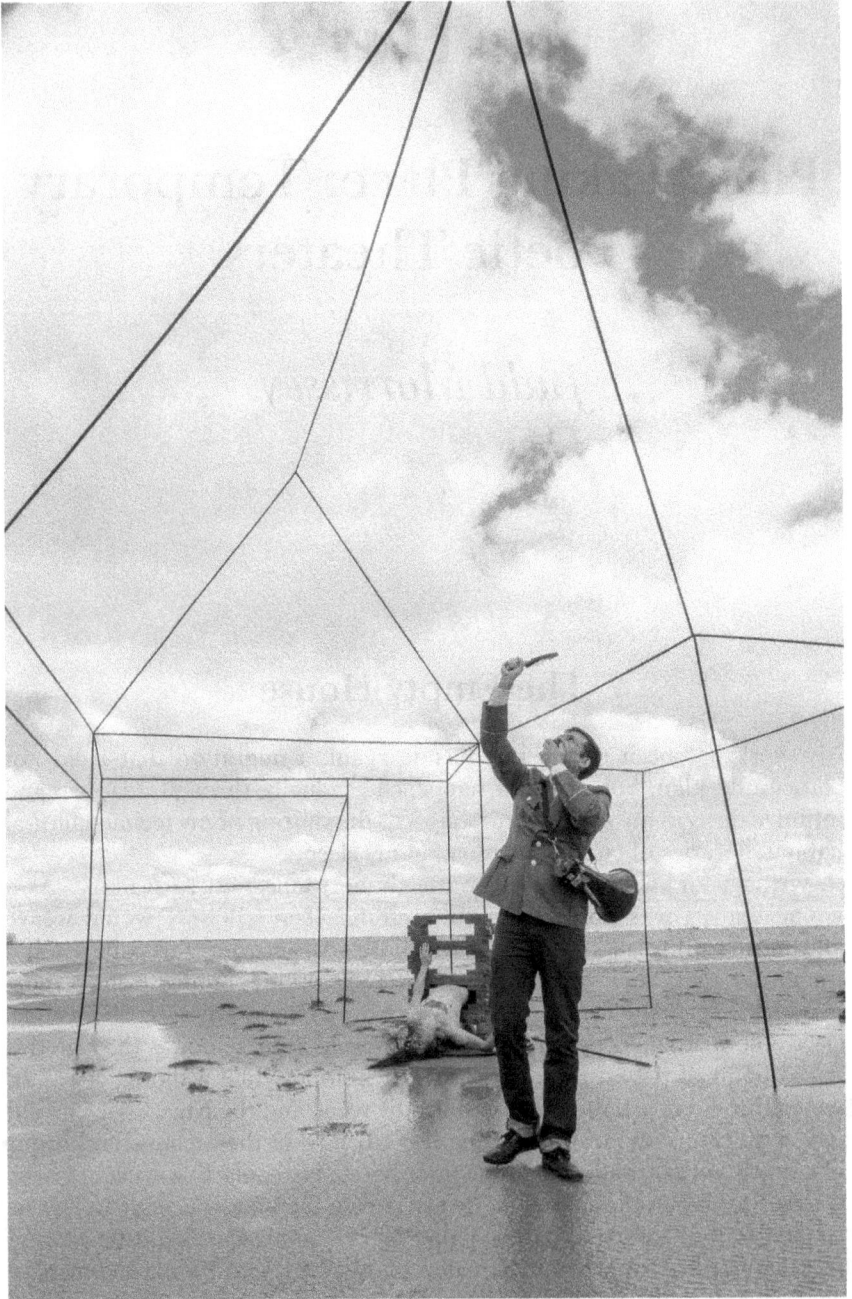

FIGURE 1 *Photo by Ji Yang.*

geophysical positions in space or from where they were originally conceived; some texts were inlaid with the floral patterns of the tattoo. When I read the word "buoy," in homage to Alan Turing, who wanted the letter u to sink under its own weight, for the word to erode into an embodiment of his desire, I was looking at an actual buoy in Lake Michigan. The words around me were mapped to specific coordinates of latitude, longitude, and altitude. I was able to measure my distance to each line: I am 2745.60.40 feet from *bays of bayesian boys*, 4382.40 feet from *Den Norske Gutt (The Norwegian Boy)*. Directly overhead, at a higher altitude and facing down, I saw these words: *his patience at the center of the zero* (see Figure 1).

I am struck by the uncanniness of performing within Sarah FitzSimons's sculpture *House*, an invitation connected to the Chicago Architecture Biennial, because the poem I read, distributed as a layer of augmented reality within the physical environment of the lakefront, once had as its working title, *The Empty House*, chosen to describe its liminal form as a virtually present performance of textual architecture. In this event of situated poetics, a nonrepresentational concept of space, created through a meticulous arrangement of language, inhabits another more concrete yet still porous expression of itself: a house within a house, a theater within a theater. Looking through the tenuous rafters, I read: *my house is sawn and double-slit*.

Installation Space

The installation of words on a page, software on a computer, and bodies in space can be transformative in nontrivial ways, unfolding as the symbolic process of a shipwreck, harmful invasion of malware, or training of military personnel. Installation can be elusive—silent, headless, self-updating in the background. Someone can be ceremonially placed into a position of authority, sometimes by being physically positioned within a spatial context: *the installation of a canon or prebendary of a cathedral consists in solemnly inducting him into his stall in the* choir. In this sense, a simple dramaturgical gesture can remodel one's world. Like the word *theater, installation* describes both a space and an activity. Installation, in art, for example, is the process of installing the installation. An installation begins to take place once it is in place, perhaps evolving over time as a growing series of installments.

Installed programs of performative code, like the stage directions in the scripts that create leaders, can transform a space and its inhabitants, and are attributed with the power to generate new mixed, augmented and virtual realities. To close this circle of code and stage, let's recall that the term virtual reality originally appears within Antonin Artaud's *The Theatre and Its Double*, housed within his formulation of a Theatre of Cruelty.

I stand in the center of the *House* installation and scan the space, reciting a found play of George Perec: *To the north, nothing. To the east, nothing. To the south, nothing. To the west, nothing. In the center, nothing.* I look down and see these words in the sand: *a satellite view of the corpus.* To begin this writing is to place myself in the text's position. Lying down, font-face up, supine, I offer my recent body of work for self-dissection, to be excavated within its own locus of operations, those performative processes that stage, between space, language, code, and bodies, a sense of *place taking place.* With this phrase, remixed from Mallarmé's spatially arranged words, *Rien n'aura eu lieu que le lieu* (Nothing will have taken place but the place), I want to inhabit *the place* as it takes place, as placement, replacement, and displacement, as what is taking our place when the body, seated under the scanner's tripod, disappears within the radius of a mechanical blindness.

Growth of a Form

Beginning in 2013, I spent a year living in Norway as a Fulbright Scholar, teaching and developing work in poetic augmented reality. While preparing for my geographical dislocation, I came across a gap that offered a connection. The gap occurs within the Sherlock Holmes stories of Sir Arthur Conan Doyle, between *The Final Problem*, when Holmes seems to die, and *The Empty House*, when he returns. In the latter story, Holmes claims that he'd spent the interim under the alias of a Norwegian explorer, Sigerson, applying his forensic talents to botany and the natural world. *The Empty House* became the working title for experiments undertaken during my own hiatus in the lushly dark and solitary landscape of Bergen, bringing together themes of limbo and exploration with the spectral in-betweenness of augmented reality.

The words *empty house* might conjure the image of Holmes in the story gazing out through a window at his double across the street, a wax decoy designed to entrap a would-be assassin. But a house is also a theatre as well as its audience, a *place of seeing*, where the one we observe on stage perhaps gazes out into a circular pattern of informational architecture as in the memory theater of Giulio Camillo.

While in Bergen, I developed a technique I'd been exploring that involved live-sampling tiles of digital satellite imagery from GPS locations and filling letterforms with the captured textures. In this system, the appearance of a text hovering in the space around a reader refers materially to a location: either where it was written, where it is being read, or a place that it is referencing (see Figure 2).

Initially, I created the texts and tracked my coordinates as I walked, sending both to a database that was visualized by the LAYAR AR browser,

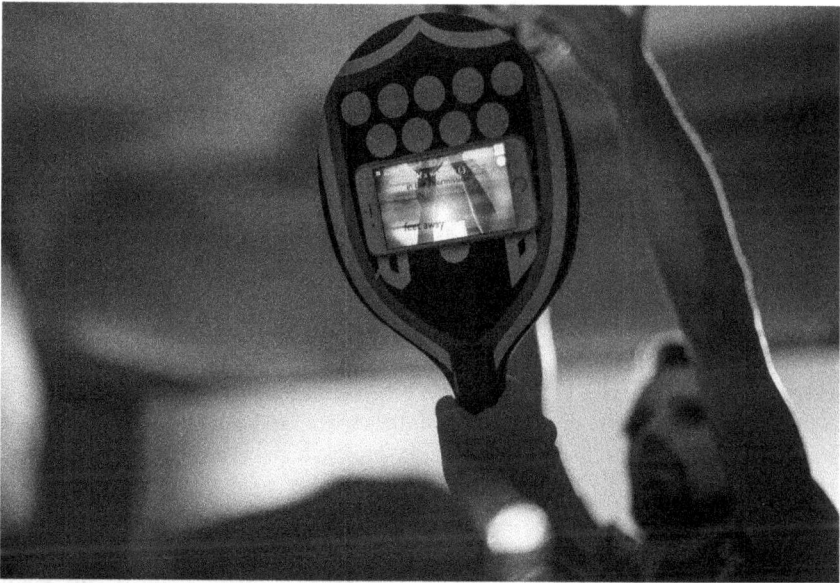

FIGURE 2 *Photo by Grace DuVal.*

so that the language was placed where it was composed, but I became more interested in visualizing the texts as aggregated visual structures constrained to locales. At this point, I began using an algorithm for great circle distances to identify GPS points in proximity to my location, and to use the discovered points to place language in circular and constellation-based patterns, plotted at different altitudes within the surrounding environment.

This technique of arranging the texts in space with an automated deliberation, rather than physically inhabiting every point, still did not eliminate the need to respond to the various physical environments inhabited by the generated text. I spent two weeks modifying *The Empty House* specifically for the atrium of the Museum of Contemporary Art in Chicago in the summer of 2014. In this case, the museum's wall text for a Simon Starling exhibition entered the poem, as an appropriate under-layer that prompted further compositional changes, his name alone generating a movement of passeriformes. Starling's project, *shedboatshed*, in which he transforms a decrepit boathouse into a boat that he then rows to Switzerland and reconstructs into a shed again, entered into the poetics, just as it now helps to imagine the geophysical course of this work from Bergen to Chicago and back to Bergen, each instance building upon and also replacing the last since until recently the poem could only exist in one place at a time.

In Bergen, as the system for composition and visualization evolved, the context of the work also shifted as I discovered a situational connection to Alan Turing. In the last two years of his life, Turing, the gay computing pioneer, visited Norway, seeking a more tolerant environment after his trial and conviction in the UK for crimes of gross indecency due to a consensual homosexual affair. During the time of his Norwegian travels, Turing was developing his theory of morphogenesis to account mathematically for patterns and forms in nature, and he seems to have named this project for a love interest. Kjell, a young man from Bergen, referred to by

(SIGHTHOLE)

Turing as *Den Norske Gutt* (the Norwegian Boy), later attempted to visit him in England but was apparently intercepted and deported by Scotland Yard, due to the intense scrutiny into Turing's private affairs. His name entered the space of my poem and also formed a new title for the work when I came across the words *Kjell Theory* in scans of handwritten notes in Turing's digital archives.

While Turing was developing his theory of morphogenesis, his own body had undergone a morphological change in that the estrogen treatments he received as a legally imposed form of chemical castration, an alternative to prison time, had resulted in gynecomastia, or the growth of male breasts.

Patients at the Center of the Zero

The anatomical theatre, in its earliest and clearest expression, the one at the University of Padua built in 1594, is a roughly circular space comprised of a central operating table surrounded by six tiered balconies where training physicians and other observers would gaze down upon the autopsy or surgical procedure being demonstrated. This structure that foregrounds the body as subject of observation, penetration, and enhancement was adopted as the symbol for the performance and technology collective, Anatomical Theatres of Mixed Reality (ATOM-r), which I co-founded in 2012.

Our first work, *The Operature* (2014), combined research into the early history of surgery with materials reanimated from the corpus of the writer, pornographer, and tattoo artist Samuel Steward. Steward was a friend and protégé of Gertrude Stein, who rather than embracing literary Modernism, located his work in direct relation to sexuality and bodily experience. Steward's *Stud File* is an autobiographical card catalogue and forensic archive tracking homosexual encounters over five decades, beginning in the 1930s when his activities were against the law. He used a self-devised alphanumeric coding system to loosely encrypt and cross-reference his accumulating experiences of sex with other men.

Steward, under the aliases, respectively, of Phil Andros and Phil Sparrow, wrote gay pulp fiction and became a tattoo artist. In Steward's tattoos, the rose, subject of Stein's Modernist resuscitation of literary language, is disseminated as an inscription in skin, carried by the bodies of thousands of sailors.

For *The Operature,* ATOM-r recreated Steward's designs as temporary tattoos and applied them to our performers' bodies as markers for augmented reality within a live performance. The set design referenced the form of anatomical theater, its central component being an operating table with an embedded interactive screen. This was part of a larger modular assemblage of five tables that, when combined, formed a stage for choreography with openings in which bodies could be placed. More abstractly, the circular form of the theater was referenced in the visual placement of geospatial texts that surrounded us in tiered radial configurations.

In the middle of the ninety-minute work, we staged an intermission in which performers laid themselves down for examination upon the assembled tables, and the audience was invited to scan their tattoos with smartphones, revealing texts, videos and 3D objects overlaid within the performance space, itself a museum of health and medicine.

Sighthole (an intermission)

In the approximate middle of this writing, I've placed a *sighthole.*

I've been using this word to name a void in the center of environmental data captured by Lidar scanners. Short for light detection and ranging, this technology swallows up whole landscapes with its rotating laser-eye, producing massive point clouds of data, while being unable to capture information in the immediate radial vicinity of the scanner's position. This creates a circular nothing in the center of the data. For this reason, if one wants to disappear, they can simply place themselves beneath the device, as in *when the station and target digital scan of images were in progress, we sat under the scanner tripod to avoid being visible in the scan.*

While the above italic text is sampled from a Google search to confirm this practice of self-deletion, I experienced the phenomenon when working on a new project with fellow poet-artists, Jennifer Scappettone and Abraham Avnisan, in which we used a Lidar scanner to capture an underground copper mine. We have since been processing the mine's virtual double in relation to a poetic system drawn from a book of nineteenth-century telegraph codes for the mining industry. In the project, SMOKEPENNY LYRICHORD HEAVENBRED, we are engaging the multivalence of these codes and relationships between apparently disembodied communication networks and the exploitation of natural and human resources. The data-voids in our mined scans are becoming portals between disparate times and places, or gaps in the cloud from which machine-poetic lament rains down to the tune of the song *Pennies from Heaven*.

Double Theaters

The double view of augmented reality took place as a particularly appropriate juxtaposition when *The Operature* was presented in 2013 at the Anatomy Theatre and Museum in London, a former anatomical theater converted into a new media performance space. Here, the work, as though returning to an origin, was contained within its own spatial metaphor. The self-similar complexity of this arrangement seemed to express itself as a persistence of unexplained technical glitches that culminated in the interruption of the work's final moments by a fire alarm with the relentless shriek of a ghost from a previous century. As soon as we finished, we all had to immediately evacuate.

A Circular Theater with Two Stages / One in the Middle and the Other Like a Ring

Kjell Theory evolved, out of the space created by *The Empty House*, as an augmented reality poem, various exhibitions, a series of solo performances, an outdoor site-specific walk, and a large-scale performance of ATOM-r (see Figure 3). The subject matter of the work expanded through doubling or juxtaposition when my mental image of Turing in Norway with gynecomastia brought to mind the blind prophet of Greek myth, Tiresias, described by T. S. Eliot as an *old man with wrinkled female breasts* who is *throbbing between two lives*. The apprehension of this hybrid figure led me to *Les Mamelles de Tirésias (The Breasts of Tiresias)*, a feminist, genderfluid play by Guillaume Apollinaire that was first produced in 1917, and for which he coined the term surrealism. In Apollinaire's play, a cisgendered woman, Theresa, wills her transformation into a man, Tiresias, while her husband gives birth to 40,049 babies. The play's performance of gender reversal in fact doubles as a plea to the men of France to replenish a population that had been ravaged by the First World War.

The comedic absurdity of Apollinaire's play is stylistically contrasted by the poetics of his visionary prologue which contains a war-time hallucination of stars in the night sky becoming the eyes of newborns. Apollinaire's astronomical vision of birthing is also impregnated with his ambition to transform theater, to infuse its tired conventions with a new spirit. While evoking this explosive generation of babies, stars, and theater, Apollinaire also specifies for the architectural design of a space in which the play should be staged: *A circular theater with two stages / One in the middle and the*

FIGURE 3 *Photo by Grace DuVal.*

other like a ring / Around the spectators permitting / The full unfolding of our modern art. This space was never built, but it notably expresses the importance of a form that can appropriately house the intricacies of work's event. This may be a complex theater with multiple vantage points and no clear center, one in which a viewer may surround or be surrounded by the spectacle of the work.

ATOM-r's *Kjell Theøry* layers Turing's theory of forms, which describes *a mathematical model of the growing embryo*, and Apollinaire's theater of gender transformations and male birthing, into a queerly embodied mixed reality that is also an expression of visionary blindness. The performance, which premiered in 2017 to coincide with the centennial of *Les Mamelles de Tirésias*, is surrounded by configurations of geospatial text merging original writing with algorithmic mutations of Apollinaire's prologue, and implements augmented costumes, props, and tattoos, to create a performative emergence of *place taking place*, a generative ephemerality impregnated with the present absence of virtuality, a nothingness that is also the birth of a new constellation.

References

Apollinaire, Guillaume (1964), "The Breasts of Tiresias," in Michael Benedikt and George E. Wellwarth (eds.), *Modern French Theatre: The Avant-Garde, Dada, and Surrealism; an Anthology of Plays*, 66, New York, NY: Dutton.

Eliot, Thomas Stearns (1922, 2011), *The Waste Land*, New York, NY: Horace Liveright; Bartleby.com, www.bartleby.com/201/1.html#218.

"installation," *Online Etymology Dictionary*, 2020. https://www.etymonline.com (accessed August 26, 2020).

Mallarmé, Stéphane (1914), "Un Coup de Dés Jamais N'Abolira Le Hasard," *La Nouvelle Review Française*, http://www.writing.upenn.edu/library/Mallarme-Stephane_Coup_1914_spread.pdf.

Perec, Georges (1997), *Species of Spaces and Other Pieces*, 7, London: Penguin Books.

Turing, A. M. (1952), "The Chemical Basis of Morphogenesis," *Philosophical Transactions of the Royal Society of London. Series B, Biological Sciences* 237 (641): 37.

Whitney, William Dwight (1906), *The Century Dictionary and Cyclopedia: A Work of Universal Reference in all Departments of Knowledge, with a New Atlas of the World*, 3122, New York, NY: Century.

15

Kinetic Poetry

Álvaro Seiça

The kinetic poem may still be in its infancy.
—MARY ELLEN SOLT (1968)

The term "kinetic" derives from the Greek verb *kinein*; that is, "to move." Therefore, action and movement infuse kinetic poetry as it describes poetic works that employ motion. Within the realm of digital poetry, where it is today mostly deployed, the composition of methods that output textual movement—such as transitions, timeouts, and intervals—incorporate temporality in the process of coding and display of writing. Yet a discussion of current works of kinetic poetry must be situated in the wider flux of aesthetic, artistic, and media antecedents that pervaded the twentieth century. These antecedents inform us about the will to move beyond the static linearity of the printed page and the notion of poetry as living in

I want to express my gratitude to Rui Torres for his contribution in suggesting and outlining this chapter from the point of view of kinetic forms. Many thanks to Scott Rettberg and C. T. Funkhouser for their revising suggestions, and to a number of poets I interviewed during 2013–16: vimeo.com/channels/setintervalconversations. I would also like to thank Richard Kostelanetz, Johanna Drucker, Chelsea Spengemann, Violette Garnier, Fondazione Bonotto's Luigi Bonotto, Patrizio Peterlini and Enrica Sampong, Eduardo Darino and Philip Steadman for insightful information; and to Stephen Bann and Brona Bronač Ferran for generously providing materials and feedback. I am aware that this brief history of kinetic poetry is mostly centered on Western Europe and the Americas. It has been difficult to access bibliographies from Asian and African literatures on this theme. I am sure there are works out there that complicate and redefine my focalization, and as such, I appreciate comments in order to complement or refute it. This chapter was made possible by funding from the University of Bergen and, in its final stage, the European Commission via the Marie Skłodowska-Curie Action ARTDEL, and the Norwegian Research Council.

a single medium. The most obvious animation medium is film, but many animation mechanisms preceded film. Kineticism can be traced back to the invention of technical apparatus such as the kinetograph and the kinetoscope, developed by Thomas Edison and William Dickson at the end of the nineteenth century.[1]

In order to create bridges between narratives from different fields and artistic movements, I will focus on five forms of time-based kinetic poetry: mechanical poetry, film poetry, videopoetry, holopoetry, and digital poetry. These five media-specific forms are better seen as media clusters with resemblances, not as groups of homogeneous media artworks, even though they all rely on temporal and spatial dimensions to achieve literary and artistic expressiveness. What they strictly have in common is the way poets and artists engage with a broader vision of "poetry in motion;" that is, kinetic poetry. They are operative insofar as they execute a set of instructions or algorithms, being that of the time slots between frames in a storyboard, or the intervals set for transitions in digital poetry. Even if this chapter offers relations and points of departure, a concise history of kinetic poetry cannot be grasped without understanding some of its immediate antecedents: Mallarmé's exploration of space in the page, Morgenstern's phono-visual poems, the Futurists' typographic quest to set "words in freedom" (Govoni and Marinetti's *parole in libertà*), Apollinaire's *calligrammes*, the Dadaist random and sound performances, the abstract films of the Modernists, and the postwar experimentalism involving sound, text, and image with spatialization, collage, montage, and other techniques unfolding with the concrete and visual poets.

Kinetic Origins

Throughout the history of writing, modes of textual inscription have been dependent on space, but rarely on time. The printing process activates text as a discrete element to be displayed on a planographic surface. In film, video, and the computer, textual inscription is presented in different outputs, and potentially acquires new forms of artistic expression—given that it allows for displacement, tridimensional space, time scheduling, and media integration. Certainly, poetry's progressive transition from static to kinetic media owes its roots to investigations and transgressions done by poets and artists working with visual text from the antiquity to the baroque period,

---

[1] These machines were envisioned upon earlier chronophotographic techniques developed by Marey, Reynaud, Demeny, Anschutz, and Muybridge, to mention but a few, in order to build stop motion devices that would set the illusion of movement: the magic lantern and the flip book (kineograph), the thaumatrope, phenakistoscope, zoetrope, praxinoscope, zoopraxiscope, electrotachyscope, and the "photographic gun." Dickson also developed the mutoscope.

via the late nineteenth century and Modernism. Stéphane Mallarmé's work is symptomatic of a quest to stretch the boundaries and conventions of words and blanks in the page. Mallarmé's poem "Un Coup de Dés Jamais N'Abolira le Hasard" (1897) is notorious for the displacement of words in space, creating voids and pauses in the free poetic line, and extending the reading area to the double-page spread. The suggestion of movement in the page was later explored by Guillaume Apollinaire in *Calligrammes* (1918), whose visual component is achieved by calligraphic elements that are syntactically and graphically arranged in relation to semantics.

It is within the Modernist period that kinetic works start to be technically activated. In the 1910s, Italian and Russian Futurist writers envisioned a world in which the machine and speed would set words free, with effect on literary expression, spatial composition, and cacophonic phonemes. During the 1910s and 1920s, painters, sculptors, architects, photographers, and filmmakers, used to material experimentation, engaged with mixed media that allowed for motion techniques. Futurist abstract films from the 1910s and Marcel Duchamp's "assisted readymade" *Bicycle Wheel* (1913) can be seen, in this sense, as some of the earliest kinetic artworks. Duchamp's piece is a sculpture that simply modifies two objects, although in 1920, with an engine, Duchamp assembled *Rotary Glass Plates (Precision Optics)*, an installation which produced both kinetic and optic rhythms. Naum Gabo's *Kinetic Construction (Standing Wave)* (1919–20) is a further step in kinetic art, insofar its mechanical motor creates four dimensions by vibrating. Gabo and Antoine Pevsner's *Realisticheskii Manifest*—where the ideas of kinetic art were introduced on August 5, 1920—paved the way not only for the establishment of an abstract constructivism, which contrasted with the political Soviet Constructivists but also for what would follow in kinetic arts: "Space and time are the only forms on which life is built and hence art must be constructed. (...) We affirm *in these arts a new element the kinetic rhythms as the basic forms of our perception of real time*" (Gabo and Pevsner 1957: 152, emphasis original).

Celebrating their hundredth anniversary, kinetic arts have traversed multifaceted experiments with artistic and literary forms in diverse media. Always connected to changes in science and technology, kineticism rapidly became a source of fascination: from László Moholy-Nagy's lumino-kinetic sculptures and abstract films, to Hans Richter, Man Ray, and Fernand Léger's movies; from Duchamp's kinetic mixed-media objects, sculptures, and films, to Alexander Calder's air stream *mobiles*. Kinetic art emerges in the 1920s and remerges in the 1950s postwar. In 1953, Yaacov Agam's solo exhibition *Peintures en Mouvement* at the Galerie Craven in Paris singles kinetic paintings out, which will resonate in the 1955 collective exhibition at Galerie Denise René. The exhibition *Le Mouvement/The Movement*, curated by René and Pontus Hultén, compiled kinetic and op(tical) works by Agam, Bury, Calder, Duchamp, Jacobsen, Soto, Tinguely, and Vasarely.

Today, it can be considered as a pivotal point in kinetic arts, signaling but also amalgamating two different branches of artistic motion: kinetic art, involving applied physical movement, and op art, suggesting movement or illusion.[2]

The post-Second World War era certainly provoked a need for artistically reimagining the world and experimental art soon blended even more media. But the effect of war, with its human cruelty and sadistic technologic development, had already shaken the artistic milieu during the twentieth century. During the First World War, Dada artists in Zürich, Berlin, and New York embraced the absurdity of human existence in face of war, and reacted, by turning chaos and meaninglessness into manifestos, literary and visual works, like Hannah Höch's photomontages, and sound performances. Sound poetry arose from the Dadaist tradition of phonetic experimentation, playful and performative randomness, in now emblematic works by Hugo Ball, Raoul Hausmann, or Tristan Tzara, which resonated in Kurt Schwitters's *Ursonate* (1922–32). Following upon innovations in electroacoustic music, such as Pierre Schaeffer's *musique concrète*, sound poetry continued as a concerted movement in France and elsewhere in the 1950s, with Henri Chopin, François Dufrêne, Ilse and Pierre Garnier, and Bernard Heidsieck placing emphasis on language's oral atomization and deconstruction via vocal techniques and reel-to-reel tape recorders. Poets also resumed research with the movement of letter shapes influenced by flows of practice that came from before the war and continued to occur during war time. But the typewriter began to be used by younger poets to establish visual patterns of linguistic signs in a new semiotic reading experience.

As the narrative usually goes, concrete poetry was initiated by Eugen Gomringer and Öyvind Fahlström in Europe, and the Noigandres group in Brazil—Augusto de Campos, Haroldo de Campos, and Décio Pignatari. According to Emmett Williams (1967: vi) and Solt (1968), Fahlström and Gomringer/Noigandres were unaware of each other's work. In fact, by 1951 Gomringer had already conceptualized some of the "constellations" collected in *Konstellationen* (1953), while Fahlström had published "Hätila Ragulpr på Fåtskliaben" (1953–4), a text that became known as the "Manifesto for Concrete Poetry" only in 1966 (Olsson 2005, 2016; Bäckström 2012). Yet E. M. de Melo e Castro (1962), in an eye-opening *TLS* letter for the United

---

[2]Future exhibitions during the 1960s—such as *Kinetische Kunst* in Zurich (1960), *Bewogen Beweging* (1961) in Amsterdam, the *Nove Tendencije* (1961–5) and *Tendencije* (1968–73) series in Zagreb, *Arte Programmata* (1962) in Milan, *The Responsive Eye* (1965) in New York, *Kinetika* (1967) in Vienna, *Cinétisme Spectacle Environnement* in Grenoble (1968), or *Cybernetic Serendipity* (1968) in London—would depart from *Le Mouvement*, or expand its scope around constructivism, concrete art, conceptual art, cybernetics, and electronic art.

Kingdom's poets, affirms that concrete poetry was born in Brazil. Franz
Mon (1988: 31), on the other hand, attributes its beginning to the work
of Italian Futurist-descendent poet Carlo Belloli in 1943, an author earlier
credited by Emmett Williams in *An Anthology of Concrete Poetry* (1967)
and by Mary Ellen Solt in *Concrete Poetry: A World View* (1968). If it is
true that Belloli's *Parole per la Guerra* (1943) follows a Futurist graphic
treatment, several poems in *Testi-Poemi Murali* (1944) and *Tavole Visuali*
(1948) already show a break that resembles what would be called "concrete
poetry" by the 1950s. According to Belloli's remarks to Solt (1968), even if
he saw his work as a precursor of concrete aesthetics, he preferred the term
poesia visiva because it conveyed an approach to visual poetry that was
semantic, not asemic.

The 1950s concrete poets absorbed creative and theoretical influences
that came from "the area of fine arts, primarily those of de Stijl, Theo
van Doesburg and Max Bill [concrete art]" (Mon 2011: 28–9). To these
references, it is important at least to mention, from the part of the Brazilian
Noigandres poets, Ernest Fenollosa's and Ezra Pound's writings about
oriental ideograms, James Joyce's and e. e. cummings's work; while from
the part of the Swedish- and German-speaking poets, the influence of Hans
Arp's concrete art, concrete and electronic music. Eduardo Kac (2015) goes
further along these lines and re-contextualizes what are, to be sure, the
multiple origins of concrete poetry: Vasilii Kamenskii's 1914 visual poems
and subtitle reference *Tango s korovami. Zhelezobetonnye poemy* [Tango
with Cows: Ferro-Concrete Poems], and importantly, for his immediate
antecedence, the less-acknowledged Brazilian poet Wlademir Dias-Pino,
whose 1940s work greatly influenced the São Paulo and Rio de Janeiro
concrete groups prior to the neoconcretism split.

These influences spread at different pace and via different networks
of friendship and collaboration. Yet the core notion to retain is that the
concrete poets pushed forward in radically transforming the disposition
of letters and words with new semantic, syntactic, phonetic, and visual
compositional strategies that aimed at reinventing poetics and breaking
away from verbose lyricism and discursiveness—what Rosmarie Waldrop
(1976: 141) called "a revolt against [the] transparency of the word." The
influence of ideogrammatic writing helped in approaching the grammar
of mass media, advertisement, and information aesthetics via typography
and industrial design. Letters, symbols and words were seen as atoms and
sequences ingrained with power—what Gomringer (1954) described as
"concentration and simplification." Furthermore, the political repression in
which some of these authors lived in, or would live in, both in Europe and
Latin America, would have an impact on works of a second wave of concrete
and visual poetics. Like Ilse Garnier, Bohumila Grögerová, Ana Hatherly, or
Salette Tavares in Europe, in the United States Mary Ellen Solt infiltrated

the male-dominated concrete poetry scene with her inventive *Flowers in Concrete* (1966). If we are to assess today's legacy of concretism, we have to necessarily address the gendered canonization at the global scale of the movement. But this fact is not new either, since there have been occasional attempts since the 1970s to claim back territory and rewrite the narrative of women's role, perhaps starting with the yearly expanding exhibition *Between Language and Image*, first organized in 1972 in Italy by Mirella Bentivoglio (Zoccoli 1976).[3]

Mechanical Poetry: Motorized Sculpture Machines

By the 1960s, compelling examples of flip books, object poems, and scroll poems, such as those made by Japanese Vou group member Takahashi Shohachiro in the *Poésieanimation* series (Toshihiko 1977; Donguy 2007: 227, 236) show that the scroll and the signifiers could create the illusion of motion. But there was more: artists were also constructing mechanical motorized sculptures with textual elements that actually moved by themselves. That was precisely what the *First International Exhibition of Concrete [Phonetic] and Kinetic Poetry* aimed at in 1964, in Cambridge, United Kingdom.[4]

The exhibition's poster includes a poem by Pierre Garnier that suggests movement due to its visual rhythm (Figure 1). Organized by Mike Weaver, with the assistance of Reg Gadney, Philip Steadman, and Stephen Bann, it recognized kinetic poetry as an expanded form of poetry, especially because Weaver was "soliciting poem-sculpture proposals" (Thomas 2019: 135). For Weaver (1964: 14), "In kinetic poetry the boundaries of the visual poem are extended in time." At this point, some poets and critics thought of "kinetic poetry" as dynamic visual poems, flip books, or book objects (artists' books) that would convey the illusion of movement, such as those by Williams or

---

[3]See also Emerson (2011), Beaulieu (2013, 2014), and Barok (2018). It is impressive the lack of women authors selected by Williams in his anthology (Ilse Garnier, Bohumila Grögerová, and Mary Ellen Solt), but even more so in Max Bense and Elisabeth Walther's *Konkrete Poesie International* (1965), Stephen Bann's *Concrete Poetry: An International Anthology* (1967), or Gomringer's anthology of German-speaking authors *Konkrete Poesie* (1972): zero! This is at odds with Solt's broader study and criteria, which is neither alphabetical nor linguistic, but rather geographical, in *Concrete Poetry: A World View* (1968). In contextualizing, Solt refers to the work of Ilse Garnier, Bohumila Grögerová, Elisabeth Walther, Salette Tavares, Blanca Calparsoro, Pilar Gómez Bedate, Louise Bogan, and her own, even though the panorama was larger. I am thinking, for instance, of Ana Hatherly.

[4]The poster and the catalog titles in fact differ (Bann 2020). The "Catalogue" (1964) included the term "phonetic."

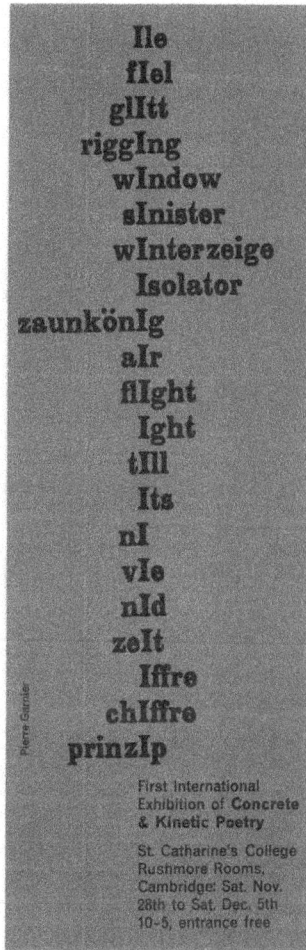

FIGURE 1 *Poster of the First* International Exhibition of Concrete and Kinetic Poetry, *St. Catharine's College, Cambridge, Nov. 28–Dec. 5, 1964. Poster designed by Philip Steadman with poster-poem "i (prinzIp)" by Pierre Garnier.* Jasia Reichardt Archive of Concrete and Sound Poetry, 1959–1977, Getty Research Institute, Los Angeles (890143B). Copyright Pierre Garnier and Philip Steadman. Courtesy of Violette Garnier and Philip Steadman.

Ian Hamilton Finlay (Solt 1968; Bann 2015), or typewriter patterns that would produce optical effects, like Timm Ulrichs's *Typotexture* (1962)—all of which seem closer to op poetry.[5]

---

[5]The "actual" and "virtual" (effect on the retina) kineticism of these works is debatable. See Vasarely's "cinétisme" (1955, 1966), Weaver's distinction (1964), Bann's unity and diversity (1966b, 2020), and Popper's historical threading (1968).

For those engaged with art, science, and technology, alongside an idea of the neo-renaissance and interdisciplinary artist, kineticism meant another possibility—kinetic art; that is, mechanical moving art. This tradition was inherited from the 1920s kinetic arts and acquired momentum in postwar arts.[6] Visual artists got also interested in exploring the potential of moving text. From 1960 onwards, Liliane Lijn created kinetic cylinders in which she would by 1962 include text in a series of mixed-media "poemcons" and "poem machines," such as *Time is Change* (1964–5), a motorized conic turning sculpture using stenciled text. Also drawing from kinetic art and under the aegis of Dom Sylvester Houédard, Ken Cox started adding letterforms to his motorized sculptures, such as *Shadow Box, Four Seasons Clock* (1965) and *Three Graces* (1966–8), which were seen as "kinetic poems" as well as "poetry machines" in a posthumous exhibition given the artist's premature death.

For the *First International Exhibition of Concrete and Kinetic Poetry*, the four members were crucial in compiling their contacts. Gadney brought news about kinetic art from Paris, such as Frank Malina's, and thus the attempt of Weaver in intersecting concrete poetry and kinetic art in order to forge an exhibition on kinetic poetry as well. This context would drive the exhibition's organizers to conceptualize kinetic poems, some of which materialized, like Weaver's motorized poem "Tempoem." John Sharkey devised the film poem *OPENWORDROBE* (1964), while the kinetic artist José María Cruxent included text, for instance, in the *Métromane* (1964) installation. Groundbreaking in scope and geography, the exhibition ended up focusing more on its concrete than kinetic dimensions (Bann 2015; Gadney 2017).

Concretism—and by extension experimentalism—seems to be of the utmost relevance for the development of kinetic poetry. Various movements that constitute the landscape of experimental poetry draw from the synthesis and compression of language in mixed-media approaches. They also tend to place an emphasis on visual materiality as a communication means, processuality, collaboration, and participation, which are decisive in the experiences with film, video, and computers enacted by many of the same poets that started in the realm of concretism. Like Fluxus and other groups or movements that populated the landscape of 1950–60s experimental arts, experimental poetry was concerned with the expanded possibilities of media and the materiality of language. At this point, poets were creating and theorizing about a proliferation of non-verbal-bound poetics: visual poetry, auditive poetry, tactile poetry, respiratory

---

[6] Besides those exhibiting at *Le Mouvement*, collectives working in this area included Group Zero, Group N, Group T, Group Y, Groupe de Recherche d'Art Visuel, and Dvizhenie. Individual artists were numerous—see the catalogues of the exhibitions indicated on footnote 2.

poetry, linguistic poetry, conceptual and mathematical poetry, synesthetic poetry, and spatial poetry (Melo e Castro 1965, 2014). This galaxy of proliferating media-oriented poetics finds echo in Adriano Spatola's "total poetry" (1969, 2008) and Dick Higgins's "intermedia" and "metapoetries" diagrams (2018 [1967, 1978]). This sense of innovative poetries led by material or media-specific dimensions—instead of psychologic content and discursive communication as the basis of poetics—operates a rupture that emphasizes media poetry as form. On the one hand, the historical thread that derives from kinetic art repurposes kinetic poetry as mechanical moving poetry. On the other hand, it is clear why poets working with film, video, or computers felt the need to name their artworks not "kinetic" but rather "film poetry," "video poetry," and "computer poetry"—not only to signal the importance of technics but also to point out what it culturally meant to shape poetry with newer media.

Antecedents: Abstract and Animated Films

Abstract films from the 1920s were singular for their unique vocabulary in relation towards moving image, shape, expressive time, spatial movement, and light. However, even if lost today, during the 1910s Futurist artists and brothers Bruno Corra and Arnaldo Ginna were already pioneering abstract films: Corra's *Musica Cromatica* (1912) and Ginna's *Vita Futurista* (1916). In the same year, their manifesto "The Futurist Cinema" called for "filmed words-in-freedom in movement" (2009: 233).

By the 1920s, the concern with film as a dense and pictorial medium with a specific visual language, as well as unconventional explorations with the camera as a mechanical apparatus and hand-painted film became primary directions for artists working in Weimar's Bauhaus, Berlin, and Paris. Walther Ruttmann's color film *Lichtspiel Opus I* (1921) acquires cinematic flow by way of organic and dancing forms. Temporal dimensions and form are clearly investigated in Richter's *Rhythmus 21* (1921), in that squares are used to reinforce and choreograph the frames' transitions like breathing organisms. Viking Eggeling, Richter's companion, created *Symphonie Diagonale* (1921–4), a silent film full of dynamism and rhythm because of shapeshifting forms that recall musical intervals. Richter's *Filmstudie* (1926), on the other hand, differs by combining abstract film with Surrealist collage in a nonlinear montage whose soft-edge forms show kinetic text. At the same time, Man Ray, who directed and collaborated in many experimental films, also signed *Le Retour à la Raison* (1923), a Dadaist film which incorporates kinetic *rayographs*, or photograms, a photographic technique used by Ray to create images without camera, that is, solely with light exposure. Léger's *Ballet Mécanique* (1924), which is a film without scenario initiated by

Dudley Murphy and Ray (later redacted from the credits by Léger himself), operates by nonlinear, but also sequential association of abstract geometric shapes and figurative depictions, in line with Léger's Cubist paintings and Ray's random shapes.

A seminal work from this period, due to the materialization of kinetic text, is Duchamp's *Anémic Cinéma* (1926). This 35 mm film uses moving *rotoreliefs*—double-sided 40 rpm disks—with hypnotic patterns that combine cinematic montage, optical tridimensional illusion, and text movement. The film's composition features absurd and whimsical lines of text that act as wordplay and turn in spiraling circles mounted on disks. Seminal contributions came as well from the author of *The New Vision* and *Vision in Motion*, Moholy-Nagy, whose experiments in lumino-kinetic sculpture would openly influence his own filmic production. In *Ein Lichtspiel: Schwarz-Weiss-Grau* (1930), likewise Richter's *Filmstudie*, Moholy-Nagy uses film techniques, such as multiple exposure and negative image, while developing a very specific vocabulary in terms of light, shades, and geometric sculptural patterns with the *Light-Space Modulator*. Early abstract films thus make evident Cubist, Dadaist, Expressionist, Surrealist, and Constructivist affiliations, which would resonate in the experimental films of the 1950–60s.[7]

Film Poetry

Besides early 1920s experimental film, authors working at the intersection of cinema, animation, and visual arts devised as well other influential pieces. These include Sergei Eisenstein's, Dziga Vertov's, and Len Lye's 1930s textual and typographic incorporations in long feature films and short animation movies; for instance, in Vertov's 1931 *Enthusiasm (Symphony of the Donbas)* and Lye's *A Colour Box* (1935) or *Trade Tattoo* (1937). One of the interesting features about Lye's work is the combination of collage techniques with cameraless hand-painted celluloid film. Lye has been credited as an important precursor of kinetic poetry (Dencker 2011; Rettberg 2011, 2019), though Vertov's specific contributions need to be further highlighted (Dencker 2011). Among the many fascinating aspects of his oeuvre, Vertov is particularly important because of his early and inventive use of animated

---

[7]The *First International Avant-Garde Film Exhibition* (1925) at the UFA Theatre in Berlin speaks to this prolific moment in experimental film production. The "Absolute Film" show included Richter's *Rhythmus 23* and *Rhythmus 25*, Eggeling's *Symphonie Diagonale*, Ruttmann's *Opus III*, Léger's *Ballet Mécanique*, Hirschfeld-Mack's live performance, and René Clair and Francis Picabia's *Entr'acte*.

typography in film, while exploring its relation with sound and rhythm, in "musical and literary word-montages" (Vertov 2011: 2).

Today, kinetic poetry can be investigated from an array of fields and lenses: literature, visual arts, media, design, film, animation, or social semiotics. To be sure, animated movies and the design of text and film opening titles are among other areas that contributed for reimagining textual movement.[8] Moreover, the titles and credits of movies started to be treated as living animations, and so semantics and semiotics gained an additional layer: motion. *Kinetic semantics* represents meaning-making not only from lexemes but also from their movement and relation in space-time—what could be described as an additional modality in semiotics.

Yet "film poetry" and "film text" appear consistently described as such with the experimental poets.[9] Experimental film poetry was influenced by Surrealist and Lettrist film, but even more so by concrete poetry.[10] Marc Adrian was one of the early inter- or transmedia artists connecting these traditions, while working with a range of media including analog film and computer-generated processes such as randomization. In the silent, and black and white 35/16 mm film poem *WO-VOR-DA-BEI* (1958), the artist creates movement by alternating close-ups and distant shots of permutated syllables (Husslein-Arco, Cabuk, and Krejci 2016). In *Schriftfilm* (1959–60), Adrian makes use of word replacement with combinatorial game at the level of substantives and verbs, whereas *Random* (1963), *Text I* (1964a) and *Text II* (1964b) are permutation films with sound developed in Berlin with a Zuse computer.

Ferdinand Kriwet, who also worked across media, composed with *Teletext* (1963) a very different collage film. *Teletext* appropriates found footage, radio, and popular music, as it intersperses signs, letters, urban symbols, and advertisement with a sharp multiplicity of sensory inputs. It reads as a critique of capitalism, the consumer society, mass media, war, and acceleration. It is a subliminal window into the 1960s, as political and pop culture events unfold at a pace that shows the contradictions of the decade.

---

[8]Lotte Reiniger, Oskar Fischinger, Berthold Bartosch, Norman McLaren, Mary Ellen Bute, Saul Bass, Pablo Ferro, or Daniel Szczechura bridged the divide between experimental and mainstream animation film, and the boundaries between art venues and the commercial industry. On Bass, see Cayley 2005.

[9]For the sake of compression, I am departing from the way authors describe their creative works, rather than opening up the discussion about what constitutes "poetry," "text," "text-based art," "language-based," or "language art."

[10]Films such as 1951 Isidore Isou's *Traité de Bave et d'Éternité*, Maurice Lemaître's *Le Film est Déjà Commencé?*, and Gil J. Wolman's *L'Anticoncept*. Surrealist and Lettrist film also influenced a different type of experimental film that developed in parallel, often using found footage and nonlinear montage: for instance, with French *cinepoésie*, or Italian *cinepoesia*, such as Gruppo '70's *Volerà nel 70* (1965).

The montage and multichannel-like simultaneous techniques impress, but more so do the visionary kinetic effects that ignite the sense of information overload and vertigo that are now commonplace in the media-polluted city, as well as in the internet. Kriwet addresses the infant television network by alternating the textual noise of the cityscape, radio and newspapers with his own newspaper collages and circular poems. The alternation of disparate images and text create a tension between legibility and readability, in such a way to destabilize perception modes (Benthien, Lau, and Marxsen 2019). The radio cut-ups are also precious: "Save now, buy later!" or "The *New York Times*: You don't have to read it all, but it's nice to know it's all there."

Gerhard Rühm, the Vienna Group co-founder, was another poet who strongly emphasized the materiality of language across media. In *3 Kinematographische Texte* (1969–70), Rühm creates a series of three black and white kinetic texts. The first silent film poem contains white shapes that progressively form the glyphs :, *i*, *!*, *o*, *a*, *q*, *d*, *b*, and *p*, while recombining as molecules via elongations and contractions. The second silent film shows dislocations of *gehe, gehen gegen, geben ruhen, eben, benen, rufen, enge, ende, dehnen, neben, nahen, gab* and, at the end, the verb *gehen* ("go, move, walk") blinks with a fade-out. The third film, with sound, involves an interplay of the "written" and "audible" words *du, durch, dich, ich, da, haus, hausmann, und, undundundundun, und, undundu*. Questioning the relation between signifier and signified, Rühm's sensual and multimodal language explores poetry aesthetics as a written, sonic, and visual art.

Adrian, Kriwet, and Rühm were not alone. In fact, there were plenty of artists working with text and film.[11] This artistic landscape was indeed diverse and spread across geographies. In Finland, filmmaker Eino Ruutsalo, who collaborated with electronic musician Erkki Kurenniemi, created the vibrant *Kineettisiä Kuvia* (1962)—the title reinforces the very nature of moving images as "kinetic pictures." In the United States, the increasing immersion of artists in collaborative computer environments contributed to another type of experimentation with moving image, text, and sound. At the Bell Telephone Laboratories, Kenneth Knowlton developed the programming languages BEFLIX and EXPLOR. He further assisted the artistic work of Lillian Schwartz and Stanley VanDerBeek in the creation of computer-generated films, as computers and microfilm recorders could process and integrate various data formats.

VanDerBeek's collaboration with Knowlton resulted in the *Poemfield* series. *Poemfield No. 2* (1966, Figures 2a and 2b) is a fascinating 16 mm

---

[11]Klaus Dencker, in the monumental study *Optische Poesie* (2011), refers to other equally influential *schriftfilme* and *textfilme* by Eric Andersen, Szczechura, Dieter Roth, Ernst Schmidt Jr., or Ulrichs.

FIGURE 2A AND 2B *Stan VanDerBeek,* Poemfield No. 2, *1966 (Film stills). 16 mm film, color, sound. Soundtrack by Paul Motian. Realized with Ken Knowlton. Copyright Stan VanDerBeek Archive.*

"study in computer graphics" produced with an IBM 7094 and BEFLIX. The film makes use of vibrant magenta and strong colors, which are woven in a textile-like dot matrix with jazz music and blinking text. This kinetic artwork impresses psychedelic and synesthetic feelings on the viewer's retina and ear with its vivid color transitions and Paul Motian's soundtrack.

Due to their multimodality, *Poemfield* and Paul Sharits's *Word-Movie (Fluxfilm #29)* (1966)—a fifty-word fast pace letter replacement in 16 mm—have been emphasized as examples that complicate the boundaries between experimental film, computer-generated animation, visual arts, and electronic literature (Gerrits 2014; Wingate 2016). It is in this prism that Wingate further refers to John Whitney's *Permutations* (1966–8), which was developed with an IBM research grant and computer coding by Jack Citron. Like many early animators, John and James Whitney created their own animation tools. John Whitney also assembled a real-time studio with a mechanical analog computer that produced *Catalog* (1961).

Arthur Layzer's *Morning Elevator* (1971), a kinetic film poem programmed in FORTRAN, further signals the entanglement of film poetry with programming languages already being used as creative platforms. Other electronic technologies were also being completely repurposed via artistic implementations. If we consider the kinetic text installations made with LEDs by Kriwet or Jenny Holzer, we have yet another avenue of exploration within media and moving text.

Videopoetry

Videopoetry is a form of kinetic poetry that directly derived from experimental film and film poetry as being time-based. However, its creation and recording relied on aspects specific to the medium of analog video. It was neither cinema nor television, even if it related to both in a critical way with regard to the use of text, the construction and representation of time, and memory. It employed not celluloid film, but magnetic videotape (VT), and it used electronic tools such as computational generators, synthesizers, and editors. Inasmuch as in film poetry, the possibility of animating letters, words, signs, and images became an exciting perspective for poets such as Melo e Castro (2007: 176), who had the chance not just to suggest movement in time and space, but rather to let letters and signs "gain actual movement of their own [and] at last be free, creating their own space."

Melo e Castro's videopoem *Roda Lume* (1968) draws on the poet's earlier experiments in film poetry, such as *Lírica do Objecto* (1958), a self-reflexive black and white 8 mm film. *Roda Lume* is also displayed in black and white, but it was already developed in the video studio of RTP. After being broadcast in a 1969 literary program, the Portuguese public

broadcasting company—which at the time was under fascist ruling—
deplorably destroyed the recorded reel. Following the 1974 Carnation
Revolution, Melo e Castro re-enacted the piece in U-matic format as *Roda
Lume Fogo* (1986), with a new soundtrack, given that he had preserved the
original storyboard. Shapes, signs, syllables, and vowels, combined with a
sound poem, construct multiple semiotic dimensions that stress the power
of art to unlock alternative worlds as paths to freedom. Its multimodality,
and the juxtaposition of sound, moving image, and kinetic text create a
particular reading experience, closer to Spatola's notion of "total poetry," in
that temporal, spatial and mnemonic dimensions are activated and evoked
in complex ways. As Kac (2004: 332) notes:

> O ponto central da criação videopoética é o tempo e suas múltiplas formas
> de manipulação, como a retenção da memória, a duração, a permanência
> breve, o corte abrupto, a compressão, a aceleração, a interrupção,
> a passagem lenta, e muitas outras formas que, conjugadas às cores
> sintéticas, ao som electrônico, aos osciladores e a outros equipamentos,
> estabelecem novos parâmetros para a arte poética.[12]

Throughout the 1970s and 1980s, many artists engaged with the medium
of video and its electronic tools to foster a dialog with other genres. Peter
Weibel's multiple "videospecific poems" from 1973 to 1975 (qtd. in Dencker
2011: 145), such as *Augentexte* (1975), or Tom Konyves's *Sympathies of War*
(1978) show how video poetry could depart from video art, concrete poetry,
or documentary film traditions. Poets collaborated as well with national
broadcasting stations and dedicated TV art hubs to produce strikingly
singular videopoems. While working at several Italian RAI studios, Gianni
Toti forged the notion of "poetronica" in often feature-length improv videos
such as *Per Una Videopoesia* (1980). Toti's sociopolitical works throughout
the 1980s and 1990s were technically activated by synthesized kinetic
lettering superimposed on an amalgam of video art, virtual worlds, and
popular TV aesthetics. At the Experimental Television Center in New York,
Richard Kostelanetz compiled several series of videofiction and videopoetry
that explore typologies of word movement and letter replacement, as well
as the electronic effects made possible by the video-editing studio and the
Amiga 500 computational lettering.[13] Kostelanetz's short videos constitute

---

[12]"The central point of videopoetic creation is time and its multiple forms of manipulation,
such as memory retention, duration, brief permanence, abrupt cutting, compression,
acceleration, interruption, slow passage, and many other forms that combined with synthetic
colors, electronic sound, oscillators and other equipment set new parameters for poetic art"
(free translation mine).
[13]The series *Video Poems* (1985–9), *Partitions* (1986), *Kinetic Writings* (1988), and *Videostrings*
(1989).

a visual encyclopedia of kinetic forms, frequently via wordplay, which parodies capitalism, bureaucracy, and sexuality.

Videopoetry is a form that has greatly evolved with digital video and still captivates contemporary poets, who do not need professional studios to work with electronic editing tools anymore. With the migration of video into digital platforms and the higher portability of cameras and editing hardware, the very conception and presentation modes have suffered a stylistic and aesthetic transformation tied to the role of video processing and editing software as creative processes.

Holopoetry

While nondocumentary videopoetry and digital poetry might render 3D spaces as 3D objects in a 2D screen, holopoetry creates a clear rupture in visual perception, as it introduces third and fourth dimensions in letters and shapes. In the late 1970s and 1980s, Richard Kostelanetz and Eduardo Kac combined visual poetry and holographic technology, thus expanding the realm of experimental poetics.[14] Kostelanetz's *On Holography* (1978)— a stereo 360-degree multiplex holographic film poem—is a spinning cylindrical sculpture that does not use laser, but rather film, by animating a self-reflexive text, frame by frame, that can be horizontally and vertically read (Kostelanetz 2017).

Kac went on to deeply explore the medium with *Holo/Olho* (1983), the first in a series of holopoems that engage with light as a medium, tridimensionality, and two important characteristics of holography: the possibility for the viewer-reader to see multiple volumes in the same spatial point, and the fact that, in a hologram, the part contains the whole and the whole contains the part. As such, *Holo/Olho* is physically, semantically, and syntactically structured with that purpose, whereas the "olho" (eye) is contained within the "hol(o)-" (*hólos*, the whole) and vice-versa, thus creating both a material and content synecdoche. In Kac's (2004: 287) words, the "holokinetics" and "lumisigns" arising from the poems establish a peculiar relation between verbal and visual signs, as well as re-envisioning kinetic forms in space.[15] In addition, *Wordsl* (1986a) is created in a curved space, using integral holography, while *Chaos* (1986b) and *Quando?* (1987–8) are computer-generated.

Holopoetry takes advantage of vertical and horizontal parallax, and the dematerialization of words in space. Kac's poems impress due to the interplay

---

[14]For further information on holography and poetry, see Funkhouser (2007: 265–70).
[15]Holopoems *Abracadabra* (1984–5), *Zyx* (1985a), and *Oco* (1985b).

between "virtual" (hologram) and "real image" (in front of the hologram), and the gradation of colors produced by the visible light spectrum. They experiment with discontinued space and the movement of letters in order to produce a new reading experience. The very movement of the viewer around the hologram transforms the text, thus implying a physical and embodied reading process. Due to its technical apparatus, the hologram does not allow for an extensive output of words. Language needs then to be worked in a compressed manner akin to concrete and visual poetry.

Digital Poetry

As we move from one medium to another, again and again we see two concurrent streams: the reimplementation of old notions in new media, but also an emphasis on how old and new differ from each other—either to specify its singularities or to claim new territory. Moreover, older and newer media tend to coexist during transition periods and that produces interesting feedback loops of artistic practice. As Philippe Bootz (2006) has stressed, digital poetry is not videopoetry. Kinetic poetry specifically written with the computer—and meant to be read and presented via a computer— is comprised of textual, visual, and aural elements. Yet it strictly depends on its underlying code to run and function. In this sense, kinetic digital poetry is algorithmically programmed animation. Furthermore, it often requires interaction or participation from the reader-user, while scheduled events can be determined by random and generative algorithms. The earliest works of kinetic digital poetry sprang from the usability offered by personal computers, though we find static poems being composed with institutional mainframe computers at least since the 1950s in Europe. In the context of Latin America, artists like Eduardo Darino and Erthos Albino de Souza, who worked in the oil industry, had access to mainframes. By 1965–6, the young film animator Eduardo Darino combined GE mainframes, BASIC, teletype printer, and a recording camera to create *Correcaminos*, "an animated visual poem" (Darino 2020; Kozak 2020). Thus, this shows how complex it was for an artist to animate an encoded sequence and how important it was for computational kineticism the digital personal computer and the popularization of simpler programming languages.

The bulk of early kinetic digital poems occurs during the 1980s. It is relevant to understand that most of the following poets were affiliated with the experimental movements of the 1960s. In 1981, Silvestre Pestana coded the first two poems of the *Computer Poetry* suite in BASIC, for a Sinclair ZX81, with white words waving on black background. The final poem (1983) was programmed in a Sinclair ZX Spectrum with more features and symbolic dimensions: color and circular movement suggested tridimensionality, and the word-shape *dor* (pain) replaced all the potential of the *new people*, Pestana's view of social reform and political freedom. Using the statement PAPER and BORDER for blue background and frame, and INK for white, yellow, green, and red squares and font, the artist represents the Portuguese and EEC flags in a critical stance to the aftermath of the Carnation Revolution and the prospect of joining the EEC.

Marco Fraticelli's *Déjà Vu: Poetry for the Computer Screen* (1983) compiles previous hand-written haikai as visual poems to be read on-screen, while Jacques Donguy and Guillaume Loizillon's *Poème Ordinateur* (1983) outputs an "endless stream of consciousness" (Donguy 2007: 331). Like Pestana, bpNichol's *First Screening: Computer Poems* (1983–4) draws from previous work with concrete poetry and novel graphic exploration of words in motion. The series of twelve poems written in AppleSoft BASIC for an Apple IIe operates with varied kinetic behavior: blinking, vertical and horizontal dislocation, letter replacement, and TV script-like scrolling transitions. Still, the greatest surprise is the fact that bpNichol annotated the source code and inserted an "Easter egg" in the last poem—that is, it contains material that is hiding in the source code. The most interesting codework appears between lines 3900 and 3935. In line 3900, the self-reflexive creative comment announces "REM ARK," whereas in line 3910 it reads: "REM AIN." (REM introduces a comment in BASIC.)

Kinetic digital poetry at this point was in many ways a re-enactment of the experimental practices of the 1960s, when poets were working in the realm of concrete poetry. bpNichol writes about "filmic effects that I hadn't the patience or skill to animate at that time" (qtd. in Huth 2008: n.p.). As Geof Huth asserts, "Earlier kinetic digital poetry tended to use the computer to illustrate the poems; Nichol used it to animate them, to make them live." This first wave can be further exemplified by Tibor Papp's *Les Très Riches Heures de l'Ordinateur n° 1* (1985), a live performance at the Polyphonix 9 festival in Paris, in which Papp, coding with an Amstrad, projected the "visual dynamic poem" onto ten screens (Donguy 2007: 314; Bootz 2014: 11). It is relevant that all these works contain the word "computer" in their titles, attesting the need to disclaim the specificity and novelty of creating poems with, and for the computer medium, but also extending the notion that all these authors perceived the computer program as a poem in itself, or as a fundamental part of the process.

The second half of the 1980s sees an intensification of authoring programs and collective gatherings.[16] In terms of publishing and distribution, the French review *alire* is launched as the first electronic journal dedicated to digital poetry. The journal, initially stored and distributed in floppy disks,

---

[16]João Coelho's *Universo* (1985), Paul Zelevansky's *SWALLOWS* (1985–6), and Huth's *Endemic Battle Collage* (1986–7) further explore poetry's kineticism in BASIC.

was published by the L.A.I.R.E. collective and included *poèmes animés*.[17] Jean-Marie Dutey's *Le Mange-Texte* (1989 [1986]) and Bootz's *Amour* (1989) demonstrate the DOS-based pixelated and flat aesthetics of the 1980s, which was rather different from the 3D virtual textual modeling investigated by artists like Jeffrey Shaw.

Throughout the 1990s and 2000s, very fast developments in technology greatly contributed for diversifying the aesthetic approaches, which are difficult to isolate in clusters. Yet the popularization of the Graphical User Interface and the World Wide Web network gave rise to ubiquitous models of presentation and dissemination that artists sought to transgress. During this period, Caterina Davinio created net poetry that addressed the noise and glitch of communication networks, while earlier experimental poets started reimplementing their concrete poems as animations—Ana María Uribe's *Tipoemas y Anipoemas* (1997) being a case in point. HTML facilitated a poetics of links, which is explored by Annie Abrahams in the multilingual and GIF-animated *understanding / comprendre* (1997–8), as well as DHTML applications such as Jim Andrews's visual poetry. But other types of time-based and trans-linguistic poetics, like "transliteral morphing," were being enacted by John Cayley's *windsound* (1999), developed for HyperCard, or *translation* (2004).

Animation software such as Flash, Director, and After Effects dictated a mainstream shift in vocabulary from *kinetic* to *animation* techniques. Unlike before, the end user was offered a software interface that did not involve coding and had a practical cinematic timeline. Flash became the 2000s most popular animation suite, being intensely used not only in industry and commercials but also by visual artists and writers with not-for-profit goals. The platform enabled works such as Brian Kim Stefans's *The Dreamlife of Letters* (2000), a self-referential and vivid catalog of moving letters and words; Young-Hae Chang Heavy Industries' prolific narrative puns with black and white graphics; David Jhave Johnston's *Sooth* (2005), an interactive and generative superimposition of text on video; or Stephanie Strickland, Cynthia Lawson Jaramillo, and Paul Ryan's *slippingglimpse* (2007), an artwork that departs from the rhythm and patterns of waves (*chreods*) to display the font's "text fields" according to the waterscape's encoded motion-capture. Collaborative endeavors also show how Flash was fit for grand-scale projects, such as David Clark's *88 Constellations for Wittgenstein* (2009). Authoring platforms became influential in terms of fostering novel ways of integrating media formats with interactive functions, but they also created homogenization. This meant that authors felt a need to understand its inner workings in order to expand the platform's possibilities

---

[17]Founded by Bootz, Frédéric Develay, Jean-Marie Dutey, Claude Maillard, and Papp in 1988. The early issues of the journal published these authors, as well as Jean-Pierre Balpe, Christophe Petchanatz, Donguy, and Philippe Castellin.

or to transgress them. Poets and artist-programmers such as Bootz, Eugenio Tisselli, and Jörg Piringer went as far as developing their own software for live audiovisual performances. Piringer's *soundpoems* (2002–8) and subsequent pieces clearly rework in digital kinetic systems the traditions of typographic, sound and concrete poetry, especially those by Rühm, Hansjörg Mayer, and others working with the alphabet's units.

Meanwhile, the dissemination of dynamic browser-based scripting languages, such as JavaScript, and open-source software for the arts, such as Processing, generated richer possibilities for animation, social coding, and the collaborative development of interfaces. Networked collaborations between writers and programmers, such as those initiated by María Mencía and Zuzana Husárová, have resulted in a synergy of skills. The two decades of the twenty-first century show a striking variety of artworks and styles that continue to redefine poetic interface, space, flow, and kineticism. Two works that explore these features, with extremely fast textual movement, are Ian Hatcher's *U* (*Total Runout*) (2015) and Montfort's "Alphabet Expanding" (2011)—to run the Perl program copy this single line into your terminal and press enter:

```
perl-e '{print$,=$"x($.+=.01),a..z;redo}'
```

Finally, María Mencía's *El Winnipeg: El Poema que Cruzó el Atlántico* (2017, Figure 3) emphasizes the importance of collective authoring and participation. This collaborative work repurposes the JavaScript library Three.js with programming by Alexandre Dupuis-Belin, while expanding the application of motion to documentary poetry. The artwork departs from testimonies of the 1939 Winnipeg boat's passengers—who were Spanish Civil War refugees helped to exile by Pablo Neruda—while it allows users to contribute with new stories and create poems out of the letters of these fragments in a zoomable planispheric ocean.

Future Movement

Kinetic poetry emerges with the historical avant-garde and it is clearly recycled with the experimental arts. The experimental poets and artists of the 1960s were galvanized by a multitude of *new tendencies* whose practices involved the critique of media and early computational systems. Moreover, these artists updated each other with letters and magazines, while exchanging works for publication and showcase. This network of contacts and collaboration, shaped at a global scale, pre-dates today's emailing lists and digital forums.

Kinetic poetry gained from one of the essential legacies of experimental poetics—its interdisciplinarity—by creating disruption in commonly

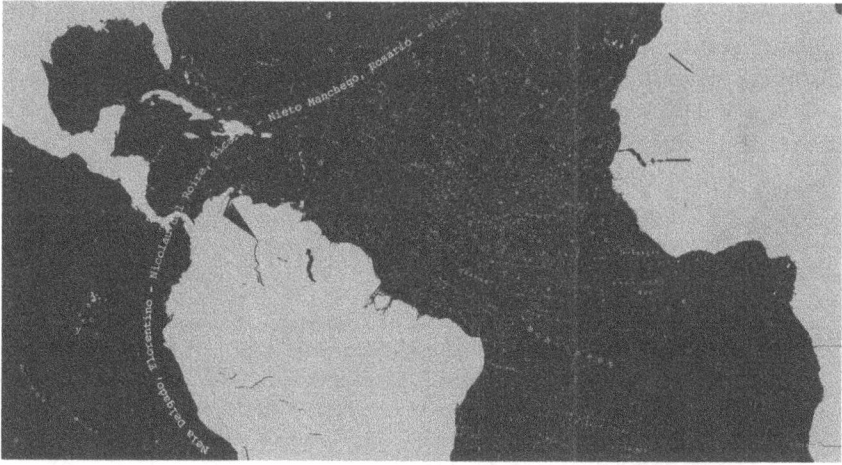

FIGURE 3 *María Mencía,* El Winnipeg: El Poema que Cruzó el Atlántico, *2017 (Screenshot). JavaScript, jQuery, HTML. Programming by Alexandre Dupuis-Belin. winnipeg.mariamencia.com/. Courtesy of the artist.*

accepted boundaries of what constitutes literature, cinema, music, live and (as in "live arts") visual arts. What is transversal to all forms of kinetic poetry is a fascination with motion, visuality, temporal modification, and how the animation of language can impact the aesthetic experience. What unfolded from the artistic experimentation with motorized mechanical sculpture, film, video, and digital media influenced current forms of site-specific mixed-media installations. In recognizing its cross-artistic form and its techno-cultural context, this narrative on how kinetic poetry has evolved and branched out in the twentieth century becomes necessarily broader and richer: not only in relation to its media but also to its artistic antecedents and other forms of kinetic writing. This transmedia approach does not locate, nor equate kinetic poetry with the beginning of the World Wide Web and animation software packages. This is why kinetic poetry is not a computational media-specific form, but rather a transmedia form. In each period, poets have, and will continue to engage with media while reacting to artistic and sociopolitical contexts, whereas embodying continuation or rupture, dialog, or radical creation.

If this broader perspective can be expanded and thoroughly researched, it is important to delineate future ways to address the relation between poetry and motion techniques. Along with the specific impact of each medium in the types of kinetic works they make possible, there is uncharted research in trying to understand how authors transgress the way each medium is supposed to be used, or what types of text behavior and meaning-making are enacted through motion. Perhaps then we can reach a satisfactory grammar or taxonomy of movement. Though practitioners and scholars in digital literature have sketched out types, it is important to understand how dialoging with film and social semiotics, animation and kinetic typography, and with non-Western and non-Roman typography may open more complex modes of moving forward.[18]

References and Further Reading

Abrahams, Annie (1997–8), *Understanding / comprendre*, accessed October 8, 2015, www.bram.org/beinghuman/converfr/conver1.html.
Adrian, Marc (1958), *WO-VOR-DA-BEI*, 35/16 mm film, 1'10", b/w, sound.
Adrian, Marc (1959–60), *Schriftfilm*, 35/16 mm film, 5'28", b/w, silent.
Adrian, Marc (1963), *Random*, 35 mm film, 4'45", b/w, sound.
Adrian, Marc (1964a), *Text I*, 35 mm film, 2'34", b/w, sound.
Adrian, Marc (1964b), *Text II*, 35 mm film, 3'40", b/w, sound.
Apollinaire, Guillaume (1918), *Calligrammes: Poèmes de la Paix et de la Guerre (1913–1916)*, Paris: Mercure de France.
Aragaki, Sayako (2007), *Agam: Beyond the Visible*, Jerusalem: Gefen.
Bäckström, Per (2012), "Words as Things: Concrete Poetry in Scandinavia," in Harri Veivo (ed.), *Transferts, Appropriations et Fonctions de l'Avant-Garde dans l'Europe Intermédiaire et du Nord, 1909–1989*, 1–16, Paris: L'Harmattan.
Balla, Giacomo, Remo Chiti, Bruno Corra, Arnaldo Ginna, F. T. Marinetti, and Emilio Settimelli (2009), "The Futurist Cinema [Sep. 11, 1916]," in Lawrence Rainey, Christine Poggi, and Laura Wittman (eds.), *Futurism: An Anthology*, 229–33, New Haven and London: Yale University Press.
Bann, Stephen (1966a), "Kinetic Art and Poetry," *Image* (winter/spring): 4–9.
Bann, Stephen (1966b), "Unity and Diversity in Kinetic Art," in Stephen Bann, Reg Gadney, Frank Popper, and Philip Steadman (eds.), *Four Essays on Kinetic Art*, 49–67, London: Motion Books.
Bann, Stephen (ed.) (1967), *Concrete Poetry: An International Anthology*, London: London Magazine Editions.
Bann, Stephen (ed.) (1974), *The Tradition of Constructivism*, New York, NY: The Viking Press.
Bann, Stephen (2020), "Pierre Garnier Poster Poem and Kinetic Poetry," email to Álvaro Seiça (February 12).
Bann, Stephen, and Gustavo Grandal Montero (2015), "From Cambridge to Brighton: Concrete Poetry in Britain. An Interview with Stephen Bann," in Sarah Bodman (ed.), *Artist's Book Yearbook 2016–2017*, 70–93, Bristol: Impact Press.

---

[18]In digital literature, see Ikonen 2003, Bootz 2006, Funkhouser 2007, Saemmer 2010, Piringer 2015, and Johnston 2016. In film and social semiotics, see Kress 2009, Leão 2013, van Leeuwen and Djonov 2015. In animation and kinetic typography, see Brownie 2015. In non-Western and non-Roman typography, see Khajavi 2019.

Barok, Dušan (2018), "Women in Concrete Poetry," *Monoskop* (December 14), accessed January 21, 2020, monoskop.org/Women_in_concrete_poetry.

Beaulieu, Derek (2013), "Concrete Poetry," *Flaunt* (April 1), accessed October 25, 2015, www.flaunt.com/content/art/concrete-poetry.

Beaulieu, Derek (2014), *Transcend Transcribe Transfigure Transform Transgress*, Ottawa: above/ground press.

Belloli, Carlo (1943), *Parole per la Guerra*, Milan: Edizioni di Futuristi in Armi.

Belloli, Carlo (1944), *Testi-Poemi Murali*, Milan: Edizioni Erre.

Belloli, Carlo (1948), *Tavole Visuali*, Rome: Edizioni di Gala.

Bense, Max, and Elisabeth Walther (eds.) (1965), *Rot 21: Konkrete Poesie International*, Stuttgart: Hansjörg Mayer.

Benthien, Claudia, Jordis Lau, and Maraike M. Marxsen (2019), *The Literariness of Media Art*, New York, NY: Routledge, doi:10.25592/literariness.

Bootz, Philippe (1989), *Amour. alire* 1, 3.5 floppy disk.

Bootz, Philippe (2006), "Les Basiques: La Littérature Numérique," *Leonardo/ OLATS*, accessed October 8, 2015, www.olats.org/livresetudes/basiques/ litteraturenumerique/12_basiquesLN.php.

Bootz, Philippe (2014), "Animated Poetry," in Marie-Laure Ryan, Lori Emerson, and Benjamin J. Robertson (eds.), *The Johns Hopkins Guide to Digital Media*, 11–13, Baltimore, MD: Johns Hopkins University Press.

bpNichol (1984), *First Screening: Computer Poems*, Toronto: Underwhich Editions; Stephanie Boluk, Leonardo Flores, Jacob Garbe, and Anastasia Salter (eds.) (2016), *Electronic Literature Collection*, Vol. 3, Cambridge: ELO, accessed December 29, 2016, collection.eliterature.org/3/work.html?work=first-screening.

Brownie, Barbara (2015), *Transforming Type: New Directions in Kinetic Typography*, London: Bloomsbury.

"Catalogue of the First International Exhibition of Concrete, Phonetic and Kinetic Poetry," *Granta*, supplement 1964, 69 (1240): i–iv.

Cayley, John (1999), *Windsound*; Katherine N. Hayles, Nick Montfort, Scott Rettberg, and Stephanie Strickland (eds.) (2006), *Electronic Literature Collection*, Vol. 1, College Park: ELO, accessed October 8, 2015. collection. eliterature.org/1/works/cayley__windsound/windsound.mov.

Cayley, John (2004), *Translation*; Katherine N. Hayles, Nick Montfort, Scott Rettberg, and Stephanie Strickland (eds.) (2006), *Electronic Literature Collection*, Vol. 1, College Park: ELO, accessed October 8, 2015, collection. eliterature.org/1/works/cayley__translation.html.

Cayley, John (2005), "Writing on Complex Surfaces," *Dichtung Digital*, 35 accessed October 25, 2015, www.dichtung-digital.org/2005/2/Cayley/.

Clark, David (2009), *88 Constellations for Wittgenstein (To Be Played with the Left Hand)*, 88constellations.net; Laura Borràs Castanyer, Talan Memmott, Rita Raley, and Brian Kim Stefans (eds.) (2011), *Electronic Literature Collection*, Vol. 2, Cambridge: ELO, accessed October 8, 2015, collection.eliterature.org/2/ works/clark_wittgenstein.html.

Coelho, João (1985), *Universo*, IBM PC, BASIC.

Cox, Ken (1965), *Four Seasons Clock*, sculpture, mixed media, accessed January 22, 2020, vimeo.com/133478858.

Cox, Ken (1965), *Shadow Box*, sculpture, mixed media, accessed January 22, 2020, www.kencox.org/artwork/4588376082.

Cox, Ken (1966–8), *Three Graces (Amor-Voluptas-Pulchritudo)*, three sculptures, brass, steel, copper, electric motors and oil paint on plywood, accessed January 22, 2020, vimeo.com/134630409 and www.lissongallery.com/exhibitions/ken-cox.

Cruxent, José María (1964), *Métromane*, painted wood, wire mesh metal (shaped for moiré effect), plastic, letraset, light bulb, electrical fitting, accessed January 21, 2020, www.william-allen.net/jos-maria-cruxent.

Darino, Eduardo (1965), *Correcaminos* aka *Caminante*, GE-235, GE DN-30, BASIC, Telex, Teletype printer, Keystone camera, accessed November 6, 2019, edicionesdelcementer.wixsite.com/ciberrrrdelia/caminante.

Darino, Eduardo (2020), "Kinetic Poetry/Text," email to Álvaro Seiça (February 10).

Dencker, Klaus Peter (2011), *Optische Poesie: Von den prähistorischen Schriftzeichen bis zu den digitalen Experimenten der Gegenwart*, Berlin and New York, NY: De Gruyter.

Donguy, Jacques (2007), *Poésies Expérimentales – Zone Numérique (1953–2007)*, Dijon: Les Presses du Réel.

Duchamp, Marcel (1951) [1913], *Bicycle Wheel*, metal wheel mounted on painted wood stool.

Duchamp, Marcel (1920), *Rotary Glass Plates (Precision Optics)* aka *Revolving Glass Machine*, painted glass, iron, electric motor, and mixed media.

Duchamp, Marcel (1926), *Anémic Cinéma*, 35 mm film, 6', b/w, silent.

Dutey, Jean-Marie (1989), *Le Mange-Texte. alire*, 1, 3.5 floppy disk.

Eggeling, Viking (1921–4), *Symphonie Diagonale*, 35 mm film, 6', b/w, silent.

Emerson, Lori (2011), "Women Dirty Concrete Poets," *loriemerson* (May 4), accessed January 22, 2020. loriemerson.net/2011/05/04/women-dirty-concrete-poets/.

Emerson, Lori (2012), "Recovering Paul Zelevansky's Literary Game 'SWALLOWS' (Apple IIe, 1985–86)," *loriemerson* (April 24), accessed May 25, 2017, loriemerson.net/2012/04/24/recovering-paul-zelevanksys-literary-game-swallows-apple-e-1985-86/.

Fahlström, Öyvind (1954), "Hätila Ragulpr på Fåtskliaben," *Odyssé* 2–3: n.p.; (1966), *Bord: Dikter 1952–55*, 57–61, Stockholm: Bonniers; (1970) [1968], "Manifesto for Concrete Poetry," trans. Karen Loevgren and Mary Ellen Solt, *Concrete Poetry: A World View*, 74–8, Bloomington, IN: Indiana University Press, available at *Ubu* (n.d.), accessed February 25, 2017, www.ubu.com/papers/fahlstrom01.html.

Ferran, Bronać (2017), "The Movement of the Poem in the 1960s: From Circle and Line to Zero and One, from Concretion to Computation," *Interdisciplinary Science Reviews* 42 (1–2): 127–43, accessed February 12, 2020, doi:10.1080/03080188.2017.1297168.

Fraticelli, Marco (1983), *Déjà Vu: Poetry for the Computer Screen*, Montreal: Guernica Editions.

Funkhouser, C. T. (2005), "A Vanguard Projected in Motion: Early Kinetic Poetry in Portuguese," *Sirena* 2: 152–64.

Funkhouser, C. T. (2007), *Prehistoric Digital Poetry: An Archaeology of Forms, 1959–1995*, Tuscaloosa, AL: University of Alabama Press.

Gabo, Naum (1957), "The Realistic Manifesto," *Constructions, Sculpture, Paintings, Drawings, Engravings*, trans. Naum Gabo, 151–2, London and Cambridge: Lund Humphries and Harvard University Press.

Gabo, Naum (1985) [1919–20], *Kinetic Construction (Standing Wave)*, metal, wood and electric motor.

Gabo, Naum, and Antoine Pevsner (1920), *Realisticheskii Manifest*, Moscow: 2nd State Printing House.

Gabo, Naum, and Antoine Pevsner (1957), "The Realistic Manifesto," *Constructions, Sculpture, Paintings, Drawings, Engravings*, trans. Naum Gabo, 151–2, London and Cambridge: Lund Humphries and Harvard UP.

Gadney, Reg (2017), "Kinetic Art," *Interdisciplinary Science Reviews* 42 (1–2): 180–92, accessed August 4, 2017, doi:10.1080/03080188.2017.1297169.

Gerrits, Jeroen (2014), "projected Poetry: From the Medium Specific to the Complex Surfaces," paper presented at ELO 2014 Conference, University of Wisconsin–Milwaukee, Milwaukee, June 21.

Gomringer, Eugen (1953), *Konstellationen*, Bern: Spiral Press.

Gomringer, Eugen (1954), "vom vers zur konstellation," *Augenblick* 2, trans. Mike Weaver (1964), "From Line to Constellation," *Image*: 12–13.

Gomringer, Eugen (ed.) (1972), *Konkrete Poesie*, Stuttgart: Reclam.

Gruppo '70 (1965), *Volerà nel 70*, 16 mm film, 7'15", color, sound, accessed February 7, 2020, www.fondazionebonotto.org/en/collection/poetry/collective/8053.html.

Hatcher, Ian (2015), *(Total Runout)*, *Imperial Matters*, accessed December 14, 2016, imperialmatters.com/ian-hatcher.

Higgins, Dick (1978), *A Dialectic of Centuries: Notes towards a Theory of the New Arts*, New York, NY: Printed Editions.

Higgins, Dick (2018), *Intermedia, Fluxus and the Something Else Press: Selected Writings by Dick Higgins*, Steven Clay and Ken Friedman (eds.), Catskill: Siglio Press.

Husslein-Arco, Agnes, Cornelia Cabuk, and Harald Krejci (eds.) (2016), *Marc Adrian. Monograph and Catalogue Raisonné*, Vienna: Belvedere and Ritter Verlag.

Huth, Geof (1986–7), *Endemic Battle Collage*; (2011), *Electronic Literature Collection*, Vol. 2, Laura Borràs Castanyer, Talan Memmott, Rita Raley, and Brian Kim Stefans (eds.), Cambridge: ELO, accessed October 8, 2015, collection.eliterature.org/2/works/huth_endemic_battle_collage.html.

Huth, Geof (2008), "First Meaning: The Digital Poetry Incunabula of bpNichol," *Open Letter* 13 (5), accessed October 8, 2015, vispo.com/bp/geof.htm.

Ikonen, Teemu (2003), "Moving Text in Avant-Garde Poetry: Towards a Poetics of Textual Motion," *Dichtung Digital* 30, accessed April 1, 2015, www.dichtung-digital.de/2003/4-ikonen.htm.

Isou, Isidore (1951), *Traité de Bave et d'Éterni*, 16 mm film, 124', color, sound.

Johnston, David Jhave (2005), *Sooth*; (2011), *Electronic Literature Collection*, Vol. 2, Laura Borràs Castanyer, Talan Memmott, Rita Raley, and Brian Kim Stefans (eds.), Cambridge: ELO, accessed October 8, 2015, collection.eliterature.org/2/works/johnston_sooth.html.

Johnston, David Jhave (2016), *Aesthetic Animism: Digital Poetry's Ontological Implications*, Cambridge, MA: MIT Press.

Kac, Eduardo (1983), *Holo/Olho*, reflection hologram.

Kac, Eduardo (1984–5), *Abracadabra*, transmission hologram.

Kac, Eduardo (1985a), *Zyx*, laser transmission hologram.

Kac, Eduardo (1985b), *Oco*, transmission hologram.

Kac, Eduardo (1986a), *Wordsl*, integral hologram.

Kac, Eduardo (1986b), *Chaos*, CG reflection hologram.

Kac, Eduardo (1987–8), *Quando?*, cylindrical CG hologram.

Kac, Eduardo (2004), *Luz & Letra: Ensaios de Arte, Literatura e Comunicação*, Rio de Janeiro: Contra Capa Livraria.

Kac, Eduardo (2015), "Dispelling Myths of Origin," presented at Concrete Poetry: International Exchanges Symposium, University of Cambridge, February 14, accessed February 17, 2020, www.spreaker.com/user/kettlesyard/part-6-eduardo-kac-dispelling-the-myths-.

Kamenskii, Vasilii (1914), *Tango s korovami. Zhelezobetonnye poemy*, Moscow: D. D. Burliuk.

Khajavi, M. Javad (2019), *Arabic Script in Motion: A Theory of Temporal Text-Based Art*, Cham: Palgrave Macmillan, doi:10.1007/978-3-030-12649-0.

Konyves, Tom (1978), *Sympathies of War*, VT, 10', color, sound.

Kostelanetz, Richard (1978), *On Holography*, stereo 360-degree multiplex hologram.

Kostelanetz, Richard (1985–9), *Video Poems*, VT.

Kostelanetz, Richard (1986), *Partitions*, VT, 27"40'.

Kostelanetz, Richard (1988), *Kinetic Writings*, VT, 22'.

Kostelanetz, Richard (1989), *Videostrings*, VT, 29"42'.

Kostelanetz, Richard (2007), "Language-based Videotapes & Audio Videotapes," in Eduardo Kac (ed.), *Media Poetry: An International Anthology*, 185–9, Bristol: Intellect Books.

Kostelanetz, Richard, and Álvaro Seiça (2017), "Syntactical Circumstances: A Conversation with Richard Kostelanetz," *Vimeo* (March 18), vimeo.com/209010609.

Kozak, Claudia (2020), "Electronic Literature Experimentalism Beyond the Great Divide: A Latin American Perspective," *Electronic Book Review*. Mar. 1. doi:10.7273/rpbk-9669

Kress, Gunther (2009), *Multimodality: A Social Semiotic Approach to Contemporary Communication*, London: Routledge.

Kriwet, Ferdinand (2011) [1963], *Teletext*, 13', 16 mm film, b/w, sound, accessed February 13, 2020, www.openaccess.uni-hamburg.de/literariness.html#kriwet.

Layzer, Arthur, and J. Miller (1971), *Morning Elevator*, 16 mm film, 4', sound.

Leão, Gisela (2013), "A Systemic Functional Approach to the Analysis of Animation in Film Opening Titles," PhD thesis, University of Technology, Sydney, accessed January 29, 2018, hdl.handle.net/10453/28015.

Léger, Fernand (1924), *Ballet Mécanique*, 35 mm film, 12', b/w and color, sound.

Lemaître, Maurice (1951), *Le Film est Déjà Commencé?*, 16 mm film, 62', color, sound.

Lijn, Liliane (1964–5), *Time is Change*, sculpture, cork, plastic, metal, paper, letraset on painted truncated cork cone, motorized turntable, accessed January 22, 2020, www.lilianelijn.com/portfolio-item/time-is-change-1966/.

Lischi, Sandra, and Silvia Moretti (eds.) (2012), *Gianni Toti O Della Poetronica*, Pisa: Edizioni ETS.

Lye, Len (1935), *A Colour Box*, 35 mm film, 4', color, sound, accessed January 17, 2020. www.lenlyefoundation.com/films/a-colour-box/21/.

Lye, Len (1937), *Trade Tattoo*, 35 mm film, 5', color, sound, accessed January 17, 2020. www.lenlyefoundation.com/films/trade-tattoo/25/.

McCaffery, Steve and bpNichol (eds.) (1978), Sound Poetry: A Catalogue, Toronto: Underwhich Editions.
Mallarmé, Stéphane (1897), "Un Coup de Dés Jamais N'Abolira le Hasard," Cosmopolis 6 (17): 417–27.
Melo e Castro, E. M. de (1958), Lírica do Objecto, 8 mm film, 3'27", b/w, silent, accessed November 23, 2016, po-ex.net/taxonomia/materialidades/ videograficas/e-m-de-melocastro-lirlca-do-objecto.
Melo e Castro, E. M. de (1962), "Letter," The Times Literary Supplement, May 25.
Melo e Castro, E. M. de (1965), A Proposição 2.01: Poesia Experimental, Lisbon: Ulisseia.
Melo e Castro, E. M. de (1966), Poemas Cinéticos, Lisbon: Galeria 111.
Melo e Castro, E. M. de (1968), Roda Lume, video, b/w, sound.
Melo e Castro, E. M. de (1986), Roda Lume Fogo, U-matic video, 2'43", b/w, sound, accessed October 8, 2015, po-ex.net/taxonomia/materialidades/ videograficas/e-m-de-melo-castro-roda-lume.
Melo e Castro, E. M. de (2007), "Videopoetry," Visible Language (1996) 30 (2): 140–9; Rep. in Eduardo Kac (ed.), Media Poetry: An International Anthology, 175–84, Bristol: Intellect Books.
Melo e Castro, E. M. de (2014), "Experimental Poetry [Excursus A]," in Rui Torres and Sandy Baldwin (eds.), PO.EX: Essays from Portugal on Cyberliterature and Intermedia, 71–6. Morgantown, WV: Center for Literary Computing.
Mencía, María (2017), El Winnipeg: El Poema que Cruzó el Atlántico, accessed August 5, 2017. winnipeg.mariamencia.com/poem/.
Moholy-Nagy, László (1930), Ein Lichtspiel: Schwarz-Weiss-Grau, 16 mm film, 6'35", b/w, silent, accessed October 8, 2015. monoskop.org/images/6/6e/ Moholy-Nagy_Laszlo_1930_Lightplay_Black-White-Grey.webm.
Moholy-Nagy, László (1930), The New Vision, New York, NY: Brewer, Warren and Putnam.
Moholy-Nagy, László (1947), Vision in Motion, Chicago, IL: Paul Theobald.
Mon, Franz (1988), "Über konkrete Poesie (1969)," in Peter Weiermair (ed.), Franz Mon: Kunsthalle Bielefeld, 2.10. –27.11.1988, 31–3, Bielefeld: Kunsthalle Bielefeld; English translation (2011), "About Concrete Poetry (1969)," in Tobi Maier (ed.), Waldemar Cordeiro and Franz Mon, 28–9, Leipzig: Spector Books.
Montfort, Nick (2011), "Alphabet Expanding," Concrete Perl, accessed October 8, 2015, nickm.com/poems/concrete_perl/.
Munari, Bruno, and Giorgio Soavi (eds.) (1962), Arte programmata: Arte Cinetica, Opere Moltiplicate, Opera Aperta, text by Umberto Eco, Milan: Olivetti.
Olsson, Jesper (2005), Alfabetets Användning: Konkret Poesi och Poetisk Artefaktion i Svenskt 1960-tal, Stockholm: OEI Editör.
Olsson, Jesper (2016), "'Hätila ragulpr på fåtskliaben' – Conceiving of Concrete Poetry," in Tania Ørum and Jesper Olsson (eds.), A Cultural History of the Avant-Garde in the Nordic Countries 1950–1975, 477–85, Leiden: Brill | Rodopi, accessed January 23, 2020, doi: 10.1163/9789004310506_052
Papp, Tibor (1985), Les Très Riches Heures de l'Ordinateur n° 1, Performance at Polyphonix 9 Festival, Centre Georges Pompidou, Paris, June.
Pestana, Silvestre (1981–3), Computer Poetry, ZX-81, ZX Spectrum, BASIC.
Pilling, Jayne (ed.) (1992), Women and Animation: A Compendium, London: British Film Institute.

Piringer, Jörg (2011), *Soundpoems* (2002–8), in Laura Borràs Castanyer, Talan Memmott, Rita Raley, and Brian Kim Stefans (eds.), *Electronic Literature Collection*, Vol. 2, Cambridge: ELO, accessed October 8, 2015, collection. eliterature.org/2/works/piringer_soundpoems.html.

Piringer, Jörg (2015), *Some Kind of Book – Prototype 1*, app, accessed April 12, 2017, vimeo.com/115223102.

Popper, Frank (1968), *Origins and Development of Kinetic Art*, trans. Stephen Bann, London: Studio Vista.

Ray, Man (1923), *Le Retour à la Raison*, 35 mm film, 2', b/w, silent.

Rettberg, Scott (2011), "Bokstaver i Bevegelse: Om Elektronisk Litteratur," trans. Kjetil Sletteland, "Letters in Space, At Play," *Vagant* 1: 12–3, accessed August 4, 2017. retts.net/index.php/2011/02/letters-in-space-at-play.

Rettberg, Scott (2019), *Electronic Literature*, London: Polity Press.

Richter, Hans (1921), *Rhythmus 21*, 35 mm film, 3', b/w, silent.

Richter, Hans (1926), *Filmstudie*, 16 mm film, 5', b/w, sound.

Russett, Robert, and Cecile Starr (1988) [1976], *Experimental Animation: Origins of a New Art*, New York, NY: Da Capo Press.

Ruttmann, Walther (1921), *Lichtspiel Opus I*, 35 mm film, 11', color, sound.

Ruttmann, Walther (1924), *Opus III*, 35 mm film, 3', color, silent.

Ruutsalo, Eino (1962), *Kineettisiä Kuvia*, 4' 49", 16 mm film, color, sound.

Rühm, Gerhard (1969–70), *3 Kinematographische Texte*, film, 4'02", b/w, sound.

Saemmer, Alexandra (2010), "Digital Literature—A Question of Style," in Roberto Simanowski, Jörgen Schäfer, and Peter Gendolla (eds.), *Reading Moving Letters: Digital Literature in Research and Teaching. A Handbook*, 163–82, Bielefeld: Transcript Verlag.

Schwitters, Kurt (1932), *Ursonate*, Recorded on South German Radio, May 5, 78 rpm.

Seitz, William C. (ed.) (1965), *The Responsive Eye*, New York, NY: MoMA.

Sharits, Paul (1966), *Word-Movie (Fluxfilm #29)*, 16 mm film, 3'50", b/w and color, sound, accessed October 8, 2015. www.ubu.com/film/fluxfilm28_sharits.html.

Sharkey, John J. (1964), *OPENWORDROBE*, 16 mm film, c. 5', b/w, silent.

Sharkey, John J. (1965), "OPENWORDROBE," *TLALOC* 8: 18.

Sharkey, John J. (ed.) (1971), *Mindplay: An Anthology of British Concrete Poetry*, London: Lorrimer.

Solt, Mary Ellen (1966), *Flowers in Concrete*, Bloomington, IN: Indiana University Press.

Solt, Mary Ellen (ed.) (1970) [1968], *Concrete Poetry: A World View*, Bloomington, IN: Indiana University Press, available at Ubu (n.d.), accessed February 25, 2017. www.ubu.com/papers/solt/.

Spatola, Adriano (1969), *Verso la Poesia Totale*, Salerno: Rumma Editore; English translation (2008), *Toward Total Poetry*, trans. Guy Bennett and Brendan Hennessey, Los Angeles, CA: Otis Books/Seismicity Editions.

Stefans, Brian Kim (2006), *The Dreamlife of Letters* (2000), in Katherine N. Hayles, Nick Montfort, Scott Rettberg, and Stephanie Strickland (eds.), *Electronic Literature Collection*, Vol. 1, College Park: ELO, accessed October 8, 2015, collection.eliterature.org/1/works/stefans__the_dreamlife_of_letters/dreamlife_index.html.

Strehovec, Janez (2003), "Text as Loop: On Visual and Kinetic Textuality," *Afterimage* 31 (1): 6–7.

Strickland, Stephanie, Cynthia Lawson Jaramillo, and Paul Ryan (2011), *slippingglimpse* (2007), in Laura Borràs Castanyer, Talan Memmott, Rita Raley, and Brian Kim Stefans (eds.), *Electronic Literature Collection*, Vol. 2, Cambridge: ELO, accessed October 8, 2015. collection.eliterature.org/2/works/strickland_slippingglimpse.html.

Thomas, Greg (2019), *Border Blurs: Concrete Poetry in England and Scotland*, Liverpool: Liverpool University Press.

Toshihiko, Shimizu (1977), "The Visual Poems of the Vou Group," in James Laughlin with Peter Glassgold and Frederick R. Martin (eds.), *New Directions 34: An International Anthology of Prose and Poetry*, 32–45, New York, NY: New Directions Books.

Toti, Gianni (1980), *Per Una Videopoesia: Concertesto e Improvvideazione per Mixer (...)*, 50', video, color, sound, accessed February 5, 2020. www.fondazionebonotto.org/it/collection/poetry/totigianni/1/10387.html.

Ulrichs, Timm (1975), *Typotexture* [1962], in Alan Riddell (ed.), *Typewriter Art*, 119, London: London Magazine Editions.

Uribe, Ana María (2016), *Tipoemas y Anipoemas* (1997, 2016 [1968–2001]), in Stephanie Boluk, Leonardo Flores, Jacob Garbe, and Anastasia Salter (eds.), *Electronic Literature Collection*, Vol. 3, Cambridge: ELO, accessed January 25, 2017, collection.eliterature.org/3/work.html?work=tipoemas-y-anipoemas.

van Leeuwen, Theo, and Emilia Djonov (2015), "Notes towards a Semiotics of Kinetic Typography," *Social Semiotics* 25 (2): 244–53, accessed January 29, 2018, doi:10.1080/10350330.2015.1010324.

VanDerBeek, Stan, and Kenneth Knowlton (1966), *Poemfield No. 2*, 16 mm film, 5'40", color, sound, accessed October 8, 2015, www.ubu.com/film/vanderbeek_poem.html.

Vasarely, Victor (1955), "Notes pour un Manifeste," *Le Mouvement*, Paris: Galerie Denise René, available in "Kinetic Art," *Monoskop*, accessed February 12, 2020, monoskop.org/Kinetic_art.

Vasarely, Victor (1966), "Victor Vasarely à Propos du Cinétisme," *Ina*, accessed February 13, 2020, www.ina.fr/video/I16144890.

Vertov, Dziga (1931), *Enthusiasm (Symphony of the Donbas)*, film, 67', b/w, sound, accessed January 17, 2020, www.dovzhenkocentre.org/eng/product/2/.

Vertov, Dziga (2011), *Enthusiasm: Symphony of the Donbas*, Kyiv: Oleksandr Dovzhenko National Centre, accessed January 17, 2020. issuu.com/dovzhenkocentre/docs/enthusiasm.

Waldrop, Rosmarie (1976), "A Basis of Concrete Poetry," *Bucknell Review* 22 (2): 141–51.

Wall-Romana, Christophe (2012), *Cinepoetry: Imaginary Cinemas in French Poetry*, New York, NY: Fordham University Press.

Weaver, Mike (1964), "Concrete and Kinetic: The Poem as Functional Object," *Image*: 14–15.

Weibel, Peter (1975), *Augentexte*, U-matic video, 1'20", b/w, sound.

Whitney, John (1961), *Catalog*, mechanical analog computer, 16 mm film, 7'21", color, sound.

Whitney, John, and Jack Citron (1966–8), *Permutations*, IBM 360, 16 mm film, 7'28", color, sound.

Williams, Emmett (ed.) (1967), *An Anthology of Concrete Poetry*, New York, NY: Something Else Press; Rep. New York, NY: Primary Information, 2013.

Wingate, Steven (2016), "Watching Textual Screens Then and Now: Text Movies, Electronic Literature, and the Continuum of Countertextual Practice," *CounterText* 2 (2): 172–90, accessed December 14, 2016, doi:10.3366/count.2016.0051.

Wolman, Gil J. (1951), *L'Anticoncept*, 35 mm film, 60', b/w, sound.

Zelevansky, Paul, and Ron Kuivila (1985–6), *SWALLOWS*, Forth-79, Apple IIe, BASIC.

Zoccoli, Franca (1976), *Arti visive. Poesia visiva / Visual Poetry by Women, An International Exhibition in Venice*, Rome: Studio d'Arte Contemporanea.

16

Kinepoeia in Animated Poetry

Dene Grigar

In their Preface to *Literary Terms*, published in 1989, Karl Beckson and Arthur Ganz argue that their Third Edition is needed because "new developments in literary criticism (with its inevitable efflorescence of fashionable terminology) have grown so alarmingly" ("Preface"). Sadly, there is no Fourth Edition of this useful text—but if there were, it would be precipitated by new developments in literary *forms* that have *grown so alarmingly* since introduction of the personal computer that occurred just five years before their Third Edition was released. One new literary form on which this chapter focuses is kinetic poetry—or poetry animated through the affordances of the computer environment, specifically programming languages like Applesoft BASIC, Visual Basic, and JavaScript and software programs like Flash, Shockwave, After Effects, and others. A particular feature of kinetic poetry is its ability to instantiate a thing, idea, or action expressed verbally through the movement of text and images, a term I identify as "kinepoeia." This chapter provides examples of kinepoeia in three kinetic poems: Rob Kendall's "Faith," Thom Swiss's "Shy Boy," and Sasha West and Robert Lavandera's "Zoology."

Computers as Cite of Poetic Expression

The introduction of the personal computer in 1984 raised awareness that writing—and even literature—could be different from print. In his book *Writing Space*, published in 1991, Jay David Bolter argues that computers provide "visual expression to our acts of conceiving and manipulating topics" (1991: 16). In fact, software programs like HyperCard and authoring systems like HyperGate, Storyspace, Intermedia, and Narrabase that emerged in the mid- to late 1980s made it possible for literary artists to experiment not only with what Bolter describes as "creat[ing] and track[ing] formal structures" (19) but also with conceiving and creating in a new medium of expression. This notion is discussed in Peter Gendolla and Jörgen Schäfer's *The Aesthetics of Net Literature* published in 2007, where they acknowledge "the decisive difference to traditional literary texts that lies in the recursive processes between humans and machines" arising out of what they call "net literature" (2007: 9). Canadian poet Barrie Phillip Nichol (bpnichol), writing in 1984 on an Apple IIe and authoring in Apple BASIC, produced what many scholars cite as the first collection of poetry that saw words moving on the computer screen—a work entitled *First Screening*. It was distributed as a signed, special edition of 100 copies on a 5.25-inch floppy disk by his publishing company, Underwhich, and later also made accessible by scholars to the public via the web.

With a few exceptions, learning to program was not a practical approach for publishing poetry for most literary authors. Robert Kendall did produce his poem, *A Life Set for Two*, with Visual Basic and published it with Eastgate Systems, Inc. in 1996. Other examples can also be cited, but it was the mainstreaming of the web browser and the software program, Flash, that heralded an extraordinary period of experimentation when literary artists were making the leap from print to the electronic medium in order to produce and publish their work on computers for the purpose of poetic expression. Argentinian poet Ana María Uribe, for example, shifted her attention from writing concrete poetry, as seen in her collection *Tipoemas* ("Typoems," or typed poems) to animated poetry, what she referred to as *Anipoemas* ("Anipoems," or animated poetry), when she discovered that she could, using Flash animation software, make her words *move*. "Equilibrio" ("Balance"), published in *The Iowa Review Web* in 2000, exemplifies the impetus toward this type of experimentation. Many other writers found the ability to express themselves with movement, sound, music, and user interaction and participation—not to mention with an expanded color palette and spatial orientation—a draw for exploring Flash, particularly, since it was fairly easy to use and prepare for dissemination on the web as a platform for creating and publishing their work. So popular was Flash by 2005 as a platform for intellectual and creative expression that theorist Lev Manovich called this new era of cultural producers "Generation Flash" (Manovich, qtd. in Salter and Murray 2004: 3). Flash's ubiquity as a platform of poetic expression contributed to kinetic poetry being referred to interchangeably as Flash poetry. If indeed there were a Fourth Edition of Beckson and Ganz's *Literary Terms* published today, kinetic or Flash poetry could be a plausible new entry.

Imagery and Kineticism

Physical phenomena expressed with words in print poetry can be expressed with movement in kinetic poetry. What I mean by this is that the phrase found in Samuel Taylor Coleridge's "Kubla Khan"—that is, "[h]uge fragments vaulted like rebounding hail"—evoking a visual picture of a volcanic explosion with the use of words—could be expressed in a kinetic poem as words literally erupting as boulders on the computer screen. That said, kinetic poetry does not rely any less on "represent[ation]" when it renders "things, actions, or even abstract ideas" (Beckson and Ganz 1989: 119) through movement; rather, it overlays kinetic imagery with the visual in a way that expresses a multisensory experience with a text. As such, imagery, which has the potential to "clarify," "express," and "externalize ... mental activity," "dispose the reader ... toward various elements in the poetic situation," and "guid[e] the reader's expectations" (363–70) is broadened, as Alex Preminger suggests in *Encyclopedia of Poetry and Poetics*, beyond the visual to harken other sensory modalities. Writing about born-digital literature in her book, *Writing Machines*, in 2002, N. Katherine Hayles complains—close to forty years after the publication of Preminger's book—that "print-centric view[s] fail to account for ... the ... signifying components of electronic texts, including sound, animation, motion, video, kinesthetic involvement, and software functionality, among others" (2002: 20). More recent, digital poet John Cayley claims in his recent book, *Grammalepsy*, that the "digitalization of typographic visuality tends to facilitate new ways of reading, especially less familiar temporalities of reading, and new relationships between reader action and what is read" (2018: 214).

Instantiating Movement: Kinepoeia

One example of a new way of reading texts stemming from born-digital literature is reading movement. Robert Kendall's "Faith," a philosophical poem published in *Cauldron & Net Volume 4* in 2002 about coming to grips with existential darkness, presents us with words that literally fall, spin, slide, expand, blink, appear, and disappear, representing kinetically the confusion and inner turmoil expressed verbally by the narrator. The word "edge," for example, sidles—or literally edges—itself into the lines, "I/edge/logic/out," thus visually representing the notion that we cling to deeply held beliefs even when they are sorely tested by logic. Later in the poem, the phrase "off the rocker" slants down out of alignment with the rest of the words next to it, reflecting the narrator's own deviation from convention (see Figure 1). At the end of the poem, the word, "leap" grows in size as if the narrator has indeed jumped and is getting closer to the object of his quest: faith.[1]

This injection of movement, a sensory modality not possible to produce in print medium, necessitates the need for an expansion of critical vocabulary. "Kinepoeia"—or movement suggested by the textual or pictorial representation of the word—is drawn from the term onomatopoeia, the rhetorical strategy that associates sound with textual representation (e.g., bam/bam) but unlike onomatopoeia, kinepoeia is indigenous to the digital medium.

Kinepoeia is frequently used in kinetic poetry. Thom Swiss's "Shy Boy," published in *Cauldon & Net* in 2002, recounts the abusive "batter[ing]" a child endures at school. Seeking to escape notice in the hallway, the boy "presses his back against the wall," "melting." The word, "melting," evaporates as it collapses down toward the bottom of the screen, much like the boy wishes he could do when confronted by the bullies (see Figure 2). The inner struggle that pits his yearning for acceptance yet fear of abuse is expressed as a visual tug of war by the phase "to be there" appearing on one side of the screen and then reappearing as the phrase "not to be there" in another. The appearance of the word, "ghost," sees the block of gray on which the word rests disappear with the word itself. In short, the use of physical movement performs the activities suggested in the words.[2]

Sasha West and Robert Lavandera's "Zoology," published in *Born Magazine* in 2009, offers yet another example of kinepoeia. As the title suggests, animals figure largely in the poem and are used to concretize the conflict the narrator feels about the death of her husband. The pure love she feels for him, despite his fading looks and narcissism, is expressed in first four lines:

> The rhino loves the camel as the camel is the color of the dying grass, the
>   muddy stream
> and the camel loves the turtle because its shell reflects a dulled sun &
>   tarnished moon
> and, O, how the turtle loves the bee
> but the bee loves only the flower & its own making of honey.
>
> (lines 1–4)

---

[1]Robert Kendall, "Faith," 2002. Cauldron & Net. Republished in 2006 and available at the *Electronic Literature Collection*, Vol. 1. Electronic Literature Organization, http://collection. eliterature.org/1/works/kendall__faith.html.

[2]Thom Swiss, "Shy Boy," 2002. Cauldron & Net. Republished in 2002 in *Born Magazine*, http://www.bornmagazine.org/projects/shyboy/.

I step to the idea edge elegantly and oh so
ultimately, not just any watered-down walking out
but a fine wine of leave-taking, a full-bodied
forgoing-going-gone upon the logic lip.

No, I just can't make the usual sense anymore so
I'll simply stride out of my mind, press my foot firmly

into the black, all-but-bottomless chasm beyond the brink,
around the bend, off the rocker (yippee!), to leave behind

only this consummate poem, this visionary, incorruptible
transcript of the deeper world's One True Word:

I step to the idea edge elegantly and oh so
red
winking
neon   logic.

No, I just can't make the sunny
side   of my mind   press

the black   button, think
around the bend   of theory   to   be

only this consummate "o,"   this visionary
"+"   of the deeper world.

hedge.   red   Oh
winking
neon   logic.

No, I just can't make the
mind press

around the bend   to

consummate   this vision   on
of the deep   "or"?

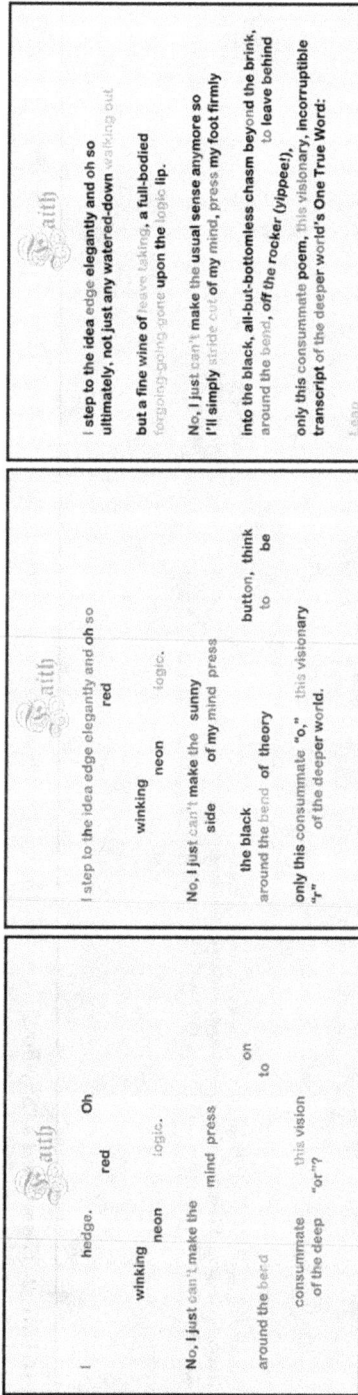

FIGURE 1 *Rob Kendall's "Faith."*

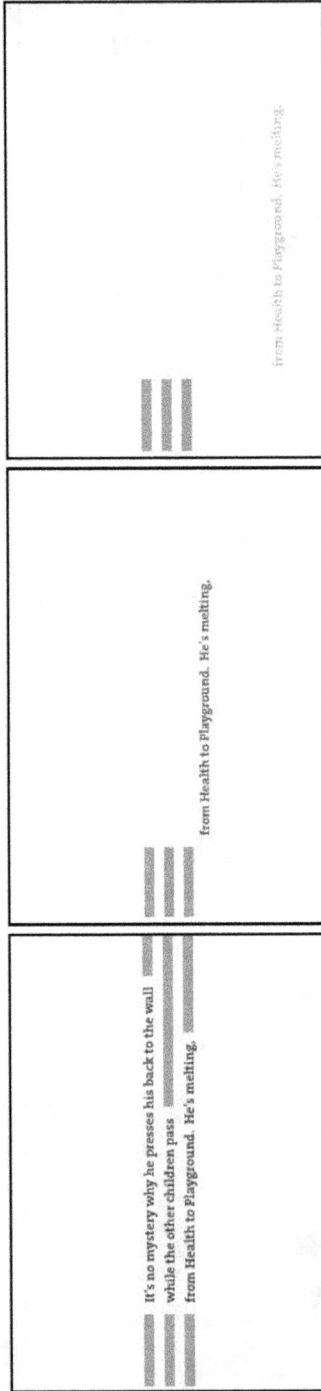

FIGURE 2 *Thom Swiss's "Shy Boy."*

FIGURE 3 *Sasha West and Robert Lavandera's "Zoology."*

Segments of each line are delivered on its own screen, highlighted pictorially with a stylized image of the animal mentioned. The images themselves are formed out of small circles moving like molecules making the animals' shapes. To advance the screens and thus complete the poem, readers move the cursor over the one circle beating physically and sonically like a heart. In line 4—"but the bee loves only the flower & its own making of honey"—is emphasized by the instantiation of the movement of honey. This movement, the golden goo dripping languidly off the flower and through the receptacle offered as the letter "Y," underscores visually the sexual selfishness of this man who carelessly wastes his abundant nectar for his own enjoyment (see Figure 3).

Words and Images in Literary Performance

In the first two kinetic poems, the words carry into effect what they promise to fulfill. The enlarging size of the word "leap," in Kendall's "Faith," effects the experience of jumping, with the earth rising to meet the narrator as he hurdles toward it. The slow disappearance of the word, "melting," in Swiss's "Shy Boy," effects the boy's desire to vanish after a humiliating beating by bullies. In the third kinetic poem, it is the images that carry into effect the promise. The provocatively dripping image of "honey," in West and Lavandera's "Zoology,"[3] effects the squandering of passion, with the honey misplaced on the flower and emptying nowhere in particular. In each example of kinepoeia we see the words and images doing as Gendolla and Schäfer suggest occurs in net literature: performing. Performance from this perspective is undertaken by objects, as Bolter observes, "danc[ing] across the screen before the reader's eyes" (1991: 145). Performance,

---

[3] Sasha West and Robert Lavandera, "Zoology," 2009. Republished in 2002 in *Born Magazine*, http://www.bornmagazine.org/projects/zoology/.

however, takes places at other locations with kinetic poetry. Talking about the "characteristics that digital media bring to specific artist statements," Roberto Simanowski sees performance as the computer-based processes occurring at the "invisible textual level" and the level where the more accessible code functions in the background of a work (2011: 31). Later, in his discussion about concrete poetry, he claims that

> [It] deals with the relation between the visible form and the intellectual substance of words. It is concrete in its vividness, in contrast to the abstraction of a term. It is visual not because it uses images, but because it adds the optical gesture of the word to its semantic meaning: as completion, expansion, or negation.
>
> (Simanowski 2011: 62)

One can extrapolate from this claim that kinetic poetry, like Kendall's, Swiss's, and West and Landavera's, that incorporates both the visual along with movement, adds a physical gesture to the "optical." Finding a vocabulary, such as kinepoeia, to describe this experience is a necessary step in discussing and understanding them more fully.

References

Beckson, Karl and Arthur Ganz (1989), *Literary Terms: A Dictionary*, 3rd edn., New York, NY: Farrar, Straus and Giroux.

Bolter, Jay David (1991), *Writing Space: The Computer, Hypertext, and the History of Writing*, Hillsdale, NJ: Lawrence Erlbaum Associates.

Cayley, John (2018), *Grammalepsy: Essays on Digital Language Arts*, New York, NY: Bloomsbury Publishing.

Gendolla, Peter and Jörgen Schäfer (2007), *The Aesthetics of Net Literature*, Bielefeld: Transcript; Piscataway, NJ.

Hayles, N. Katherine (2002), *Writing Machines*, Cambridge, MA: The MIT Press.

Preminger Alex (ed.) (1965), *Encyclopedia of Poetry and Poetics*, Princeton, NJ: Princeton University Press.

Salter, Anastasia and John Murray (2014), *Flash: Building the Interactive Web*. Cambridge, MA: The MIT Press.

Simanowski, Robert (2011), *Digital Art and Meaning: Reading Kinetic Poetry, Text Machines, Mapping Art, and Interactive Installations*. Minneapolis, MN: University of Minnesota Press.

17

Mobile Electronic Literature

Jeneen Naji

The thing to remember about mobile electronic literature (or e-lit as it is also known) is that it is not Flash. Let me clarify—mobile e-lit is electronic literature that is experienced through mobile devices such as iPhones, iPads, Google nexus, Samsung phones, etc. Before mobile devices electronic literature authors and readers were having quite a merry time of it using Adobe Flash[1] on standalone machines[2] and then came the "big falling out." For a myriad of reasons both technical and commercial, Flash was not compatible with iOS[3] (Keizer 2013). Given this meant that electronic authors could not use Flash to make content for the most talked about and used platform around, a steep learning curve was in order for existing e-lit authors and an exciting opportunity for new authors. HTML5, JavaScript, XCode, and Objective C became the new tools of trade; however, they often frustrated those used to the visual affordances of vector graphics software Flash. Donna Leishman (2015: 149) reminds us that the Flash community had more to do with creative concerns and experimentation that with a shared platform. Therefore, mobile devices' lack of compatibility with Flash had huge ramifications for electronic literature authors and artists as it effectively shut down a stream of years of vibrant artistic expression and communication. Nonetheless, over time electronic literature authors found ways to utilize mobile media for producing new forms of expression. The

---

[1]Adobe Flash: multimedia authoring and viewing software. Previously known as Macromedia Flash prior to Adobe's purchase of Macromedia.
[2]Standalone machines: desktop computers.
[3]iOS is Apple's operating system on mobile devices such as the iPad or iPhone.

lack of access to older electronic literature works is still a loss due to the limited provision of backward compatibility[4] on desktop computers; we are unlikely to be able to view older works designed for desktops on today's desktops. This is why projects such as Dene Grigar and Stuart Moulthrop's *Pathfinders* (that documents electronic literature pieces) and Dene Grigar's *Electronic Literature Lab* (that curates and preserves electronic literature) are so valuable. These projects allow us to replicate and even experience older works of electronic literature as the authors envisaged and designed them. While acknowledging the past we must nonetheless look forward and recognize the many new exciting and interesting examples of mobile electronic literature currently accessible; this chapter will discuss a small few toward giving a brief overview of the current state of play of mobile electronic literature.

Pry by Samantha Gorman and Danny Cannizzaro is an exciting example of a mobile e-lit app that exhibits considerable fluency in the mobile medium. *Pry* is made by Tender Claws, an art collective and studio founded by Cannizzaro and Gorman. This app tells the story of James, a demolition expert who has returned from the Gulf War (*Pry* 2015). Part 1 of *Pry* (Gorman and Cannizzaro 2014) was released in 2014 and is available for both iOS and android mobile platforms. What is particularly engaging about *Pry* is the way the haptic gestures of tap, swipe, and pinch are also imbued with meaning. The haptic gestures are not simply there as replacements of a mouse click; they are part of the storytelling experience. To see through the main character James' eyes you spread and hold open your fingers, using a "reverse pinch" gesture similar to one that could be used to actually open someone's eyes. We are also offered the option to enter James' subconscious by pinching and holding closed again in a gesture similar to one that might be used to close something. The further association of the title of the work *Pry*, meaning to peer in or pull apart, further reinforces the haptic gesture of the reverse pinch. This tactile illusion of the act of opening and closing someone's eyes automatically imbues the reader[5] with a feeling of intimacy with the main character. Sight—internal, external, and lack thereof—is a theme of the piece with the reader even being required to move his/her finger across braille on the screen as the character reads aloud. The ability provided in *Pry* (Gorman and Cannizzaro 2014) to switch between the conscious and subconscious provides what seems to be the missing link between movies and books. In films, we often lose the insight into a main character's internal dialog that is so integral to books; however, in *Pry* (Gorman and Cannizzaro 2014), an application that relies quite heavily on video, the reader is

---

[4]Backward compatibility is the capability of newer forms of technology to run older versions.
[5]In this chapter I use the word reader to refer to what Barthes terms the active reader since electronic literature may require reading, viewing, listening, playing, or using.

afforded agency to choose between either mode. In between the video-based photorealistic mode of the conscious and the surreal text and images of the subconscious we find text, telling the story simply and plainly but in a nonlinear associative manner evocative of our internal recollections of people and events. The storytelling narrative is not straightforward and the full sequence of events is constructed accretively but this seems appropriate for such a distracted medium as a mobile app in which a reader quite potentially dips in and out of content as they simultaneously perform many different tasks on their device. Despite this capacity for a fragmented reading experience the piece nonetheless still maintains a sense of momentum for the reader, a desire to reveal the full story, to find out what really happened, which is key for any good story electronic or analogue.

It is significant that *Pry*'s (Gorman and Cannizzaro 2014) capacity to allow a reader to dip in and out of a mobile e-lit piece exactly corresponds to Funkhouser's advice for electronic literature authors. In *New Directions in Digital Poetry* (2012: 245) Christopher Funkhouser states that, given the attention span and temporal constraints of the average mobile device user, authors could benefit from making electronic literature pieces that can be read in small chunks. Furthermore, Funkhouser notes it is important to remember that many readers will be reading e-lit on smaller mobile screens. *Pry* (Gorman and Cannizzaro 2014) in fact does all of this at it has both iPhone and iPad versions and the fact that it saves your progress and its' disjointed story arc allows the reader to dip in and out of the app as suits.

David O'Reilly's *Mountain* is another app that offers a similarly ambient experience for the reader but with less of a plot-driven story. *Mountain* (O'Reilly 2014) is an app that once launched asks the reader some questions; the answers to these questions are then used to construct a unique mountain. Time passes, weather happens, and the mountain continues, and once the reader gets over their initial automatic instinct to click relentlessly in order to make something happen a Zen-like state for both reader and app descends as the app continues to run in the background as the reader goes about their business. *Mountain* isn't as obviously a piece of electronic literature as *Pry*, but neither can it be described as a game. It does, however, provide the same reader freedom for intermittent interaction and engagement that Funkhouser (2012: 245) advises for electronic literature authors and artists. Furthermore, both *Pry* (Gorman and Cannizzaro 2014) and *Mountain* succeed in prompting the reader to reflect on her world, *Pry* (Gorman and Cannizzaro 2014) by raising contemporary issues of real-life impact of war on the individual, and *Mountain* on the modern instinct and desire for continuous actions and control. The evocation of reflections such as this is in fact exactly what we have come to expect from literature, so rather than trying to categorize electronic literature by specific content, modalities, or forms perhaps instead it might be more useful to identify electronic literature instead by its capacity to transform or evoke critical reflections or connections. Funkhouser (2012: 246) suggests that desktop content designed to provoke viewer transformations may not effectively transfer over to a mobile device, that does not mean, however, that content designed and developed specifically for mobile cannot transform. Funkhouser (2012: 246) cites Talan Memmott's Artaudian-inspired test for a digital poem as to whether it has the ability to transform or cause thinking (303). In these instances, however, Funkhouser and Memmott are referring to digital poetry. The examples we have looked at so far do not fall into this category and it is important to remember that the poetic experience is often much more intangible and fragile to achieve than in other narrative forms.

Abra! by Amaranth Borsuk, Kate Durbin, and Ian Hatcher (& You) describes itself as a poetry instrument/spellbook that responds to touch. It is a playful piece that allows you to modify the text on screen in a myriad of different ways through touch. You can select specific spells from the top of the screen such as mutate, graft, prune, erase, and cadabra. The reader can modify the poem by selecting one of these "spells" and then touching the text with her finger. At the bottom of the screen there is a rainbow dial that you can use to navigate the poems in the *Abra* cycle. A limited edition clothbound book is also available; however, I only experienced the app, which is colorful and playful, but I wouldn't classify it as a transformative literary experience and I was disappointed by the lack of audio. Nonetheless, the strengths of the piece lie in the colorful aesthetics and enormous scope for reader interaction in the variety of spells and settings the reader can access to modify the text and even include her own. The reader can then share her own creation easily on Facebook and Twitter or simply save a photo. This potential social media connectivity in *Abra* (Borsuk et al. 2015) is an aspect of the work that draws on the affordances of the mobile medium, which thrives on and even demands at times a social media connection. Unusually, however, the app doesn't include audio which given the playful nature of the work potentially could have added an extra dimension to the experience—without it the reader's focus is retained on the written words.

Jason Edwards Lewis and Bruno Nadeau's P.o.E.M.M. (Poetry for Excitable [Mobile] Media) project is an example of digital poetry on a mobile device that maintains a strong focus not only on written words but also very evocative (sometimes overpowering even) audio. P.o.E.M.M. is a series of eight mobile iOS apps that deals with themes of belonging, identity, youth, and multiculturalism among others. The touchscreen interactivity of the apps uses the pinch and swipe gestures we have come to associate with iOS technology. The pieces also allow for the creation of your own version as well as connecting with online social media such as Twitter. Interestingly, the reader can even register their own version of an app in a similar fashion to limited edition print artworks; that is, one of one hundred. The P.o.E.M.M. website describes the apps as "making sense of crazy talk & kid talk, the meanings of different shades of purple, the conundrums of being a Cherokee boy adopted by Anglos and raised in northern California mountain country, and the importance of calling a sundae a sundae." There are eight apps available, entitled *What They Speak When They Speak to Me, Buzz Aldrin Doesn't Know Any Better, The Great Migration, Smooth Second Bastard, No Choice About the Terminology, The Summer the Rattlesnakes Came, The World Was White, The World That Surrounds You Wants Your Death* (P.o.E.M.M.).

For brevity's sake, I will focus on one of these apps; namely, *The World Was White* (Lewis et al. 2015). The reason for this selection is due to the striking visual impact of the work when the reader is initially presented with a pure white screen on launching the app, which remains that way until the reader touches the screen at which point text appears. If the reader maintains a finger press on-screen and swipes she hears audio, so this very immediate reaction to her touch evokes a strong engagement with the piece by providing a sense of agency for the reader. The P.o.E.M.M. website describes *The World Was White* (Lewis et al. 2015) as a "homage to the many, many road trips—short and long—I took across northern California with friends while a teenager. Now, much later, I have come to realize that it is also about growing up one of the few brown kids in white, rural mountain country." So perhaps here we find another reason for my specific engagement with this piece as this theme speaks to me as an Irish woman with Arabic heritage who grew up in Ireland's countryside. So once again this is an aspect that we find also in nonelectronic literature, the connection or relatability of a piece that quite often depends on your own codex of memories and experiences.

It is interesting to note that it is the more "poetic" apps that seem to be more suited to the mobile medium as there is less of a sense of a beginning, middle, and end to them. Poetry's nonreliance on linearity means that it is easier for a reader to dip in and out of each of these apps whereas when reading *Pry* (Gorman and Cannizzaro 2014) for example, a story, the reader is more likely to seek an end, a resolution as such, whereas in *Mountain* (O'Reilly 2014), *The World Was White* (Lewis et al. 2015), and *Abra* (Borsuk et al. 2015) a reader is more likely to be open to a nonlinear experiential literary engagement.

There are many, many more examples of mobile electronic literature available for download on the App Store and Google Play and online. This chapter has only skimmed the surface of just a few; however, it has I hope shown how many of the same aspects, problems, strengths, and weaknesses of analog literature can be also found to be at play in mobile e-lit, thereby proving its value as a medium for artistic communication, experimentation, and expression despite the challenging and sometimes limiting technical aspects of developing/writing for this medium. As the P.o.E.M.M. website suggests, mobile devices can offer a more intimate interaction experience for a reader and the higher-resolution screen can provide an aesthetically pleasing visual and textual experience. Perhaps it is the haptic nature of interaction, the ability to literally make things happen through touch that provides the reader with a potentially more intimate experience than in other mediums. Larissa Hjorth refers to haptic screens in relation to the touch screens of mobile media—she suggests that the screen is no longer about visuality but about touch (2011: 440). This is ideal for the purposes of e-lit as the more intuitive nature of these technologies mean that the spell of the piece need not be broken by the need for the reader to lift his/her head and look for the mouse; a simple hand movement will be enough to proceed within the piece. As Hjorth (2011: 444) proposes when discussing mobile media "it is the touch of the device, the intimacy of the object, that makes it so meaningful"—the tactile process of the analog is recreated in the digital through the haptic screen.

References

Barthes, R. (1970), *S/Z*, Paris: Editions de Seuil.
Borsuk, Amaranth, Kate Durbin, and Ian Hatcher (2015), *Abra!* [Computer software], Apple App Store, Vers. 1.0. Ian Hatcher.
Electronic Literature Lab. Web Tender Claws Studio, https://tenderclaws.com/pry (accessed November 30, 2015).
Funkhouser, C. T. (2012), *New Directions in Digital Poetry*, London: Continuum.
Gorman, Samantha and Danny Cannizzaro (2014), *Pry* [Computer software], Apple App Store, Vers. 1.1.0. Tender Claws LLC, Web (accessed November 30, 2015).
Hjorth, L. (2011), "Domesticating New Media: A Discussion on Locating Mobile Media," in S. Giddings and M. Lister (eds.), 437–48, London and New York: Routledge.
Keizer, Greg (2013), "Apple Dumps Flash from Mac OS X," *Computerworld*, October 22 Web (accessed November 30, 2015).
Leishman, Donna (2015), "The Flash Community: Implications for Post-Conceptualism," in Scott Rettberg, Patricia Tomaszek and Sandy Baldwin (eds.), *Electronic Literature Communities*, 131–50, Morgantown, WV: Center for Literary Computing.
Lewis, Jason Edward, S. Maheau, C. Gratton, and B. Nadeau (2015), *The World Was White* [Computer software], Apple App Store, Vers. 1.0.2. Jason Lewis, Web (accessed November 30, 2015).
Memmott, Talan (2009), "Beyond Taxonomy: Digital Poetics and the Problem of Reading," in Adalaide Morris and Thomas Swiss (eds.), *New Media Poetics: Contexts, Technotexts, and Theories*, 293–305, Cambridge, MA: MIT Press.
O'Reilly, David (2014), *Mountain* [Computer software], Apple App Store, Vers. 1.2, David O'Reilly, Web (accessed November 30, 2015).
Pathfinders, Web Tender Claws Studio, https://tenderclaws.com/pry (accessed November 30, 2015).
P.o.E.M.M., Web Tender Claws Studio, https://tenderclaws.com/pry (accessed November 30, 2015).
Pry, Web Tender Claws Studio, https://tenderclaws.com/pry (accessed November 30, 2015).

18

The Voice of the Polyrhetor: Physical Computing and the (e-)Literature of Things

Helen J. Burgess

Sustaining a deep engagement with the machines we use has long been assumed to be fundamental to what we do as authors and scholars. In technical communication and writing studies, the term "digital literacy" is frequently used to signify the process of learning how to "read" (and occasionally write) digital texts. A growing cohort of digital humanists asserts that both writing and coding are crucial components. For practitioners of electronic literature, it is the process of what N. Katherine Hayles calls "concealing and revealing" (2005: 54) that guides our work with platforms, authoring systems, and code. Matt Kirschenbaum, for example, argues that "the distinction between what's on the screen (or page) and what lies beneath is beginning to disappear, as computer languages seep into the visible, legible spaces in which we read" ("Hello Worlds"). Cathy Davidson suggests that "[d]igital literacy means not rote learning but experimentation, process, creativity, not just technology but multimedia imagination, expression–and principles too" ("Digital Literacy"). And Ian Bogost has argued for what he calls "procedural literacy" (2012: 32) in which we learn not only how to code but also learn how the disciplinary nature of code itself encourages structured thinking and facilitates an understanding of the world as a series of interrelated systems.

One question we might like to ask in the pursuit of procedural literacy is how far we can extend what it means "to write." Jody Shipka, for example, warns of the conflation of "multimodality" (composition processes that cross multiple tactile, visual, and oral media) with the more traditionally understood electronic "multimedia," arguing that digital composition often substitutes one "narrow range of practices" (2011: 5), such as writing, for another, such as hypertext, while undermining the "complex relationship between writing and other modes of representation" that might include the spatial, oral, and haptic (12). Thus, if we are to realize the full promise of Bogost's call for procedural literacy, we might like to consider the proposition that working with technology to promote digital literacy should go beyond the manipulation of digital objects using software, and even beyond the manipulation of software itself.

One fruitful series of digital literacy practices involves looking not just at the surfaces and screens of the computer as a writing tool, but looking inside and under the hood. This field, known as "humanities physical computing," or "critical making," emphasizes the role of student and scholar as builder and maker, as well as critic. Thus, "writing the machine" includes learning how to assemble it from the ground up, and understanding its physical components, how they connect, and how they function. In this chapter, I'll look at the role of physical computing (that is, the practice of creating electronic objects and circuits using microprocessors, servos, and other small pieces of electronics) as a potential component in "digital literacy" practices, and suggest that studying physical computing can offer us insights into the way communication is moving from the screen to a much more complex world of 3D electronic objects. These objects, I'll suggest, expose the innards of writing as a practice that is embedded much more deeply in layers of encoding and staging than we might initially think, and offer a fertile space for the creation of an electronic literature of Things.

Rhetoricians are by their training fond of an apposite piece of classical terminology, so let's find something to suit. If one can speak of "discursive" or "rhetorical engineering" when discussing the composition of technical writing, the appropriate term for the compositional processes involved in physical computing might be "skenic engineering." The *skene* in classical Greek theater was the building behind the main stage area (the proscenium) where props and materials (and actors) were kept in waiting for use in productions; sometimes for dramatic purposes action happened "off-stage" in the skene area. Such a model rings intuitively true with writing, in which we draw upon historical references, metaphors, and argumentative turns. But because of its origins in the physical spaces of the theater, the term *skene* also calls to mind the "stage-setting" intent of physical computing, and the productive potential of a space from which electronic items and objects might be drawn or manipulated in fruitful ways and multiple combinations.

To show you the potential range of skenic engineering, let's look at examples of a couple of historical "writing machines," which employ physical technologies in very different ways.

Machine #1: The Futurama

In 1939, visitors to the New York World's Fair were introduced to the Futurama diorama exhibit: a "ride into the future" built by industrial designer Norman Bel Geddes for General Motors. Geddes' model of 1960s America, at over 35,000 square feet and housed in the Fair's GM Pavilion, was a showcase for futuristic design with its streamlined, unornamented walls and sweeping highway-like entrance. The "Futurama ride" was the highlight of the Fair, attracting up to 28,000 people a day over the two-year duration of the exhibit.

The Futurama exhibit employed the genre of the ride, which was popular in other amusement parks at the fair: visitors were seated in a "carry-go-round" consisting of 552 plush blue mohair chairs that moved slowly around the sides of the diorama as simulated night fell and the sun rose again. The carry-go-round or "mobilounge" was "... a combination conveyor-elevator-escalator," designed by Westinghouse Elevator Company, with a piped-in soundtrack generated by the Polyrhetor, an audio soundtrack delivery device created by Electrical Research Products, Inc. The winged easy chairs, upholstered in blue mohair fabric, were six-feet high "to suggest a private, traveling opera box" (Geddes "For Release"). The chairs' "wings" were designed to limit the spectator's view to the front. According to Bel Geddes' description,

> The spectator is seated in a comfortable chair on the conveyor platform and is moved through semi-darkness while a quiet authoritative voice at his shoulder explains what he is about to see ... It will be viewed through a continuous window directly before him and the voice at his shoulder will personally bring to his attention and describe to him the various features and points of interest which he is to see.
>
> ("Description")

The "quiet authoritative voice" Bel Geddes referred to consisted of a recorded voice issuing from a sound-box in each pair of chairs. The soundtrack, which was triggered as each set of chairs rolled over predetermined points in the ride, was controlled and coordinated by a centralized machine called "the Polyrhetor." This machine, also known as the "spectator sound system," and "Twenty-Tons-of-Voice," delivered guided narration (voiced by Edgar Barrier of Orson Welles' Mercury Theater) to the 552 armchairs carrying visitors through the ride.

The Polyrhetor contained 150 individual amplifiers, each playing a part of the guided tour through the exhibit. Because magnetic tape was in the early stages of development, the machine relied on motion picture film as a medium on which to record the audio guide. A contemporary image caption reads: "This huge automaton, machined to a precision rivaling the world's great telescopes, serves as a corps of 150 'private guides' to visitors ... 150 equally spaced photoelectric cell devices scan a motion picture film at the same time throughout its length [to give] visitors a perfectly synchronized description of the treats awaiting the motorist of the future" (GM Heritage Center).

When I first started writing about the Futurama ride many years ago I was primarily interested in it only for the message it was conveying: that the highways of the future were coming, that they would unite the pastoral natural world with technological convenience and speed, and that the landscape would be rationalized into a productive, pleasant, driving experience (Burgess and Hamming 2015). And certainly, the ride achieved this successfully: the designer Norman Bel Geddes would be a key voice in postwar thinking about American superhighways. But over time it became clear that what was most interesting about the ride was not the diorama, with its "half-million buildings and houses – thousands of miles of multilane highways – [and] more than a million trees" (*Highways and Horizons*). But what kept me coming back was the giant Polyrhetor machine, with its film canisters (without vision) and its radio star voice (without radio). I started thinking about the Polyrhetor as a kind of throwback to oral culture in the midst of literate culture.

Of all the components of the Futurama ride, the Polyrhetor device is particularly interesting because of the way it speaks to us. It straddles the communications divide between orality and literacy: where orality is characterized by Ong, McLuhan, and others as an aural, enveloping exchange featuring spoken word and shared experience, while literacy consists of the organization of information in the visual register, encouraging distance and discipline of the eye. On the one hand, the Futurama ride was a shared, "oral" experience. Edgar Barrier's recorded voice spoke to each person, customized to their position above the diorama, while the intimacy of the ride was magnified by soft chairs and dim lighting. People emerged from the ride bearing a pin ("I have seen the future") proclaiming their participation in a shared experience. The Polyrhetor's voice, chosen by designers for its smooth, authoritative, but comforting tone, provided guidance via the trusted medium of a radio professional's familiar-sounding narration.

On the other hand, though, the couches with their wing-backed dividers separated travelers from each other, and the distance from the diorama separated each viewer from the landscape. Rather than walking through an exhibition hallway, the visitors were placed above the diorama. There were no customized movements: once you were on the ride, there was no getting off. The voice was prerecorded and did not talk back. Indeed, the picture the diorama presented of the future was of a rationalized network across the landscape, the individual vehicles encouraging an atomistic vision of transportation.

Thus, even while the Polyrhetor provided the comfort and individual attention of the oral tradition, it disciplined its visitors into accepting what John Brinckerhoff Jackson would later call the "new odology," saying

> We do not always give credit to how the motorized American – commuter, tourist, truck driver – has accepted the new odology, how docile we have been in complying with the scientific definition of the highway as a managed, authoritarian system of steady, uninterrupted flow for economic benefits.
>
> (1994: 192)

In short, the Futurama ride was a persuasive space, with the ride itself mirroring the physical pathways of the highways being traced onto the landscape, while the Polyrhetor provided the narrative scaffolding.

Machine #2: The Universal Turing Machine

Let's step back out for a moment, and look at a skenic machine from another perspective. At the same time as the Polyrhetor and the Futurama were being conceived and staged with the help of industrial designers and engineers for the purpose of selling cars, the British mathematician Alan Turing was publishing *On Computable Numbers* (1936), in which he posited a thought experiment we know popularly as "the Turing machine."

The Turing machine features a tape of infinite length and a probe head. The tape is fed through the machine. The probe head can read and write to the tape: ones and zeroes or some analog. The tape can move back and forth, being marked and remarked to carry out computations and data processing. An "action table" contained mathematical instructions for processing the tape. A key feature of a Turing machine was that it consisted of what he called "discrete states"—for example, the number of switches turned on and off. Given enough space, a complete description of every single state in the machine could be stored. The machine could thus be described completely using a limited symbolic set.

The most significant version of the Turing machine is the universal Turing machine (UTM), which can be *programmed to behave like other Turing machines* by feeding in instructions through the tape. This means that in order to get the UTM to do something different, you just need to feed it new instructions on the tape, rather than building another machine. This was a radical new idea, coming out of Turing's realization that you didn't need to know what the physical build of a machine was; all you needed to know was its informational state at any particular point. By 1950, after having the chance to work on the earliest computers, Turing was able to state with confidence that "digital computers … can mimic any discrete state machine" (1950: 441).

Reading and Writing Machines

You would think that there is a world between these two machines, the Futurama exhibit and the machine in Turing's brain. But they have one thing in common: they're both reading and writing machines. In fact, if we want to get a little fanciful, the Futurama ride starts to look eerily like a physical manifestation of a Turing machine: the Polyrhetor and chair triggering mechanism is the probe head. The people are fed in like a ribbon. This gives us a picture of a kind of human Turing tape passing through a massive capitalist programming machine. The Polyrhetor and the Futurama ride between them created a specific context for "programming" humans: in addition to literally "reading" the script from film canisters, the Polyrhetor provides a "reading" of the landscape and "writes" on the visitors by impressing the story on them. The Futurama exhibit is, thus, both a computer and a kind of giant book to be read, built on the skenic technologies of the Polyrhetor's sensors and film voice recordings.

The idea of these kinds of large-scale technologies as reading and writing machines enables us to think about the relationship between technology and written artifacts—for example, books—with some fruitful results. First, is a book more like a UTM—infinitely programmable, regardless of form—or more like a Polyrhetor—pre-programmed specific to its circumstances? Let's map it out:

- The purpose of a Turing machine was that there only needed to be one machine, which could simulate all other machines through programming. Nothing was single-purpose any more.

- A book could be thought of as a Turing machine in the sense that the machine is programmable—the technology of the book remains stable while the programming changes.

- Nominally, the Polyrhetor is the same; its "voice" is recorded and stored.

- But at the same time, the Polyrhetor is a single-use system reliant on other parts of the Futurama—for example, the sensor system that triggered parts of the audio track as chairs passed through the ride. The Polyrhetor was created in order to provide context for one specific text: the diorama. It wasn't portable, and used technologies that were quickly outdated (in particular, film as an audio device).

The Polyrhetor and the Turing machine thus offer us two boundary scenarios for what it means to read and write. On one end of the spectrum, the Futurama ride employs tools drawn from the theater, film, and engineering. It is a profoundly physical experience that makes very little use of textual elements beyond the occasional sign. On the other end, Turing's discrete state machines are reliant exclusively on encoding and decoding: the primary function of text. Both are thus simulation machines, but in different ways. The Turing machine simulates the literate environment, encoding and recoding. The Polyrhetor simulates the place-and-time-bound environment, guiding, persuading, enveloping. Between these two machines lies a fruitful space for creative play: between material and virtual, presence and absence, speaking and writing.

An (e-)Literature of Things

Once we start thinking about reading and writing environments as potential sites for designing, staging, and engineering digital texts, many new modalities open up for electronic literature. Physical computing, with its easy access to hobbyist-level electronic components, offers us some interesting alternative directions in the creation of digital texts that respond to light, sound, movement, or the press of a button. And skenic engineering—the "staging" of code and material objects to create specific effects—can help us to create interesting literary artifacts that emphasize, like artists' books, exploration and idiosyncracy, rather than rationalism and regularity.

To create a skenic literary object, let us consider the process of electronic staging. Much work being done in physical computing right now is concerned with the non-"writing" parts of the Polyrhetor experience. As Carla Diana notes, "we're entering a time when sound, light and movement are equally important parts of the creative palette. Everyday objects whose expressive elements have long been static will now glow, sing, vibrate and change position at the drop of a hat" ("Talking, Walking Objects"). We're surrounded by such objects: our Google Nests regulate our HVAC systems. Our cars are stuffed to bursting with sensors. Our refrigerators are internet-connected. Our home networks are doling out local IP addresses to our televisions, set-top boxes. The Amazon one-click button is a physical button you can use to re-order laundry detergent. The Internet of Things is in our homes, eating our electricity.

In the midst of this cacophony of movable screens, motion-detectable bodies and electronic signals, the idea of using an IoT network to produce *actual text*—physical, printed pieces of paper with static marks on them—seems quaint. And yet I'm fascinated by the process of using IoT-era technologies (manufactured hardware object, microprocessors, communication networks, commercial and open-source APIs) to produce such old-fashioned literary artifacts. As we've already established, the Polyrhetor was a kind of reading/writing machine in the sense that like a Turing machine, the Polyrhetor "writes" its narrative onto us. Thus, my plan was to stage a skenic machine that would produce writing: an electronic literary device embedded in the Internet of Things. But unlike the Turing machine, which is purely concerned with symbols, the Futurama ride and the Polyrhetor drew upon the strengths of the oral and physical environment: the coming together of people into the same space. We still value the face-to-face experience, the closeness of flesh, the shared temporary habitat. And so, my skenic object would need to embrace both the near and the far: electronic, distant writing and physical, face-to-face writing.

The inspiration to create such an object began for me when I saw a tiny thermal printer show up in an online store for electronics. It was a small, somewhat clunky version of the many different types of thermal receipt printers that are used ubiquitously to document the moment we swipe our credit cards or pass over paper notes in exchange for food, services, and objects. It wasn't internet-connected, but various tutorial links promised me that I could hook it up to an Arduino or Raspberry Pi, and use those components to connect to the internet. Most importantly, though, the continuous paper scrolling out of the printer feed reminded me of the Turing machine tape and the human "tape" passing through the Futurama ride.

Receipt printers themselves are interesting little producers of everyday text. A receipt is what David Levy so evocatively calls "a witness" (2001: 7)—it is an object that is generated on the spot as proof of a transaction in a place and time. It stands in for a person, testifying to an interaction. In my first attempt to create a little skenic object, which I called "MashBOT," I wanted to mess with the perception that a piece of writing can either be local (produced as a kind of one-time event, like the Polyrhetor) or global (produced in a broadcast environment built for replication and repetition, like the encodings and decodings of the Turing machine), oral or literate—but not both. Thus, MashBOT was created as a writing machine that did two things: produce a piece of writing posted to that great global writing space, Twitter, while simultaneously crafting an unique, local physical copy, printed out on a little thermal printer paired with an Arduino microcontroller.

MashBOT writes love notes, generated using Markov chains and a very simple corpus of quotes from Bruno Latour's *Aramis* and Roland Barthes' *A Lover's Discourse: Fragments*. Interacting with the project can be done in two ways: by going to the Twitter handle @mashomatic and reading the generated tweets from anywhere, or entering the physical space where MashBOT is exhibited, waiting for the "ready" light to turn green, and pressing a tiny button, which prints a copy of the latest tweet queued up on the printer. In the process, the human "reader" crosses repeatedly between the generality of computing and the physical particularity of that embedded moment in which the button is pressed. The tweet is the same, but it can be torn or cut from the stream of "receipts" slowly being produced by the printer, placed in the pocket, and taken away like a little talisman of the written word. The technology is simultaneously broadcast and narrowcast: the love notes are broadcast by Twitter, but the note that appears on the printer is narrowcast for the person in front of it. The snippet of Markov-generated literature twins, duplicates: the same note appears online (in many places at once) and on paper (in one place).

None of this would be possible without staging principles that make use of connectivity and another nice old rhetorical borrowing, *dispositio* (arrangement). Taking as his model a typical Parisian day, Latour suggests that a comprehensive sociology must account for not merely separate people and things, but the ways they are wired together through multiple control and observation technologies: traffic lights, cameras, and so forth: "sensors, counters, radio signals, computers, listings, formulae, scales, circuit-breakers, servo-mechanisms need to be added in; it is these that permit the link to be made between one place and another, distant, one … You can't make a social structure without this compilation work" (1996: 240). The *skene* of MashBOT is bound up in this "compilation work"—the hooking together of multiple technologies, from Twitter, to backend server scripting, to Arduino coding and assembly, all the way to the moment the user presses the button to print the text. Python scripts hook together the Twitter API with text-generating scripts; my fingers assemble the delicate components into a configuration on the breadboard that will allow a flow of bits and electrons to become an inscription on thermal paper.

Of course, all literature is the result of a transaction or collaboration between multiple actors and actants—with the book acting as a physical index of the wide network of "publishing" as a means of conveying meaning. Historically, it seems that literature relies on, or leverages, this transparency to make claims about the universality of "the literary." With few exceptions (such as the artist's book), we are meant to look through books, not at them. Twitter does likewise, by making the act of tweeting and reading tweets as seamless as possible. We look straight through the browser window as though it does not exist. An e-literature of Things upends this process by introducing physical technologies into the equation, so that they are less "transparent" than the disappearing book or browser window. These objects exist to remind us that "literariness" is not universal, or virtual, but the result of a mess of interactions with materiality: the body, the object, the manufacturing process that produced that object, the specific physical circumstances in which one interacts with the object.

Finally, Latour suggests the deep wrong-headedness of a sociological model that "imagined that at root we were monkeys to which had been added by a simple prosthesis, buildings, computers, formulae or steam engines. … objects are not means, but rather mediators-just as all other actants are. They do not transmit our force faithfully, any more then we are faithful messengers of theirs" (1996: 240). MashBOT is an example of this unfaithfulness. At the time of this writing, he's been tweeting for about eighteen months, and he occasionally gets a retweet. But not all are from humans—indeed, some are from other Twitterbots, triggering on a word that MashBOT has generated and using it to produce their own response. This is the true moment of conception for my version of the giant Polyrhetor—my desktop machine of "many voices"—as an example of the (e-)literature of Things.

References

Bogost, Ian (2005), *Telemedium* 52 (1 & 2): 32–6.

Burgess, Helen J. and Jeanne Hamming (2015), *Highways of the Mind*, Philadelphia, PA: University of Pennsylvania Press.

Davidson, Cathy (2012), "Digital Literacy: An Agenda for the 21st Century," Blog posting, January 29, http://www.cathydavidson.com/2012/01/digital-literacy-an-agenda-for-the-21st-century/.

Diana, Carla (2013), "Talking, Walking Objects," *New York Times*, January 26.

Geddes, Norman Bel (1939), *Description of the Conveyor Ride Models in the General Motors Exhibit*, February, Austin, TX: Harry Ransom Center, University of Texas.

Geddes, Norman Bel (1939), "For Release in Saturday Afternoon and Sunday Papers, April 15-16, 1939," April, GM Press release draft, Austin, TX: Harry Ransom Center, University of Texas.

General Motors Corp. (1939), *General Motors Highways and Horizons*.

GM Heritage Center (n.d.), "Twenty-Tons-of-Voice. This huge automaton, machined to a precision rivaling the world's great telescopes, serves as a corps of 150 'private guides' to visitors touring the General Motors' 'Highways and Horizons' exhibit at the New York World's Fair," Stirling Heights, MI: GM Heritage Center Image Archive.

Hayles, N. Katherine (2005), *My Mother Was a Computer: Digital Subjects and Literary Texts*, Chicago, IL: University of Chicago Press.

Jackson, John Brinckerhoff (1994), *A Sense of Place, a Sense of Time*, New Haven, CT: Yale University Press.

Kirschenbaum, Matthew (2009), "Hello Worlds," *Chronicle of Higher Education*, January 23.

Latour, Bruno (1996), "On Interobjectivity," *Mind, Culture, and Activity* 3 (4): 228–45.

Levy, David (2001), *Scrolling Forward: Making Sense of Documents in the Digital Age*, New York, NY: Arcade Publishing.

Shipka, Jody (2011), *Toward a Composition Made Whole*, Pittsburgh, PA: University of Pittsburgh Press.

Turing, A. M. (1950), "Computing Machinery and Intelligence," *Mind* 59: 433–60.

19

Having Your Story and Eating It Too: Affect and Narrative in Recombinant Fiction

Will Luers

It is night. A car drives up to a house and parks. The headlights go out. The next morning, the car starts again and drives away.

The above sentences barely make up a narrative, yet they describe a direction of events in time and space (a before and after) that conjure a network of possible (and possibly meaningful) narratives. Is the driver an investigator or a jealous stalker? A tired or homeless traveler? Is the car robotic? What is the narration concealing? It is within this explosion of possibilities that the simplest sequence can, not only hold attention and create expectation but also activate an array of the reader's own mental operations beyond just decoding the text's denotative meaning. Decoding a work of fiction might involve, simultaneously, visualization, imagination, associative thinking, perceptual and emotional identification, memory searches, inductive and deductive reasoning, comparative analysis, calculation, and prediction. The distinct pleasure of "getting lost" in a story might occur because these conscious and unconscious processes are tuned to the text and simultaneously involved in complex processes that are outside the text; that is, in the reader's mind. Narrative artists work hard to embed signals (useful information) so that texts can be decoded, but they also work at ambiguity and semantic noise to engage a reader's cognitive participation. Life is a daily encounter with contingencies and random events that don't fit plans and strategies. Narrative fiction, being itself a model of life experience, must wrestle with the unexpected and the random to incorporate as affect the things, events, and processes outside human meaning-structures.

One of the challenges of computational fiction, of which there are many genres and flavors of practice, is how to manage narrative coherency while exploring the more interesting and affective possibilities of noise, indeterminacy, and randomness. These qualities, foundational practices in computational arts, are also a part of lived experience. We manage to make some sense of experience, despite the noise, indeterminacy, and randomness. But we also use these very chaotic processes to our advantage in making sense. At a preconscious level, we incorporate random processes into decision-making, abstract-modeling, problem-solving, and narrative construction. We create a kind of disorder to discover higher levels of order. Paradoxically, this complexity with which we negotiate experience remains inaccessible to us. Joseph Tabbi, in *Cognitive Fictions*, writes that digital artists may enact "cognitive calisthenics that usefully defamiliarize experience at the higher level," but the deeper brain basis of cognition "escapes all signification, the level at or below which no narrative, language or social construction can go (xiii)."

> ... nowhere –not in print nor electronic media–is it possible to *represent* the actual transitions from perceptuo-motor behavior to human thought that arise, presumably, from the subpopulation of neurons at multiple locations throughout the brain.
>
> (Tabbi 2002: xiii)

Whether we use narrative fiction to "make sense," prop up beliefs, or just for amusement and play, it is a technique for modeling the chaos and flux of lived experience. To enter a story is to enter a dance (or battle) between order and chaos. Narrative artists have always worked with computational thinking to achieve the right balance between these forces. The very act of creative narrative composition involves combinatorics, conditionals, random and parallel processing, variables, and abstractions to build meaningful, affective, and believable representations of contingent experience within a narrative framework. In modern and postmodern literatures, authors use parataxis, meta-narrative, and stream-of-consciousness techniques to disrupt narrative logic in order to produce affect in the reader/viewer—a meaningful confusion that stimulates a search for other patterns outside or beyond linear cause–effect chains.

Recombinant poetics, a term coined by artist/scholar Bill Seaman, refers to a techo-poetic practice in which the display and juxtaposition of media elements are partially or fully generated by computer algorithms, rather than through an author's predetermined composition. Much like the labyrinthine and rhizomatic structures of hypertext fiction and poetry, there is a semantic instability to recombinant works. As texts, they change with each reading. Espen Aarseth labels such indeterministic texts *metamorphic*, a subset of ergodic cybertext that transform "endlessly with no final (and repeatable) state to be reached" (1997: 181). If recombinant texts inscribe change and indeterminacy as affective qualities or as metaphorical frames for negotiating fragmented content, how is meaning generated? Seaman draws on Roy Ascott's notion of "field theory" in art to describe how semantic elements "work together to form an emergent outcome through recombination and *interaction<–>intra-action* by an engaged participant" (Vesna 2007: 135). Recombinant authors program discrete media elements to display and behave through deterministic and variable processes, often in conjunction with user interaction. Although inspired by traditions of combinatorial literature and the use of constraints to generate narrative or poetic forms, recombinant works of art produce "fields of meaning" where "a finite set of media-elements is entertained through a vast set of potential combinatorial abstractions"(Rieser 2002: 242).

If narrative art is about representing causal relationships between agents and events, how can a work be received as narrative without these relationships being made explicit? Examples of recombinant poetics in works of digital poetry and art are abundant. Digital narratives that foreground recombinant processes are less common, because they tend to dismantle or dissolve themselves as narrative in favor of more nonlinear, affective, and emergent meanings. Narrative may be present in recombinant works, often within discrete semantic fragments, but narrative structure collapses when the intentions of anything like a narrator are noticeably absent.

In recent years, works of digital fiction have opened up new possibilities for incorporating recombinant poetics *inside* narratives. *Pry* and *Strange Rain* employ touch gestures that are metaphorically tied to narration; the user uncovers narrative meaning by exploring and learning about how the work operates. While both are only partially recombinant, in different ways, they are also both committed to telling psychological stories. Like much digital fiction, the works model subjective points-of-view and cognitive, nonlinear processes, but within a framework that is organized as a sequence of events. *Toxi•City*, a more overtly recombinant work, incorporates audio and video, fictional and nonfictional elements into a collage documentary about an ecological catastrophe that takes place both in the future and in the recent past. The work conflates the real with the imagined in a dense, semi-random (toxic) montage of audio and video fragments. What is interesting about these works is how their authors seek narrative coherence. Whether it is through user gestures, visual and auditory motifs, or narrative voice, the narrative elements resonate with the indeterminacy of the works' combinatorics. Neither narrative nor recombinant processes are dominant. They are rather in a productive and meaningful relationship that the reader negotiates.

Recombinant fiction seems to work when both narrative structure and stochastic processes are explicitly in play at the level of signification. That is, narrative sequence and indeterminacy or randomness are evident and recognizable in the text. *The Korsakow System*, created by Florian Thalhofer, is a downloadable program for creating database films. The software makes it quite easy for authors to create and display interactive, recombinant systems out of a collection of video fragments. These fragments, called smallest-narrative-units or "snus," are authored to display through a semi-random selection process. Inside the player of a *Korsakow* work, each video comes with a menu of other videos to play next in sequence. The random selection for these displayed choices are constrained by the author's in and out tags for each video in the work. The more a video is tagged to other videos, the more likely it will appear. It is also possible to assign probabilities with a rating scale to control the frequency or precedence of videos appearing in any given traversal. Many of the best *Korsakow* films are nonfiction—meandering, generative, and gently interactive database documentaries that explore topics, ideas, or places. *Korsakow* narrative fiction has proven to be more of a challenge. The problem is not the degree to which an author can either restrict or open the possibilities of selection, nor is it a problem of the user recognizing narrative structure and randomness. It is that, in a *Korsakow* film, there is no explicit relationship between the two processes, no visible dance between chaos and order. The user has access to narrative content in each video, but the random processes have no resonance or tension with the content.

How does resonance and tension between order and chaos emerge in a work of fiction? From my own experience of authoring both linear and recombinant fictions, I believe it starts in the act of composition. Working at the micro and macro levels of constructing a narrative fiction, drafting arrangements and patterns, testing and iterating, a fictional work grows in complexity in the mind of the author. The author's challenge is to translate this complex network into a compact system of signs that can be unpacked by a user back into a complex cognitive network. In two recent recombinant fiction projects, *Fingerbend* and *Phantom Agents*, my aim was to create narrative sequence within an interface that was indeterminately arranged with image selections, text fragments, and design elements. I wanted a "book" that always changed, but reflected that change within a static narrative sequence. *Fingerbend* is the multimedia diary of a character in a futile effort to make a story out of something radically distributed and networked. *Phantom Agents* is the episodic adventures of two characters inside a corrupted augmented reality game. In the composition of both works, the narrative and generative system emerged symbiotically. It is not only that there are narrative rationales for stochastic systems, but that these systems produce affect that complicate, disrupt, and infect the narration. Built with HTML5 and JavaScript, I used a few simple programming principles to compose, simultaneously and iteratively, the narrative structure and the recombinant system. Random text fragments, images, and design elements are held in arrays, as are the narrative fragments. With each swipe or movement forward in the work, the interface displays from the narrative arrays in increments (i++) and from the random arrays based on a random JavaScript-generated number (*Math.random()*). Once that basic system is built, my role as a narrative artist takes over and I fill the arrays with the media fragments that play with each other, that resonate and mirror what the system is doing, as well as impart the necessary information about plot, character, and setting. Interface design is another layer of refining edges between the process and structure. As recombinant fictions, the works are playfully about themselves and not the simulations of a believable inner or outer reality. They are *metamorphic* texts that simulate the murky and liminal realms of lived experience; the paradox of being both in the world and dreaming the world.

Peter Schwenger in his book *At the Borders of Sleep: On Liminal Literature*, adopts the music term "obbligato," the sometimes improvised musical line around a main theme, to refer to the barely conscious mental wanderings of a reader trying to follow a text.

> ... when we read, we are aware not only of the shapes of phrases and sentences, not only of the particular "meaning" that words like nets enmesh, but also of the associative reticulations of our own minds.
>
> (2012: 32)

Conventional narrative suppresses the possibilities of the reader's *obbligato*, keeping attention on the causal chain of events, mostly by raising questions and then answering them. Much of experimental literary fiction and electronic fiction, decidedly unconventional in its approach to narrative, uses opacity (Lamarque 2014), absence, and noise (Paulson 1988) to encourage and stimulate private, dream-like associations as an affective counterpoint to narrative cause and effect. The roots of narrative art, how a reader's affect and cognition are stimulated in a text, is less about making meaning clear than about multiplying networks of semantic connections within and outside the text. While noise in a text, the disruption and decomposition of semantic coherence, seems to work against narrative's temporal logic, it is in fact essential to embedding appropriate complexity of character, situation, and/or theme.

There are many authors of electronic literature that resist narrative, because narrative as a structuring device is just one of many that we now use to navigate contemporary life. Our environments, social interactions, diversions, and expressive tools increasingly follow database rather than narrative logic (Manovich 2002: 216). For digital artists, computation and database aesthetics offers so many possibilities for exploring networked life, that narrative is just one formal device among many. It is optional. Also, as Galen Strawson point out, there are harmful limitations to relying on causal-chains to explain nonlinear forces in lived experience (2004: 428–52). Narratives, as frameworks for "human self-understanding, close down important avenues of thought, impoverish our grasp of ethical possibilities, needlessly and wrongly distress those who do not fit their model, and are potentially destructive in psychotherapeutic contexts" (447). However, some of the most lasting works of literary fiction—*Don Quixote*, *Tristram Shandy*, *Ulysses*—probe not only the complexities of lived experience but also the viability of narrative for representing such complexity. Literary fiction is an ever-elusive quest to model, map, and harness contingency and indeterminacy on the level of signification. The possibilities for computational fiction are immense in furthering this effort. It is unlikely that recombinant poetics in electronic literature will produce fictions made of well-formed characters living explicitly in a believable external reality, but it might produce fictions that evoke the murky and liminal symbolic systems that make up our internal and external networks.

References

Aarseth, Espen J. (1997), *Cybertext: Perspectives on Ergodic Literature*, Baltimore, MD: Johns Hopkins University Press.

Cannizzaro, Danny and Samantha Gorman (2015), *Pry*, Tender Claws, January 7, App Store, https://itunes.apple.com/app/id846195114.

Korsakow (2011), "The Korsakow System," n.p., Web (accessed April 30, 2015), http://korsakow.org/.

Lamarque, Peter (2014), *The Opacity of Narrative*, London and New York: Rowman & Littlefield International.

Loyer, Erik (2013), *Strange Rain*, Opertoon, January 7, App Store, https://itunes.apple.com/us/app/strange-rain/id400446789?mt=8.

Luers, Will (2013), *Fingerbend*, December, Web, http://will-luers.com/fingerbend/.

Luers, Will (2013), *Phantom Agents*, December, Web, http://will-luers.com/phantomagents/.

Manovich, Lev (2002), *The Language of New Media*, reprint edn., Cambridge, MA: The MIT Press.

Paulson, William (1988), *The Noise of Culture: Literary Texts in a World of Information*, Ithaca, NY: Cornell University Press.

Paulson, William (1994), "Chance, Complexity, and Narrative Explanation," *SubStance* 74: 5–21.

Rettberg, Scott and Roderick Coover (2014), *Toxi•City* [Installation], exhibited at Electronic Literature Organization 2014: Hold the Light, http://elmcip.net/creative-work/toxicity.

Rieser, Martin and Andrea Zapp (eds.) (2002), *The New Screen Media: Cinema/ Art/Narrative*, Pap/DVD edn., London: British Film Institute.
Schwenger, Peter (2012), *At the Borders of Sleep: On Liminal Literature*, Minneapolis, MN: University of Minnesota Press.
Strawson, Galen (2004), "Against Narrativity," *Ratio* 17 (4): 428–52.
Tabbi, Joseph (2002), *Cognitive Fictions*, 1st edn., Minneapolis, MN: University of Minnesota Press.
Vesna, Victoria (ed.) (2007), *Database Aesthetics: Art in the Age of Information Overflow*, 1st edn., Minneapolis, MN: University of Minnesota Press.

SECTION III

Practices

20

Challenges to Archiving and Documenting Born-Digital Literature: What Scholars, Archivists, and Librarians Need to Know

Dene Grigar

In October 2015 I visited the David M. Rubenstein Rare Book & Manuscript Library at Duke University to conduct research into *Uncle Roger*, the first commercial work of electronic literature by pioneering artist Judy Malloy. There, among the twenty-seven boxes comprising the Judy Malloy Papers, I sifted through notebooks, computer readouts of code of her works, images she took of her many works, correspondence with other artists, and exhibition papers. The materials associated with *Uncle Roger* were contained primarily in Box 3. Despite the fact that *Uncle Roger* was limited to one box and was organized in folders, it was still difficult to determine what constituted the serial novel, what can be called Version 1.0, or where the information was located for the database narrative created in GW-BASIC, or Version 4.0. Furthermore, the hand-made artist's box with hand-designed inserts, Version 3.3, was dispersed in different folders in the box, and the floppy disks themselves were archived separately and were, understandably, inaccessible for use. Unless someone knew exactly what they were looking for among the materials in the archive, they would have not known that the item entitled "Topic 14: A Party in Woodside, as first told on WELL, 1986 December" represented *Uncle Roger, Version 1.0*, or known to look for *four* inserts for Version 3.3.

My experience with *Uncle Roger* is not unique. Electronic literature scholars can point to many examples of works where digital and analog materials are packaged together as "the work." Some, like John McDaid's *Uncle Buddy's Phantom Funhouse*, include music cassettes, among many other items. Others, like Kate Pullinger, Stefan Schemat, and Chris Joseph's *The Breathing Wall*, were packaged peripheral technology; this particular example came with a headset with a microphone along with the work on CD-ROM. Even web-based works like Amaranth Borsuk and Brad Bouse's *Between Page and Screen*, may necessitate a print book. In light of current archival practices, how does one make such works available for study so that they retain their integrity? Herein lies the challenge.

I have attempted to address this challenge with the lab I have built for works of electronic literature at Washington State University Vancouver and the documentation I've been doing in that lab for these works. The lab is called Electronic Literature Lab, or "ELL," and the documentation project is called *Pathfinders*.

Enter ELL

ELL consists of forty-five vintage Macintosh computers and two PCs, all representing various operating systems and media affordances dating back to 1977 and collected for the purpose of accessing electronic literature and documenting it for future generations. It also consists of a personal library of over 200 works of e-lit catalogued in an online database.

I began putting together my library when I was a graduate student in the early 1990s. As time passed, I became acutely aware that works produced a mere twenty years ago were quickly becoming forgotten and overlooked. After the introduction of the Apple iPhone in 2007, which has rendered works produced in Flash obsolete, I shared my collection through exhibits, which I have done at the Library of Congress in the United States and the Modern Language Association conferences in 2012, 2013, and 2014, among other venues.

While preservationists are able to make some electronic literature works available via emulation and migration, copyright laws prevent many early works from undergoing these processes. But beyond the legal issues, something is indelibly lost in moving electronic literature from its original source material into a new format. ELL, instead, follows the model of preservation called "collection," by making it possible for scholars to study works on the device on which they have been originally produced or for which they were originally accessed. Anyone reading the emulated version of Malloy's *Uncle Roger* instead of the Apple IIe, for example, can understand the different experience the migrated version affords. Lost in translation are the sounds—the beeps and whirring—the computer makes when interacting with the work that let us know we were on or off the right track, the heaviness of the computer keys when striking them with one's fingertips, and bright green words flickering slightly against the dark green background of the interface itself, sensory modalities indelibly inscribed as the shared cultural experience at the time of the work's publication.

Enter *Pathfinders*

Since its inception, ELL has drawn one post-doctoral student a year and has seen many visitors at its lectures, artists talks, workshops, and courses, but like any specialized lab or library it is challenged when making the work in the library widely available to scholars. Enter *Pathfinders*, a project funded by the National Endowment for the Humanities' Office of Digital Humanities and led by Stuart Moulthrop and me.

In 2013 we harnessed ELL to document four seminal works of early digital literature: Judy Malloy's *Uncle Roger* (1986–8), John McDaid's *Uncle Buddy's Phantom Funhouse* (1993), Shelley Jackson's *Patchwork Girl* (1995), and Bill Bly's *We Descend* (1997). We chose these four because they are examples of long-form writing that represent a specific individual contribution unique to the field as well as reflect a wide range of experimentation taking place during this period. As mentioned previously, Malloy's *Uncle Roger* was first published in 1986 as a serial novel delivered to an online audience on the *Whole Earth 'Lectric Link* (*WELL*). Later iterations were migrated to platforms by re-programming it in UNIX Shell Scripts, Applesoft BASIC, GW-BASIC, and HTML. John McDaid's *Uncle Buddy's Phantom Funhouse* was produced with HyperCard, a software application available on Apple computers for creating hypermedia. Like Malloy's *Uncle Roger*, *Funhouse* is a novel, but one that includes sound and printed elements as part of its storytelling strategy. Shelley Jackson's *Patchwork Girl*, produced with Storyspace—a hypertext authoring system created and sold by Eastgate Systems, Inc.—is viewed by many as the high point of hypertext literature in the pre-web period of the early digital age. Finally, Bill Bly's hypertext novel *We Descend*, also created with Storyspace, takes advantage of the affordances of this tool to experiment successfully with the multitemporal narrative and intricate narrative structure.

The method innovated for documenting these is called the Traversal, which Stuart and I define as audio and video recordings of demonstrations performed on historically appropriate platforms (Grigar and Moulthrop 2015). It involves a reflective encounter with a digital text in which the possibilities of that text are explored in a way that indicates its key features, capabilities, and themes. The term is borrowed from Michael Joyce who used it in his essay, "Nonce Upon Some Times: Rereading Hypertext Fiction," to refer to any particular reading of a hypertext (Joyce 1997: 581). His use of the term was adopted from directed-graph theory and influenced by Espen Aarseth's notion, from *Cybertext*, of the "traversal function," the mechanism by which certain elements of a systematic text are presented in a specific encounter or reading (Aarseth 1997: 62). For Stuart and me the Traversal always involves human agency, even though it may be strongly inflected by program logic or machine operations.

As we have conceived it, a Traversal must take place on equipment as close as possible in configuration to the system used to create the work, or on which the work might have been expected to reach its initial audience. This specification has two important consequences. First, it ensures fidelity to the original product. Some of the works we considered (and as we discovered, more than we first thought) have been re-engineered and even rewritten for later platforms. Using historically appropriate equipment allowed us to recognize these changes, in some cases bringing them to light for the first time. The use of early equipment also manifests subtle but important aspects of the original user experience. Malloy's Narrabase works begin with the whirring and clicking of the floppy drive, a kind of computational theme song, before resolving to the words, "Bad Information." *Patchwork Girl* and *We Descend* both open with initialization screens that enumerate the works' complement of spaces and links—a display that flashes by all too quickly on newer, faster machines. The physical context of the work is most salient in *Uncle Buddy's Phantom Funhouse*, which consists of a box of artifacts, including the backup disks from a vanished writer's hard drive. These objects literally demand an antique Macintosh and, as mentioned previously, a vintage cassette player.

Our Traversals method requires the author and then two readers to perform the work, talking through choices they encounter. Passages are read aloud, hyperlinks are selected and announced, and experiences with the words and media elements are expressed.

Stuart and I videotaped the Traversals and photographed the floppy disks, the containers with which they were sold, and other materials in the package. In some cases we included sound files of works. We provided detailed textual descriptions of the liners of the jewel cases and the notes packaged along with the folios so that if someone in the future wished to recreate the ephemera that accompany the work itself, he or she could. The result of our effort is a multimedia book published in June 2015 on the open-source Scalar platform, containing 173 screens of content, including 53,857 words, 104 video clips, 204 color photos, and 3 audio files. To date we have had over 10,000 scholars from close to 250 universities, centers, libraries, and schools (Grigar and Moulthrop 2015).

What *Pathfinders* Means for the Literary Archival Experience

Imagine with me, if you will, another type of experience with the Judy Malloy Papers at the Rubenstein. This time the scholar is carrying her laptop and has accessed the section on Malloy at the *Pathfinders* book. She is interested in looking at Version 3.3 of *Uncle Roger*, so she opens Box 3. She knows from *Pathfinders* that she should probably study the materials in the folder marked "A Party in Woodside, Apple II version written in BASIC, 1987." She also knows that she probably needs to look for the folder, "The Blue Notebook, Apple II+ version, written in BASIC, 1988," and should also consult "Terminals, stand alone copy (disk removed)," as well as "Packaging, disk components" and "Packaging, disk versions, Apple II (disks removed)." In fact, any folder that alludes to a disk for an Apple computer is more than likely related to *Uncle Roger*, Versions 3.1, 3.2, and 3.3. And because she cannot access the floppy disks, she can watch *Pathfinders'* videos of Malloy traversing through a section of the work and hear the author talk about the production of all three parts of it in videotaped interviews with her. She can compare the materials she is examining in the boxes with the images of Version 3.3 used for the *Pathfinders* project. Doing so provides her with an understanding of the variances between the different artist boxes hand-made by Malloy, thus coming to see the level of material practice involved the digital production of this work.

My lab continues the Pathfinders project as *Rebooting Electronic Literature*, publishing an annual open-source, multimedia book by the same name. Thus far, we have documented 16 additional born-digital works, including Michael Joyce's *afternoon, a story*; M. D. Coverley's *Califia*, Deena Larsen's *Samplers*, Rob Kendall's *A Life Set for Two*, J Yellowlees Douglas' "I Have Said Nothing," and Stuart's own *Victory Garden*. Since these works are inaccessible to readers due to technological obsolescence, the work that Stuart and I are undertaking with *Pathfinders* in labs like ELL to make them available for study, even at the level we are doing, contributes to long-term study of them.

Back to the Challenges at Hand

My work with documenting electronic literature raises five important questions about digital preservation:

1. For what kinds of digital objects is one approach to preservation more desirable than another?

I have been, for example, experimenting with preserving literary apps and have found that each new version of an app's operating system can result in a new version of the work, or that the work itself is updated to fix bugs and include new and better features. It is not possible to save multiple versions because the new one overwrites the old. Thus, in order to preserve versions of an app, I have to have as many smart devices as updates—an expensive endeavor. Also, beta versions I am sent to review are limited in terms of the time frame in which I get to access them. This limitation makes it impossible to compare the commercially published version with the beta version or to collect betas for long-term study.

2. How can differing approaches be combined or coordinated to best serve the interests of future scholars?

   In the case of Judy Malloy's *Uncle Roger*, the 1995 migrated web version (Version 5.0) and 2012 DOSBox emulated version (Version 6.0) both provide readers ongoing and ready access to the authorized version of the work. The latter especially works to recreate some of the visceral experience of interacting with a 1980s computer in that it simulates whirs and clicks that one associates with a 1980s computer though it should be noted that these sounds are not the same as found on the original Apple and PC computers. Additionally, the original computer provides a frame that timestamps the work, thus adding to our knowledge of the work much in the same way an original frame provides context for a painting it was constructed to hold into place and showcase. Thus, what Stuart and I learned through *Pathfinders* is that one method of preservation may not be enough; multiple approaches may be necessary to preserve electronic literature effectively for long-term study and understanding of a work.

3. What can researchers working on one sort of digital production (electronic literature, for instance) learn from those concerned with different but related areas (e.g., video games, digital writing more broadly conceived, or social-network discourse)?

   I have been guiding students in my academic program with documenting video games with the Traversal method. One such project, *Chronicles: Documenting the Articulation of Culture in Video Games* by Madeleine Brookman, documents the iconic Japanese Role-Playing-Game, *Chrono Trigger*, released originally in 1972. This publication made an excellent case study for the application of the Traversal method to other media forms and showed that it does lend itself to documenting games.

4. How, can researchers approaching the posterity of digital texts from diverse directions benefit from exchange of perspectives and results?

A little over a month ago, I received an email message from Michael Joyce, author of *afternoon, a story*, considered one of the most important works of American electronic literature today. It was published in 1990 on 3 ½-inch floppy disks and later migrated to CD-ROM technology. However, Apple computers running the El Capitan operating system cannot read the work. Curators of the *Paraules Pixelades* exhibit at the Art Santa Monica in Barcelona wanted to show *afternoon* but could not. Michael wanted to know if Stuart and I had produced a video of a Traversal of it. We had not. But within a week we produced one—James O'Sullivan, an electronic literature scholar at University of Sheffield who was visiting my lab, served as a reader for a Traversal, which we videotaped. We were able to send the video to the curators, and the work could be exhibited as a video Traversal. We fully understand that what the audience saw was not the work itself; but what they were able to experience was a performance of it, and it allowed the work to live on to a new audience.

5. How best to make works available in a way that keeps the work in tact so that it retains its integrity?

The answer to this question is twofold: it is helpful if the work is well cataloged in a database, and it is situated in the archive with all of its accompanying materials as one cohesive work.

In regards to the database, identifying and cataloging all elements of a work allows scholars to call up information about the work easily and coherently. For example, if one searches for M. D. Coverley's *Califia* in the ELL catalog, he or she will find five copies: two of the versions sold by Eastgate Systems, Inc., and three different beta versions. Upon closer inspection of any of these copies, a viewer would see the work's logo found on the CD-ROM jewel case cover on which the work was published; publication information; a brief description of the work; a link to the official website of the work; information about the format, copies, and notes; and information about the operating systems for which the work can be accessed.

In regards to the physical archive, a work like Stephanie Strickland's *True North* is both a hypertext published by Eastgate Systems, Inc. in 1997 and a book of poetry published by the University of Notre Dame Press that same year. Because the hypertext is a rearticulation of the book, it is important to keep them together in the archive. Along with these items are the review I wrote of the hypertext, *True North*, for the *American Book Review* in 1997. Thus, electronic and print components are retained together with the work itself available to be viewed and experienced by scholars visiting the lab and collection.

Final Argument

The approach to archiving the physical materials of a work of electronic literature that I am suggesting runs against the mandate to preserve, for if these items—including the floppy disks, diskettes, and CDs—are made available for scholars to use, then they are in danger of being corrupted and/ or ruined in the present time, and if the physical materials are not separated by type, then the specific practices needed to safeguard paper, plastic, and the rest against environmental and natural dangers indigenous to them will not be correctly followed. However, the goal of ELL, unlike some other collections, is *not* to preserve individual works for posterity but rather long enough to allow for the works to be properly *documented* for posterity. Truth be told, if I indeed held floppy disks of *Uncle Roger* in an environmental-friendly vault for the next 100 years, there is still no guarantee that the work would survive and lend itself to be read. All it guarantees is that it is still a physical artifact in a sleeve marked *A Party in Woodside*. Without documenting it via *Pathfinders*, Malloy's own diligent efforts, and scholarship of researchers interested in the work because it is available today to be experienced, few if anyone would know what *A Party in Woodside* means.

References

Aarseth, Espen. (1997), *Cybertext: Perspectives on Ergodic Literature*. Baltimore, MD: The Johns Hopkins University Press.

Bly, Bill. (1997), *We Descend*. Watertown, MA: Eastgate Systems, Inc.

Borsuk, Amaranth and Brad Bouse's *Between Page and Screen*.

Coverley, M. D. (2000), *Califia*. Watertown, MA: Eastgate Systems, Inc.

Douglas, J Yellowlees. (1993), "I Have Said Nothing." Watertown, MA: Eastgate Systems, Inc.

Grigar, Dene and Stuart Moulthrop. (2015), *Pathfinders: Documenting the Experience of Early Digital Literature*. DOI: 10.7273/WF0B-TQ14.

Jackson, Shelley. (1995), *Patchwork Girl*. Watertown, MA: Eastgate Systems, Inc.

Joyce, Michael. (1987, 1990), *afternoon, a story*. Watertown, MA: Eastgate Systems, Inc.

Joyce, Michael. (1997), "Nonce Upon Some Times: Rereading Hypertext Fiction." *Modern Fiction Studies*, 43 (3), 579–597. Retrieved September 28, 2020, from http://www.jstor.org/stable/26285652

Kendall, Rob. (1996), *A Life Set for Two*. Watertown, MA: Eastgate Systems, Inc.

Larsen, Deena. (1997), *Samplers*, Watertown, MA: Eastgate Systems, Inc.

Malloy, Judy. (1986–2014), *Uncle Roger Versions 1.0–6.0*. Judy Malloy Papers. David M. Rubenstein Rare Book and Manuscript Library. Duke University. https://archives.lib.duke.edu/catalog/malloyjudy.

McDaid, John. (1992), *Uncle Buddy's Phantom Funhouse*. Watertown, MA: Eastgate Systems, Inc.

Moulthrop, Stuart (1991), *Victory Garden*. Watertown, MA: Eastgate Systems, Inc.

Pullinger, Kate, Stefan Schemat, and Chris Joseph. (2004), *The Breathing Wall*. http://www.thebreathingwall.com.

21

Holes as a Collaborative Project

Graham Allen

Holes started off as something of a poetic exercise. I have written elsewhere about the theoretical context of its inception, and how it was first designed as a poetic contribution to a deconstructive conference celebrating, among other things, the 40th anniversary of Derrida's essay "Structure, Sign, and Play in the Discourses of the Human Sciences."[1] However, as far as my own practice as a poet was concerned there was a far more down-to-earth motivation for embarking on such a project, which was, in short, to practice daily my ability to write a ten-syllable line of poetry. I've always admired modern poets who can keep such a regularity of line without it affecting the vernacular effect of their poetry. I wanted to learn how to do that with ten syllables. More than that, I was interested in developing an ongoing poetic practice which I might then be able to mine for larger more publishable work. The practice of the one ten-syllable line a day was, then, supposed to be a temporary strategy, one which would quickly be replaced by other strategies. Poets should be fans of Brian Eno.

The practice of *Holes*, however, soon grew into something a little more permanent than an evanescent *oblique strategy*. Once it began it didn't appear to want to end. There was something within it that seemed immensely applicable to my own feelings about the way poets should exploit their own lives while never offering up themselves as the literal and direct

---

[1]See Graham Allen and James O'Sullivan (2016), "Collapsing Generation and Reception: *Holes* as Electronic Literary Impermanence," in Helen J. Burgess (ed.), *Hyperrhiz: New Media Cultures*, no. 15, doi:10.20415/hyp/015.e01. See also "365 Holes" essay in *Theory & Event* 2009 12 (1), http://muse.jhu.edu/journals/theory_and_event/v012/12.1.allen.html.

subject of confession or simply as subject matter. That is to say, I believe in the Romantic idea of the exercise of personal Imagination and I also take Eliot's (which is really Keats's) dictum about *impersonality*. Shelley once said that didactic poetry was his abhorrence. Contemporary poets' greatest abhorrence should be confessional poetry. No question. That leaves us with a paradox of a kind of experiential poetry which explores and mines but also avoids and transcends the self. One route is clearly a kind of contemporary version of the dramatic monologue. But there need to be other ways which meet the paradox of "self" head on. *Holes*, as it developed, seemed, in its wonderful brevity and rigid occasionalism, and most of all in its exploitation of the linguistic shifter and its refusal to offer clarifying contexts, to offer up something of a solution to the paradox of "self." It had one authorial origin, that is true; but that author was (and still is) frequently at a loss to explain the meaning of a reference or the structure of a set on lines.

Let me pursue this theme, which is in a sense a counterweight to the idea of collaboration, a little further. The vast majority of what gets published as poetry in Ireland, where I live and work, presents itself as confessional. One might talk about the Heaney-inspired dominance of something we might call *autobiographical naturalism*: a poetry that presents the "self" responding, normally in the rhetoric of the immediate present, to a natural or social scene and/or event. What I sometimes describe as "looking out of the window and seeing the hills or the seagulls or the removal van" poetry. There is, of course, much in contemporary Irish poetry which resists this *autobiographical naturalism*, some of it through subtlety and rhetorical intelligence, some of it through various Modernist pathways. The fact remains, that the poetries which embrace rather than resist this style of poetry are frequently depressingly unconscious both of the mythic nature of the autobiographical self and of the possibilities for aesthetic immediacy. Any poet worth their salt surely knows that (i) "I" am a network, an intertextual node (rather than an individual subject), and (ii) the dream of "immediacy" goes back to the Wordsworthian re-evaulation of all poetic values in *The Prelude* and other odes and lyrics. That one can show the reliance of Wordsworth's poetic self on Milton's construction of "self" in *Paradise Lost* only goes to strengthen our understanding of the paradoxical (because intertextual) poetic "self." The myth of immediacy is less frequently discussed, however, and is, I would assert, a stronger pull on serious poetic expression. I am reminded here of the brilliantly succinct title of Geoffrey H. Hartman's early book on Romantic and post-Romantic poetry, *The Unmediated Vision*.[2] It is relatively easy to accept that one has gained one's voice through long immersion in the literature

---

[2]Geoffrey H. Hartman (1954), *The Unmediated Vision: An Interpretation of Wordsworth, Hopkins, Rilke, and Valery*, New Haven: Yale University Press.

of the past. If you don't accept that then you are probably not writing anything worthy of the name of poetry. It is easy, then, to admit that one is, as a "self" and a "poetic voice," mediated. How would you know how to write poetry if you were not? The dream that one might draw one's art and one's life into a closer temporal and experiential relation than is normally possible appears, however, to be an aesthetic desire which it is less easy to cast aside.

Anyone who has ever studied Wordsworth, or any other poet for that matter, will have had the experience of registering, as something of a shock, how mediated their poetry actually is. The fact seems commonsensical and hardly worth airing; however, when one experiences the temporal delays, the revisionary and editing processes, the sheer muddy writing and rewriting of the poems which first articulated the modern poetic dream of immediacy, then it still arrives as something of a shock.[3] This is perhaps before one begins to factor in the poet's relations with their publishers, the processes involved in publishing, and then the modernizing, textual work required before we re-encounter the poet in the modern scholarly edition. There is more water under the bridge than we like to think about before we read or even Wordsworth's first readers read the words: "Five years have passed; five summers, with the length / Of five long winters! ... "[4]

Holes can be said to have grown on me as a project and a practice, partly because within its brevity of expression there seemed something like an answer to the insistent call of immediacy. Despite this, however, as I found 2007 bleeding into 2008 and then 2009, 2010, and 2011, the fact remained that some of the best poetry I had managed to compose in this period languished, not only unpublished, but as far as I could see unpublishable. As *Holes* grew the question of its publishability intensified. I imagined I might, at year ten, after a decade of work, be in a position to persuade some kind of publisher (some kind publisher) to publish the work. I imagined that serial publication every decade might be achievable.[5] But such an approach, monumentalizing the lines into something like an achieved, stable work,

---

[3]See, for example, *From Goslar to Grasmere. William Wordsworth: Electronic Manuscripts*, http://collections.wordsworth.org.uk/GtoG/.
[4]I cite Wordsworth's "Lines Written a Few Miles above Tintern Abbey, On Revisiting the Banks of the Wye during a Tour, July 13, 1798" from Jerome J. McGann (ed.) (1994), *The New Oxford Book of Period Verse*, Oxford and New York: 178–81. This edition anthologizes Romantic poetry in the year of its first publication, thus attempting to give the reader a less mediated experience. Whether that manner of arranging such editions works in the way McGann wished his to work is obviously a debatable question.
[5]This ambition was partly realised when *Holes: Decade I* was published by New Binary Press. See Graham Allen 2017, *Holes: Decade I*, Cork City: New Binary Press.

seemed destined to eradicate all the topicality, all the day-by-day relevance, and a good deal of the basic reference of the lines. The experience of writing and reading *Holes*, in its first six years, became an increasingly peculiar one. Filled with topicality, with references to political and social events, cultural moments of interest, biographical stages of my life, the text grew and grew and the absurdity of its lack of any public readership grew with it. I remember one Professor of Classics, concerned in his own work with the topicality of the texts of Greek drama, on learning of *Holes*, saying to me: "So you write topical poetry, but you don't publish it? You wally!"

I am not a blogger. Nor perhaps was I meant to be. I am interested in digital art and culture made possible by the internet and by electronics and computers and hypertextual systems, but I do not naturally place my practice as a poet in those contexts; I lack the basics skills to do so even if I wanted to. So although what I was doing with *Holes* seemed in many ways to cry out for a blog or some other e-lit format, I did not imagine I was ever going to be able to manage that solution myself. I hardly think I am unique in this mind set. Computers dominate my life, just like they dominate most people's lives. I write, compose, edit, communicate, even think on keyboards and screens. But I stay resolutely, stubbornly, helplessly on the outside of those screens. A digital artist, the author of e-lit, call them what you will, exists, of course, on both sides of the screen. What *Holes* needed, to become something that could exploit digital technology in order to really come into being, was, then, someone else, someone other than me. It was when I talked all this through with my collaborator, James O'Sullivan, who was if I remember looking for a project to demonstrate a number of then still rather unacknowledged facts about e-lit, that we realized that between us we could create something rather unique.

One of the major points that James found was confirmed in *Holes* was that digital literature (e-lit if you prefer) does not (à la George P. Landow et al.) have to be interactive to be classed as such. Certainly, there is a minimum interactivity in the various divisions (About, Responses, Gallery) and the availability of the selection of starting date, but compared to the claims concerning kinds of intense hypertextuality which dominated, until recently, discussions of digital literature, *Holes* is stubbornly static. It is important, therefore, to be clear about what we mean when we describe *Holes* as *born-digital*. Or to put things another way, to describe *Holes*, accurately, as *born-digital*, is to clarify that that often-used phrase refers not to a certain type of interactive, forking, nonnarrative art, but rather to anything, *Holes* included, that could not exist in its essential state, without being presented through a digital (electronic) medium. The essential features of *Holes* are its ongoing creation (day by day, week by week, month by month, year by year), and its conflation, in its digital form, of the activities of creation and reception. *Holes* is born-digital because its outstanding feature is that it is read as it grows, and for this to happen it requires a medium of communication which transcends the monumentalizing features of codex publication, and allows for the equivalent of weekly replenishment ("Five years and one week have passed …"). If the lines of *Holes* can be figured as a form of liquid poetry, then they require not a glass to fill but, rather, a flowing stream to join. There is a noteworthy way, therefore, in which digital media are, ironically perhaps, more authentically organic than print books. The Romantic dream of immediacy is bound up, of course, with a figure of organic form which has its most dramatic expression in Coleridge's Eolian harp, a musical instrument which creates "natural music" by responding, in an unmediated and thus immediate fashion, to the wind.[6] Coleridge's figure presents us with a dream of poetry which responds "naturally"/immediately with the world outside the poet. A dream, that is, in which experience and poetic expression are coincident. I am not, for a moment, suggesting that such an immediacy is actually attainable in digital literature, or anywhere else, for that matter. What I am suggesting is that *Holes* finds ways of using its digital platform to radically decrease the temporal gulf between conception and expression, on the one hand, and publication and reception on the other. It brings us closer to that unfulfillable dream which is also, of course, the dream of modern news media and perhaps telecommunications more generally. I am writing these lines, for example, on the morning of November 14, 2015, one day after the terrorist atrocities in Paris on November 13, 2015, and I am able as a poet to know that my response to this event will be published in *Holes* within the next 24 to 48 hours. I can even be certain of this before that response is composed and committed to written form. One of the ironies which emerges here, of course, if we take up this post-Romantic perspective on unmediatedness and immediacy, is that the gains achieved by *Holes* on the temporal level are made on the basis of a collaborative understanding of authorship which cuts across other apparently foundational Romantic principles.

In the basic sense of the word, *Holes* is collaborative. Without that collaboration it wouldn't publically exist. The mechanics of this collaboration have been described in the essay I have already mentioned (Allen and O'Sullivan, see note 1), but from my own side they consist of a combination of composition, regular delivery (every Sunday), and then trust (that James will code correctly, that he will do so promptly, and that he will more generally attend to the upkeep of the site, including periodic improvements to design, storage, etc.). The collaboration is rather like that of the old relationship between author and publisher, in that James takes control of all aspects of publication, including advertising the work on a number of social platforms and, crucially, the look of the text on the numerous kinds of computer screens it is now read on. If *Holes* is a poetic

---

[6]See S. T. Coleridge, "The Eolian Harp," in McGann (ed.) 1994: 119–20.

artifact, then it has two creators, two authors. Always generous and always eager to involve me in decision making, it is nonetheless true that much of the reader's experience of reading, looking at, and interacting with *Holes* is the work of James O'Sullivan.

Much of the emphasis on this collaboration for me lies in the issue of trust I have already raised. Invoking the figure of the Georgian publisher-bookseller like, for example, John Murray II (publisher of Byron, until Byron broke with him), or Joseph Johnson (publisher of Wollstonecraft and other radicals), or Charles Ollier (Shelley's publisher, until a breakdown of trust), is illuminating in many respects. These Georgian figures were in control of every step of the process of publication from commissioning work through to advertising, branding, and rebranding works and authors. The relationship they had with their artists was neither subservient nor impersonally bureaucratic, but surprisingly equal even to the point of debates over content and artistic direction.[7] The literary marketplace was beginning to open up into something unprecedented, offering a mass audience to those who possessed the wit and the courage (along with the talent) to seize and shape the new medium of the mass-produced and mass-distributed book. Authors we associate with Romantic ideas of originality and solitary creation (Byron, Shelley, Keats, Wordsworth) were surprisingly dependent upon the friendship, guidance, and wisdom of their bookseller-publishers.

A similar equality exists in the kind of digital collaboration represented by *Holes* where the relationship is between two poets, one of whom is also the publisher. The seven lines (or "holes" as I tend to call them) are delivered each week from a notebook application on an iPad through a direct email communication. Which is to say, the author figure (Graham Allen) is not involved directly in the overall shape of the text (as it develops), and unless he makes a special point to do so does not check how these seven lines he is sending to his publisher (James) relate to those which have preceded them. It would be interesting, because of this, to ask: who is in fact responsible for the overall shape and design of the text (as it develops)? I have invoked the issue of trust, in the way I have, since the answer to this question is in no way clear. If conventional thinking gives responsibility for the overall text to the author, then with *Holes* this responsibility has to be understood as a shared responsibility between someone who is acting the part of author and someone who is acting the part of publisher but who has taken on numerous functions previously associated with the author. The trust involved is not then principally about content, but rather about a dual custodianship for a text that has long ago escaped the confines of one single creative mind.

[7]For a recent discussion of Byron's crucial, career-defining relationship with his publisher, see Mary O'Connell (2014), *Byron and John Murray: A Poet and His Publisher*, Liverpool: Liverpool University Press.

Holes, in this last sense, is not a text, not a poem, and not a book. It is best described as a textual project, an auto-bio-graphy, where each one of those activities (self-reflection and self-expression, living/existing, writing/designing) is done by more than one person, and normally by at least two. It might be best to understand this project in its collaborative dimensions through the issue of mutual interest. Not in the usual semi-military, brinkmanship sense of that phrase, but rather in terms of the required criteria for continuation. If either of the authors of *Holes* (the author-poet and the author-publisher, if you will) decided it was a project that no longer interested them then the project and thus the text itself would finish. In fact without the author-publisher's maintenance of the site on which the text is presented the text would cease to exist, whatever decisions the author-poet wanted to effect. So *Holes* exists and will continue to exist so long as it remains interesting to at least two people. The original and originating number of existence for *Holes* is not the Romantic and post-Romantic "One" but a wholly more dialogic and collaborative "two." If the reader, that "third" party of interest, if such a person exists, were ever able to look through the text as if it were a hole through which one could see then it would be two faces, two lives, two origins that they saw.

Yet, of course, this doubleness is not strictly dependent upon its literal manifestation in two authors, two origins. Doubleness is not to be understood in a crass literalism. If there were only *one* then as we have already said there would still, necessarily, be a minimum of *two*. The poem in various ways, but particularly through a rigorous adherence to the logic of the linguistic shifter seeks to perform this fact again and again. It is worth, perhaps, elaborating on this point. The model for this utilization of the shifter in *Holes* lies in the entries on this subject in Roland Barthes's own auto-bio-graphical text. Barthes remarks on receiving a postcard signed "Jean-Louis" and dated "Monday." As he states the receiver of such a message, "must instantly choose between more than one Jean-Louis and several Mondays."[8] Barthes finds this tendency of the linguistic shifter to multiple possible referents joyful and even erotic, and a liberating basis for his own auto-bio-graphical writing. He says:

I speak … but I wrap myself in the mist of an enunciatory situation which is unknown to you; I insert into my discourse certain *leaks of interlocution* (is this not, in fact, what always happens when we utilize that shifter *par excellence*, the pronoun *I*?).

(166)

---

[8]Roland Barthes (1977), *Roland Barthes by Roland Barthes*, trans. Richard Howard, 165–6, London: Macmillan.

The answer to the rhetorical question is, of course, yes, but a yes which we always seem to forget. We forget it, that is, whenever we think that the personal pronoun has an uncomplicated coincidence with the one who is enunciating, and we forget it whenever we speak or write about ourselves as authors. In its use of the shifter, *Holes* attempts to perform this forgetting. Which is to say it attempts to remain self-conscious about processes of inevitable and yet unpredictable forgetting. There are pronouns in *Holes* to which "I" can assign referents and there are pronouns for which "I" have lost the referent or to which "I" can wager a number of competing referents. But things are more complicated even than that. One might take a line like the following, from May 26, 2007: "When the lights are off, who am I to you."[9] I have no idea, writing now on November 16, 2015, whether that line was meant to capture something personal to my life or something generic and thus impersonal. Because of that I am not sure what the referent of "you" is but also whether there is any specific referent for "I." Was the line referring to something personal or did it arrive as a poetic fiction? I have no idea. The context for this line is quite forgotten and irretrievable. The line is, therefore, both autobiographical and yet unrecoverable, and in that sense beyond all possibility of autobiographical writing or understanding. It is the potential expression of a moment of personal affect along with the eradication of the possibility of that affect.

Holes is an auto-bio-graphy which strives to include forgetting (and thus the contingencies of memory) within its own *graphy* and its own *auto*. It presents itself as a text by an author, but it constantly exploits the fact that the idea of a text by an author depends upon quite common and perhaps unavoidable modes of forgetting. Through this process of remembering its reliance on forgetting *Holes* hopes to re-member, to re-member differently. Barthes, if we can return to him, goes further and notes the socio-legal prohibitions which militate or at least would like to militate against the logic of the shifter. Through this he draws this issue back toward the question of the author's lack of singularity. He ventures, through this use of the shifter, toward a liberation from the socio-legalistic restrictions which normatively revolve around the inscription of a singularity of dating and of naming:

> Can we even imagine the freedom and, so to speak, the erotic fluidity of a collectivity which would speak only in pronouns and shifters, each person never saying anything legal whatsoever, and in which the *vagueness of difference* (the only fashion of respecting its subtlety, its infinite repercussion) would be language's most precious value?
>
> (166)

---

[9]*HolesbyGrahamAllen*, May 26, 2007, http://holesbygrahamallen.org/.

One is reminded of Michel Foucault's analysis of the history of the name and notion of "the author" as a legal construct.[10] The proper name of the author, Foucault says there, and the study of book history has shown again and again since, is on one sense a legal construct designed to attribute and fix meaning. More generally, we tend to want to restrict meaning-making to individuals, so that we talk (rather absurdly if one thinks about it) of Michelangelo's *The Last Judgement* or of Shakespeare's plays or of Stanley Kubrick's films, or even of Henry Ford's motor cars and James Dyson's vacuum-cleaner. The proper name occludes the collective and collaborative processes that are involved in the creation of works of art and of technology. So that *HolesbyGrahamAllen*, like any other title bearing a singular proper name, works to erase its own relation to the logic of the shifter. The exploitation and celebration within the work itself of the logic of the shifter, however, hopefully foregrounds such a mode of forgetting and by doing so figuratively at least refers to the *more than one* that lies behind the text on the screen. Through its performance of the shifter, or the logic of the shifter, and through the associated, analogous leading metaphor of the hole (that cannot be successfully looked through), *Holes* is a text and a project which seeks to remember again and again the inevitability of its forgetting of the complex doubles (the various versions of *more than one*) that help to create it.

---

[10]Michel Foucault, "What is an Author," in Donald F. Bouchard (ed.), trans. Donald F. Bouchard and Sherry Simon, *Language, Counter-Memory, Practice: Selected Essays and Interviews*, 113–38, Oxford: Blackwell.

22

Publishing Electronic Literature

James O'Sullivan

If publishing is the set of activities which achieves the dissemination of literature, then what can publishers offer work which can quite readily attend to its own dissemination? The creators of electronic literature often act as artist, producer, and distributor, removing the relationship between writer and publisher which has persisted since the earliest days of the literary market. Those who wish to find readers for their writing have long relied on publishers as "useful middlemen" (Bhaskar 2013: 1). Informed by my own experiences running a publishing house which publishes born-digital electronic literature,[1] this short chapter explores the extent to which electronic literature needs such middlemen, whether electronic literature has any need for publishers in the traditional sense. As just noted, why seek a publisher for something which publishes itself?

The practice of publishing is often unkind to itself, driven by a need to make literature happen but in a manner that can be economically sustained; a good publisher knows that good literature does not necessarily find good readers.[2] As Bhaskar so eloquently contends: "Publishing isn't like most industries. It busies itself with questions of intangible value and moral worth" (2013: 2). Publishers, then, often find themselves torn between the desire to turn a manuscript into a book, into something ready to seek an

---

[1]My definition of electronic literature has been detailed elsewhere (Heckman and O'Sullivan).
[2]Those interested in this problem from an Irish perspective might enjoy "The Realities of Independent Publishing in Ireland," published in the online edition of *The Irish Times* on June 9, 2017 (O'Sullivan).

audience, and the need to make a manuscript into a product, something which can survive capitalism as it seeks. As difficult as this might often seem, many publishers do find a way, utilitarian things are made of manuscripts, and readers are found.

The relationship between authors and readers has not always been facilitated by publishers. Charting the history of publishing in Britain, Feather reminds us of something which we tend to forget: "there was publishing before there were publishers" (2006: 3). The current state of e-lit publishing is perhaps the natural order restored, before writing as creative practice became writing as a commercial concern.[3] Some print authors still self-publish, and the digital economy has given rise to a whole range of platforms designed to empower writers taking this path. Services like Amazon's CreateSpace[4] have allowed authors to accomplish activities once the reserve of publishing houses. There are legitimate reasons for self-publishing, there are reasons why a talented author might have no alternative but to self-publish, and there are authors who must self-publish because their work will never be of sufficient quality to find a place under a reputable imprint. Whatever the reason for print literature being self-published, contemporary writing does stand out as other when shared without a publisher. We assume, rightly or wrongly, something about self-published literature because of the position which publishers have long held within the market.

Electronic literature has no such lineage, no historical frame from which a tension between publishing and self-publishing emanates. Publishers exist because writers are not necessarily producers, they are storytellers and artists, but their medium is language, and for language to find an audience it needs to face material realities dealt with by the practices of publishing. Publishers facilitate transactions between authors and readers, accounting for the many editorial, material, and economic matters embedded within the contemporary process of literary making. We can consider the act of publishing to entail three essential elements: production, distribution, and prestige. Such a troika can be problematized—particularly in the digital age—but this is publishing in the most fundamental sense, the selection, creation, and sharing of words deemed culturally, aesthetically, and economically worthy.

---

[3]This is not a negative appraisal of publishers: whatever the role of publishing in the emergence of culture as industry, it is too late to separate the purity of expressive writing from the contemporary situation. All we can do at this point is keep faith that we will always have at least a few publishers who go about their business with the moral worth to which Bhaskar refers firmly instilled.

[4]I have, on a previous occasion, referred to the demand for content on Kindle and iTunes as a "dangerous axis of power" (Horgan 2017: 21).

My contention that there are no e-lit publishers is, of course, rhetorical.[5] Indeed, in the late-1980s and 1990s, Mark Bernstein's fabled Eastgate Systems, Inc., an ongoing concern based in Watertown, Massachusetts, published and sold works of electronic literature as packaged disks, becoming central to the emerging e-lit community. Eastgate maintains Storyspace, one of the first intuitive hypertext authoring systems to be adapted by authors for literary purposes.[6] Based on the success of titles like *afternoon: a story* (Joyce 1990) and *Victory Garden* (Moulthrop 1991), Eastgate became known as *the* publisher of electronic literature, drawing appreciation from articles published in popular venues like the *New York Times Book Review* (Coover 1993) and *Chicago Tribune* (Gutermann 1999) throughout the 1990s. Many critics now credit the Eastgate School with first bringing electronic literature into public consciousness. Whatever the contemporary situation, Bernstein's contributions to the e-lit community demonstrate that much of this form's first generation did rely on publishing houses: Eastgate invested in its writers, providing a means of production and distribution through which its carefully curated hypertextual stories could be brought to screens before downloading became a thing. It was figures like Michael Joyce and Stuart Moulthrop who had the aesthetic vision, but it was Eastgate, with Storyspace, with the finance that purchased and packaged the diskettes, with the network of distribution that brought it to e-lit's earliest readers, which saw that vision find an audience.

And then downloading, with the spread of the internet throughout domestic spheres, became the dominant form of cultural transmission, and everything changed.

Such change has one essential consequence in this context: "new authoring and distribution channels opened up" (Walker Rettberg 2012). Floppies were no longer needed to connect authors with readers, and it made little aesthetic sense for authors and publishers to persist in committing digital fictions to physical media. This shift brought about the rise of the Flash Moment and platform poetics wherein artists co-opt prevalent systems (Flores 2018), it brought about the present-day model of up-and-down distribution now considered standard. Electronic literature went from being shared as something bookish—a corporeal structure containing literary content—to

---

[5]For a comprehensive account of relevant publishing activities in Europe, see "Electronic literature publishing and distribution in Europe" (Eskelinen et al. 2014).

[6]While I am unaware of a comprehensive history of the Eastgate School, there are some sources from which readers interested in this particular aspect of e lit's origin story might benefit (Barnet 2013: 131–2; Bernstein 2010; Walker Rettberg 2012). It is also important to acknowledge the contributions of Jay David Bolter and Michael Joyce to the emergence of e-lit authoring and publishing: while Storyspace is now maintained by Mark Bernstein's Eastgate, it was first developed by Bolter and Joyce back in 1987 (Bolter and Joyce 1987).

predominantly web-based content. Even contemporary works of electronic literature that are considered post-internet, operating on individual devices as local instances, are largely disseminated via web-based platforms like Steam.[7] Where once publishers were needed to make something of a digital fiction, present-day authors need only click "upload" and wait for readers to hit "download" in turn. We find evidence for this in the fleeting nature of Eastgate's ascendency as an e-lit publisher, coupled with a glaring absence of many successors. Multimodal authors are seemingly unconvinced of the need for publishers.

It is not just authors that need convincing. Every summer, the Electronic Literature Organization (ELO) announces the recipient of the Coover Award, an annual prize which acknowledges the work of electronic literature considered by the scholarly body's judiciary panel to be the year's best. In August 2018, the prestigious accolade went to Will Luers, Hazel Smith, and Roger Dean for *novelling*, published by New Binary Press.[8] Despite being the Founding Editor of New Binary Press, I heard news of this achievement after it had become widespread on social media: the authors had been informed, whereas the press had not been contacted. The official announcement posted on the ELO's web page (eliterature.org) found space to acknowledge Bloomsbury for their part in publishing a volume which won the Hayles Award, the Coover prize's critical counterpart. This is not intended as a criticism of the ELO, but it is telling nonetheless: as a community, we *see* scholarly publishers, while we efface their creative counterparts. This is possibly a consequence of the ELO's status as an academic organization with a largely—though certainly not entirely—scholastic culture: publishers remain an active part of how we appraise scholarship, and so it would have been seen as more important to include the critical book's publisher in the announcement. The publisher of the creative work was seen as less noteworthy.

Furthermore, the achievement was not acknowledged by any of the state-funded bodies in Ireland tasked with the promotion of literature, despite quite explicit efforts on my part to achieve some small token of recognition, even a congratulatory tweet. Whether or not these bodies appreciate what it is that the e-lit movement is seeking to achieve with computational aesthetics, the reality of the situation is that this stuff is happening, and those agencies in receipt of public funding have a responsibility to support and amplify *all* literatures. This situation chimes with the wider situation in Europe, where e-lit tends to be isolated from the mainstream (Eskelinen

---

[7] Examples of such works would include *Dear Esther* (Pinchbeck 2017) and *All the Delicate Duplicates* (Breeze and Campbell 2017).
[8] It is "my" press in name to the extent that I am its Founding Editor, but it of course belongs to my collaborators, its authors, and readers.

et al. 2014: 235). Independent publishers who engage with the precarious economic conditions of their industry thrive on validation, and it would have been a small but meaningful gesture for some of these organizations to recognize the first ever Irish press to be involved in winning one of the major international awards for electronic literature. If such disregard continues, it will not simply be e-lit authors who question the need for publishers in this domain, but publishers themselves will ask, why bother?

Using New Binary Press as a case study is ideal in that I am positioned to articulate why it is that works like *novelling* may or may not require a publisher. Founded in 2012, New Binary Press publishes literature across a variety of media, including born-digital electronic literature. In fact, the press has been built on e-lit, with one of its first titles, Graham Allen's one-line-a-day *Holes* (Allen and O'Sullivan 2016; Karhio 2017; O'Sullivan, "Publishing *Holes*"), remaining one of the imprint's flagship projects, and the publishing house includes leading figures such as Nick Montfort, Stephanie Strickland, John Barber, and Jason Nelson among its authors. New Binary Press is reflective of the culture of assemblage that one encounters in the space occupied by new media artists and writers; its catalog is somewhat dissonant, functioning as something of a laboratory designed to facilitate literary experiments, a sandbox for wilder things without a home. While I have not really fulfilled what I set out to accomplish with my press, its founding purpose remains clear in that it is an experiment in the production and publication of all kinds of literature, print, electronic, and whatever else might seem interesting.[9]

Holes is a useful staging point from which to embark on a discussion of e-lit publishing, as it demonstrates one of the key differences between print and screen forms. As I wrote in *Holes: Decade I*, a special anniversary edition print volume of the work's first ten years' worth of lines:

> *Holes* isn't something I've published, it's something I *publish*, and as such, it is a work with which I hold a very strange relationship. A manuscript is proposed and submitted, given form and sold—that is the usual order of things. The publishing process doesn't end with that first act of dissemination, publishers must always retain something of a stake in the works they have taken charge of, but the relationship does change once a manuscript is a book. There are many activities post-production—promotion, interaction with booksellers, the realisation of subsequent editions—but a publisher's intervention usually declines over time. Once a publisher has made a book of a manuscript, they release it to the wild—books live and die in public, far from the guarded confines of their press.

---

[9]It may be that the experiment will soon come to an end, but whatever its future, it has been a worthwhile endeavor.

Even with born-digital literature, aside from the odd bit of file and server maintenance, the publisher will fade to the periphery as their ability to contribute to a title's critical and commercial success slowly starts to diminish.

(O'Sullivan, "Publishing Holes": 109)

As an iterative piece of organic, autobiographical writing which grows every day,[10] *Holes* does not fit into the natural order of publishing: it is a work with which I am, as the person who brings it to the public, perennially engaged.

This particular characteristic, the need for *Holes* be regularly updated, is common across many forms of electronic literature, particularly contemporary literary games which rely on complex engines that need to be maintained so as to remain compatible with operating systems. To commit to such long-term work in a precarious market makes little economic sense for publishers, and if it is the author doing the maintenance, perhaps it is the presence of the publisher which makes no sense.

If the future of electronic literature is one which will include publishers, then it should be possible to isolate aspects of the literary process that genuinely benefit from the intervention of such. Certain ideological positions will hold that publishers have assumed a less than benevolent role within the literary market, but these criticisms are typically directed at the wider publishing *industry* of late capitalism, whereas this chapter is presented on the basis of my own critical assumption that yes, some publishers are "bad," and some are "good." The aim here is not to assess the motivations or validity of specific publishers, but rather, acknowledge that publishers do exist, and that many have made significant contributions to worthwhile artistic projects. In the age of contemporary e-lit, can such contributions continue. In other words, will there ever be another Eastgate?

Eastgate is the ideal exemplar as Bernstein's press came to prominence before the material culture of e-lit was transformed by the web. The history of Eastgate and thus of electronic literature is one of "floppies, diskettes, and compact discs" (O'Sullivan and Grigar 2019: 429), a culture which partly persisted after Eastgate in the circulation of thumb drives containing ELO collections, and in the Blu-ray disks used for special editions of literary games like *Dear Esther*. But the majority of contemporary works are now digital downloads of some sort. Previous to this turn, when diskettes of all shapes, sizes, and formats were the dominant means of sharing, projects like Eastgate had a very clear purpose: they took on the task of committing

---

[10]New lines tend to be added on a weekly basis, as Graham sends me the previous seven days' worth of writing every Sunday.

hypertextual fictions to disk and getting those disks to readers. But the task of Eastgate was not simply to provide some storage medium upon which titles could be sold; Bernstein's publishing house also provided the Storyspace platform within which many of the iconic texts were authored.[11] In this sense, Eastgate's role in the process of production was not only about making a thing of the manuscript, it also provided the tools necessary for the manuscript to be written.

It is in this latter regard, the facilitation of writing as production, as opposed to just production post-writing, that publishers have a potential role to play. Gaming engines like Unity and Unreal are, quite arguably, the future of contemporary ludoliterary works.[12] While not developed for the sake of literature, these engines are the contemporary equivalent of Storyspace, open to authors using them to achieve literary intentions. Eastgate's founder once remarked that "[a]ny hypertext system will, sooner or later, be used to make art" (Bernstein 2010), and yet, how wonderful it would be if we had more systems—such as Twine[13]—which have been *designed* for such practice. Perhaps such design would only serve to constrain, to map literary structures to pre-defined schematics and templates, and so we are better off as we are?[14] Either way, while many authors are turning to game engines like Unity and Unreal to realize their aesthetic ambitions, the dynamic is not quite the same: there is no publisher behind such systems, supporting authors in their pursuit of some act of literary expression that has no explicit commercial value—there are platforms, but these platforms do not necessarily have an Eastgate.

Herein lies the potential for publishers to contribute to e-lit from the perspective of production: as the potential for making literature through computation expands, so too will the skills required to achieve such acts of

---

[11]Eastgate still maintains Storyspace as a hypertext authoring system for MacOS ("Storyspace 3 for MacOS").

[12]I am borrowing here from Astrid Ensslin, who categorizes the "various degrees of hybridity" represented by electronic literature and literary games in terms of her literary-ludic spectrum (Ensslin 2014: 43–5).

[13]One could argue that Twine has usurped Storyspace as the field's most popular system for authoring hypertextual fiction (Friedhoff 2013).

[14]To give an example of what I mean by this, one might consider looking at works of electronic literature developed in Twine: structurally, these are all essentially *the same*. Their content varies, but users of this intuitive platform—and it has many because it is robustly and intuitively crafted—all tend to stick to the same limited out-of-the-box narrative frame offered by Twine. Hypertextual fiction, in an era where immersion matters, should be about more than just text-based forking paths. But if text-based forking paths is what the dominant authoring system offers, then text-based forking paths is what we will get, again and again. It is a wonderful platform and its creators deserve credit, but if *everyone* is using Twine, the advancement of new forms of electronic literature will suffer.

expression. E-lit's contemporary moment is *now*, in that story-driven literary games are finally being embraced by a popular audience.[15] It would be to the detriment of literature for this trend to continue alongside the alienation of authors who, while recognizing the value of computational aesthetics, are unable to realize artistic visions due to a lack of digital literacy. At present, e-lit practices usually constitute community-centered activities (Eskelinen et al. 2014: 235), confining the aesthetic affordances of this space to those who are a part of it already. Publishers can play an active and vital role in the production of electronic literature by pairing authors with technical collaborators, by supporting the development of intuitive authoring systems like Storyspace and Twine, and by generally encouraging opportunities for those who can write to do so for interactive screens.[16]

But what of distribution? When Eastgate titles were completed, they found their way onto floppies that were packaged and sold. The publisher made this happen, they managed the transaction, and so they took their cut, and the author theirs, functioning in much the same way as the print industry. Now, everything is either published freely to the web or downloaded through some Steam-like catalog, a place where *all the readers will be*. This is where the case for publishers becomes tricky. Take an artist who has set about creating a piece of multimodal writing, producing the work entirely out of their own labor and expertise: the thing is digital, if it is done then it is done and does not need to be made bookish, packaged in a way that is suited to distribution. The artist can simply take the thing they have made and bring it to the market themselves because the channels are abundantly clear and largely dictated by the platform for which the work has been created. At no point do they really need a publisher, because unlike the print trade, pretty much anyone can access these digital distribution channels—the Steams and app stores—without capital or experience.[17]

And yet, publishers can still play an active role in the task of bringing the work of writers to readers. While the app-store model of distribution is suited to certain types of projects, the time of web-based works of multimodal fiction has not yet passed. In the case of *novelling*, New Binary Press made no contribution to the production, which came from the authors readymade for dissemination. They needed a publisher with a server capable of hosting

---

[15]I have written about such titles elsewhere—see, for example, "Electronic Literature's Contemporary Moment: Breeze and Campbell's 'All the Delicate Duplicates'" (O'Sullivan) and "The Dream of an Island: *Dear Esther* and the Digital Sublime" (O'Sullivan).

[16]I appreciate that such ambitions are not so easy to accomplish, but the market for digital fictions packaged as games is thriving, and so publishers should be excited by the pursuit of any title that can bring them into such a creative and potentially lucrative space with such a diverse, global audience.

[17]The scale of the projects we are discussing needs to be considered here: we are not considering ambitious AAA video games designed for the mass-market.

the piece, and this is the part that New Binary Press plays in the work: it is public because of the technical infrastructure provided by its publisher. The challenge is that it is freely available online, and a model that allows authors and publishers to benefit from such an arrangement is not readily apparent.

It would not have been overly difficult for the author of *novelling* to set up or purchase some hosting themselves, once more removing the need for a publisher to distribute their work to its audience, but a good publisher thinks about distribution in the context of longevity; a good publisher will ensure the work persists for as long as possible. If we consider the Eastgate School to be electronic literature's first generation, then our community has already lost a generation to obsolescence. The great myth of the digital is that it persists, that data lives somewhere forever, ready to be reclaimed in its ideal state should some media archeologist come looking—the truth is that data dies all time, or is simply left to rot, unable to voice itself through systems which speak entirely different languages.

The antidote to such loss might be projects like *Pathfinders*, established to document the experience of early electronic literature (Grigar and Moulthrop 2015). But the thing and the experience of the thing are not equivalent, and while *Pathfinders* is a hugely important act of cultural preservation, such endeavors will always be playing catch-up—and only capable of capturing a very small part of the canon—if authors and publishers do not think more intentionally about the life of a work. Perhaps this is how it should be, perhaps, to quote Simon Biggs, to preserve works of electronic literature is to "fix them in time and space, like an insect in amber ... alienating the work from its context and rendering it senseless" (2010: 201). Perhaps authors have a right to create electronic literature designed for ephemerality, to establish the paratextuality of their works without concern for acts of recreation which often distort esthetics? Publishers have a responsibility to document literary history before it is erased, but there are instances where documentation is all we will have. As digital ecologies evolve, many born-digital works will be lost, and perhaps—just perhaps—the authors of such pieces are fine with future generations knowing these things existed without being positioned to actively engage with them in a more tangible sense?

Of course, there is a marked difference between ephemerality as artistic intention and loss "from a simple lack of care" (Biggs 2010: 201). Publishers can provide such care, ensuring that works and their contexts are documented, if only as a bibliographic record intended to carry the existence of a piece into the future. Artists tend not to think about legacy, partly because many assume their work might achieve this independent of their efforts, but for the critical reception necessary for preservation to be achieved through attention, work usually needs a publisher, an entity dedicated to finding some place for its wards within the cultural record.

Such a cultural record can only be so big: we cannot publish everything nor should we seek to do so. The community of practice which surrounds e-lit has suffered from an absence of publishers acting as filters. We are seeing at present an increasing number of writers do very trivial and esthetically uninteresting things with computers and calling it electronic/ digital/multimodal literature, representing the field of practice in a way that makes me, as a scholar and practitioner who has invested their intellectual time and professional labor in this space, deeply uncomfortable. Without publishers, this influx will continue, making it difficult for the uninitiated to see through the noise to the quality works of electronic literature. "Gatekeeper" is typically drawn upon as an ugly word, but when publishers act as gatekeepers—when we have enough of them and they are sufficiently dissonant in their perspective—they can play an essential role in the protection of cultural spaces.

Turning the ideals of publishing as production, distribution, and prestige into a viable model for the publishing of electronic literature may prove an insurmountable challenge for most smaller, independent operations. Electronic literature "is not a market-driven literary phenomenon, but a community-driven scene with an accompanying set of aesthetic, social, and cultural values and practices" (Eskelinen et al. 2014: 235), and so the few commercial successes that one might point to will probably remain the exception rather than the norm. But that does not mean the community should not continue to consider the role that publishing can play in the advancement of electronic literature.

Without wanting to end on a pessimistic note, my realization that publishing electronic literature is currently quite futile came in the guise of *All the Delicate Duplicates* (Breeze and Campbell 2017), an exemplary piece of e-lit which *I could not have published*. It was produced by its contributors, and released to the wild via Steam, the same marketplace where one can find all of the titles created by studios like The Chinese Room. Such works are the best that contemporary e-lit as a form has to offer, and they have been offered without a publisher. I cannot think of one thing which New Binary Press might have offered these titles. Publishers have been described as "merchants of culture," as hybrid creatures, "one part star gazer, one part gambler, one part businessman, one part midwife and three parts optimist" (Bhaskar 1). As far as publishing electronic literature is concerned, at the time of writing, I am no longer an optimist.

References

Allen, Graham and James O'Sullivan (2016), "Collapsing Generation and Reception: Holes as Electronic Literary Impermanence," in Helen J. Burgess (ed.), *Hyperrhiz: New Media Cultures*, no. 15, doi:10.20415/hyp/015.e01.
Barnet, Belinda (2013), *Memory Machines: The Evolution of Hypertext*, New York, NY: Anthem Press.

Bernstein, Mark (2010), *The History of Hypertext Authoring Software and Beyond: Interview with Mark Bernstein* [Interview by Judy Malloy], https://people.well.com/user/jmalloy/elit/bernstein.html.

Bhaskar, Michael (2013), *The Content Machine: Towards a Theory of Publishing from the Printing Press to the Digital Network*, New York, NY: Anthem Press.

Biggs, Simon (2010), "Publish and Die: The Preservation of Digital Literature within the UK," *SPIEL: Siegener Periodicum Zur Internationalen Empirischen Literaturwissenschaft*, 29: 191–202.

Bolter, Jay David and Michael Joyce (1987), "Hypertext and Creative Writing." *HYPERTEXT '87: Proceedings of the ACM Conference on Hypertext*, 41–50, ACM Digital Library, doi:10.1145/317426.317431.

Breeze, Mez and Andy Campbell (2017), *All the Delicate Duplicates*. The Space, One to One Development Trust, Dreaming Methods, & Mez Breeze Design, https://allthedelicateduplicat.es/.

Coover, Robert (1993), "And Hypertext Is Only the Beginning. Watch Out!" *New York Times Book Review*, August, pp. 8–9.

eliterature.org (2018), "Announcing the Winners of the 2018 ELO Prize," Electronic Literature Organization, https://eliterature.org/2018/08/announcing-the-winners-of-the-2018-elo-prize/.

Ensslin, Astrid (2014), *Literary Gaming*, Cambridge, MA: The MIT Press.

Eskelinen, Markku et al. (2014), "Electronic Literature Publishing and Distribution in Europe," in Scott Rettberg and Sandy Baldwin (eds.), *Electronic Literature as a Model of Creativity and Innovation in Practice: A Report from the HERA Joint Research Project*, 187–242, Center for Literary Computing and ELMCIP, http://bora.uib.no/bitstream/handle/1956/8939/rettberg_baldwin_elmcip.pdf?sequence=3.

Feather, John (2006), *A History of British Publishing*, 2nd edn., Abingdon, Oxfordshire: Routledge.

Flores, Leonardo (2018), *Third Generation Electronic Literature*, Bergen: University of Bergen, https://vimeo.com/256143153.

Friedhoff, Jane (2013), *Untangling Twine: A Platform Study*, http://www.digra.org/wp-content/uploads/digital-library/paper_67.compressed.pdf.

Grigar, Dene, and Stuart Moulthrop (2015), *Pathfinders: Documenting the Experience of Early Digital Literature*, Vancouver, WA: Nouspace Publications, http://scalar.usc.edu/works/pathfinders/index.

Gutermann, Jimmy (1999), "Hypertext before the Web," *Chicago Tribune*, April 8.

Heckman, Davin and James O'Sullivan (2018), "Electronic Literature: Contexts and Poetics," in Kenneth M. Price and Ray Siemens (eds.), *Literary Studies in a Digital Age*, Modern Language Association, doi:10.1632/lsda.2018.14.

Horgan, Joseph (2017), "Keep Going despite the Prophets of Doom," *Books Ireland*, February 2017, pp. 20–1.

Joyce, Michael (1990), "afternoon: a story," *Electronic Literature as a Model of Creativity and Innovation in Practice (ELMCIP)*, https://elmcip.net/creative-work/afternoon-story.

Karhio, Anne (2017), "The End of Landscape: Holes by Graham Allen," *Electronic Book Review*, http://www.electronicbookreview.com/thread/electropoetics/landscaped.

Moulthrop, Stuart (1991), "Victory Garden," *Electronic Literature as a Model of Creativity and Innovation in Practice (ELMCIP)*, https://elmcip.net/creative-work/victory-garden.

O'Sullivan, James (2017), "Electronic Literature's Contemporary Moment: Breeze and Campbell's 'All the Delicate Duplicates,'" *Los Angeles Review of Books*, November, https://lareviewofbooks.org/article/electronic-literature-turns-a-new-page-breeze-and-campbells-all-the-delicate-duplicates/.

O'Sullivan, James (2017), "Publishing Holes," in Graham Allen, *Holes: Decade I*, 109–13, Cork: New Binary Press.

O'Sullivan, James (2017), "'The Dream of an Island': Dear Esther and the Digital Sublime," in Astrid Ensslin et al. (eds.), *Paradoxa*, 29: 313–26.

O'Sullivan, James (2017), "The Realities of Independent Publishing in Ireland," *The Irish Times*, June 9, http://www.irishtimes.com/culture/books/the-realities-of-independent-publishing-in-ireland-1.3113708.

O'Sullivan, James and Dene Grigar (2019), "The Origins of Electronic Literature as Net/Web Art," in Niels Brügger and Ian Milligan (eds.), *The SAGE Handbook of Web History*, 428–40, Thousand Oaks, CA: SAGE.

Pinchbeck, Dan (2017), *Dear Esther: Landmark Edition*, The Chinese Room, http://store.steampowered.com/app/520720/.

"Storyspace 3 for MacOS" (2018), *Eastgate.Com*, http://www.eastgate.com/storyspace/index.html.

Walker Rettberg, Jill (2012), "Electronic Literature Seen from a Distance: The Beginnings of a Field," *Dichtung Digital: A Journal of Art and Culture in Digital Media*, 41, http://www.dichtung-digital.de/en/journal/archiv/?postID=278.

23

E-Lit after Flash: The Rise (and Fall) of a "Universal" Language

Anastasia Salter and John Murray

While electronic literature has always been found across a diverse range of platforms, some technologies emerge that define an era. Flash is one such platform: a tool for creating interactive content to run outside the restrictions of the browser through an embedded media player that at one point was so ubiquitous its controlling company, Adobe, claimed it was found on "99.9%" of computers—a percentage subject to dispute but still impressive even when challenged (Arah 2009). The appeal of this platform to electronic literature creators was clear: by breaking the rules of the browser, Flash offered an artist an author-friendly space for experimenting with interactivity and animation at the limits of potential web media. Flash rose as a form so influential among the practitioners using it that Lev Manovich coined the term "Generation Flash" to describe the artists using Flash to transform practices of animation in 2002 (Manovich 2002). Even more essentially, Flash offered the allure of a "universal language" for the interactive web, promising creators they could build once and run their work anywhere, without needing a middle-man distributor or interpreter. This "write once" philosophy suggested that Flash could reach audiences on any device. While the Flash editor remains primarily a proprietary and expensive software program, the Flash viewer and specification is both ubiquitous and free for use. Yet Flash has been declared dead, thanks in large part to Apple's decision to exclude Flash from iOS and thus from meaningful participation in the mobile web, and a world with increasing emphasis on mobile platforms and native web technologies is bringing about transformations in the landscape of electronic literature. Over its twenty-plus year lifespan as a dominant platform, Flash has had a considerable impact on the aesthetics and development of electronic literature, with its groundbreaking emphasis on visuals and accessible metaphors for coding enabling the construction of work. From the influence of a popular and powerful technology arise a number of challenges that face the field today: we consider here only a few of the problems of archiving, preserving, and analyzing Flash works given the nature of Flash as a platform. Finally, we survey the changes that electronic literature is now undergoing in what may soon be a post-Flash world, with multiple platforms vying to replace Flash but none offering the same promise of a cohesive standard.

While Flash as a platform is too broad to examine fully here (see our study on the platform, *Flash: Building the Web*, MIT Press 2014), it has a few primary characteristics that make it notable as part of the history of the web. Flash was first launched in 1996 as FutureSplash, a tool for animation: it was acquired first by Macromedia and later by Adobe, both of whom expanded the platform to include a number of new options. Flash started as a tool for timeline animation, resembling the process of working on traditional frame-by-frame animation but with the assistance of "tweening," or the algorithmic filling-in of gaps between frames. Over time programming was integrated into the interface through ActionScript, a scripting language that started as a way to add basic behaviors and interactivity and eventually became a full object-oriented language. Flash's style is founded on the use of vector graphics, scaling algorithmic images, which is part of what gave Flash works the distinctive style that Manovich notes as essential to "Flashimation" (Manovich 2002). The platform also makes use of two file-types that have become ubiquitous on the web:.SWF, the released files in Flash that are played within the browser using the Flash Player, and.FLA, the source document of Flash prior to compilation. Most works are only released as.SWF files, and the Flash Player has maintained an emphasis on supporting backwards compatibility, allowing old released works to be viewed in later versions of the Flash Player. This backwards compatibility has been essential to the popularity of Flash with web arcade game creators and electronic literature artists alike, as it means that creators working in older versions of the Flash editor (which were less complicated and inexpensive to acquire) can still distribute new works built in those unsupported versions and expect the work to be playable across modern browsers.

The Platform Studies series has explored several hardware platforms (Atari Video Computer System, *Racing the Beam*, Bogost and Montfort 2009; Commodore Amiga, *The Future Was Here*, Jimmy Maher; Nintendo Wii, *Codename Revolution*, Steven E. Jones and George K. Thiruvathukal) that provide constraints and affordances that shape from the lowest levels of hardware to software to user networks the context and potential for an experience. Montfort and Bogost define the platform as "the abstraction level beneath the code ... the humanistic parallel of computing systems and computer architecture, connecting the fundamentals of new media work to the cultures in which they were produced and the cultures in which coding, forms, interfaces, and eventual use are layered upon them" (Bogost and Montfort 2009). This attention to the layers of hardware and architecture serves as a reminder that works (including works of electronic literature) are first experienced in a material context, and removing that context divorces the work from some of its layers of meaning. While emulators allow us an entry point into the original experience of a work, they often strip away layers of the platform and interface that the lens of platform studies reminds us are essential. Flash works thus far have benefited from the status of Flash as a software platform, which to some extent means an independence from hardware constraints that makes Flash works easier to preserve and distribute than hardware-dependent works. However, Flash itself has a set of architecture and material constraints that have made a powerful impact on the aesthetics and poetics of works produced using its affordances, and thus platforms like Flash are essential to our understanding of practices within electronic literature. In many cases the dependence of a software work on a particular hardware configuration is only revealed with the progression of time, and there are a number of assumptions implicit in most Flash works of hardware configurations and interfaces—particularly the use of the keyboard and mouse for controls. And the dependence on a particular set of software constraints (particularly when rendered invisible through the act of compiling, as in the transformation from.FLA to.SWF in creating a distributed Flash work) can go even more unexamined.

Flash in Electronic Literature

How can we measure Flash's significance in the history of electronic literature thus far? One possibility lies in quantifying Flash's visibility within the landscape of electronic literature is particularly clear when exploring all of the works collecting in the ELMCIP database, a cluster Scott Rettberg describes: "one of the identifiable clusters is a set of works developed in Flash or similar software that emphasizes motion graphics, kinetic texts, or innovative interface and interaction," noting further that Flash is one of the top-ten tags within the most-cited works in the database (Rettberg 2014). Rettberg's analysis of the clustering around two dominant platforms, Flash and hypertext, further suggests Flash's influence in types of works: animation, visual poetry, audio, and music are all clustered around Flash in visualizing the database. There are nearly 400 works tagged as Flash in the ELMCIP, and not quite 500 tagged as hypertext as of March 2015. While the ELMCIP is not a comprehensive survey of all works that might be considered as electronic literature, it is inclusive of the collections, major international venues, and self-archived work by practitioners who identify with the field. The dominance of "Flash" and "Hypertext" as tags demonstrates the significance of the tool of practice to categorizing and understanding any particular work within the field.

Flash's popularity as a platform for electronic literature can be attributed both to the sustained community around it and the ease with which Flash works are created and distributed. There are other platforms with similar affordances, and solutions for animation in particular abound. Matt Kirschenbaum's observations on Flash note that its popularity as a platform (as opposed to competitors with similar affordances, such as VRML or QuickTimeVR) is thanks to distribution and flexibility in cross-platform performance, and thus "the salient question is not whether one can produce 'better' content with VRML or with Flash, but rather the extent to which the kind of content we create for environments like the Web is determined by various social histories, histories that are often corporate, but always situated within absolute zones of material and ideological circumstance" (Kirschenbaum 2003). Flash's dominance is likewise not a proof that it is a "better" platform for electronic literature, but rather a reflection of the modalities, architecture, and distribution it offers despite its position as a corporate-controlled platform.

The choice of platform is thus a definitive choice when creating any work of electronic literature, as it is a choice of both affordances within the platform and the positioning of the work in the landscape of the web. Working with a platform can even be a core influence on how the work is constructed. Geniwate's description of her process of working with co-author Deena Larsen on *The Princess Murderer* (Flash, 2003) notes how the work in ActionScript influenced the construction of the text itself: "Increasingly we found the programming code seeping into the surface text. It couldn't be helped: the world we created simultaneously existed on two levels: a surface narrative about an insatiable Bluebeard and his ferocious princesses, and a semi-subliminal narrative about performative textuality and world-creation. The act of writing code infested the act of writing narrative and vice-versa" (Geniwate 2005). Such collusion of text and code through a platform can even become central to the experience of a work: "Loseby's installation uses Flash actionscripting to 'domesticate' the monstrosity of code and directly thematize the fear of invisible and unknowable code, disturbing because she considers it to be 'a language that is both hidden and alien to me'" (Raley 2005). Such work demonstrates how the platform can become both tool and subject, embedded as it is in the process and reflexive work of electronic literature.

One of the greatest challenges in electronic literature is the preservation of a canon. Several projects have made specific efforts at archiving and preserving works: notably, the *Electronic Literature Collection*, Vols. 1 and 2 have served as an entry point into electronic literature. Likewise, the Pathfinders project has taken on the task of documenting works of electronic literature that might otherwise have become unreadable (Grigar and Moulthrop 2015). Flash works represent a particular problem for the archivist: they are a form of compiled work, and deconstructing the decompiled file into reusable or analyzable parts is difficult. Flash's own software configuration warranted particular mention in Nick Montfort and Noah Wardrip-Fruin's recommendations for creating works of electronic literature that can easily be preserved: "Flash has been widely adopted, and it does offer capabilities that are essential to some electronic literature authors' practice. However, we recommend against using Flash to create elements that can be easily created in open, non-proprietary systems … formats that are more preservable than Flash may emerge as useful in the future" (Montfort and Wardrip-Fruin 2004). Again, Montfort and Wardrip-Fruin's recommendations present a reminder that while Flash as a platform empowered many amateur creators in large part thanks to an active and sharing community of developers, it remains a closed system.

The recommendation against Flash is in accordance with Montfort and Wardrip-Fruin's overall recommendation to avoid corporate-driven systems in favor of community-driven systems, a recommendation made in 2004 a decade before Adobe's support of Flash has mostly diminished. Likewise, writing in 2007 about the first volume of the ELC, John Zuern observes his concerns for the future:

> The dates that would in some ways be even more useful but that cannot be listed with any precision are the expiration dates of the works included in the ELC. How far out are we, for example, from the end of Flash, a technology abundantly represented in this volume? What provisions have we, as a culture or as individual artists, made to ensure that something of the work remains accessible even after the program that created it and the platform on which it runs are obsolete?.
>
> (Zuern 2007)

Now that Flash's spiral of diminishing influence has clearly begun, such questions have become even more pressing—and should inform our evaluation of emerging platforms for electronic literature, which will likely suffer from the same problems in the future.

Post-Flash Electronic Literature

Our recent study of Flash represented the first major academic work on Flash as a platform, as all other books devoted to Flash have focused on it from an applied or technical lens. This omission of Flash as an object of primary study is a significant reminder that the platforms we rely upon are often less visible than the works they enable, and the post-Flash era of electronic literature will likewise rely on platforms that may or may not be sustainable and easily preserved. More attention than ever is being paid to Flash's structures following its apparent death. The subject of post-Flash electronic literature warranted a panel at the 2014 Modern Language Association conference, following on Mark Sample's assertion in the proposal:

> Unlike the avant garde art and experimental poetry that is its direct forebear, e-lit has been dominated for much of its existence by a single, proprietary technology: Adobe's Flash. For fifteen years, many e-lit authors have relied on Flash—and its earlier iteration, Macromedia Shockwave—to develop their multimedia works. And for fifteen years, readers of e-lit have relied on Flash running in their web browsers to engage with these works.
>
> (Sample 2013)

One of the panel's participants, Christopher Funkhouser, drew particular attention to how the affordances of Flash became the affordances of digital poetics, emphasizing the correlation between platform and practice (Sample 2013).

Just as Flash played a role in shaping a generation of work in electronic literature, so too are the platforms that are displacing it playing a part now in changing discourse. The apparent deciding moment for the death of Flash as a universal language came hand in hand with the rise of mobile and touchscreen chameleon interfaces, with their new demands and affordances. Mobile platforms have brought new attention to electronic literature works, particularly with the category of interactive book rising in commercial viability, but such texts are often tied to a particular ecosystem (iOS or Android) and cannot be easily archived or accessed outside those native hardware devices (Salter 2015). In this new ecosystem, the apparent new ruling universal language is a throwback to the early days of the web. HTML (now HTML5), augmented with CSS and JavaScript, can span across PC, Mac, iOS, and Android architectures easily, but without the flexibility or distribution networks available for native applications on those same platforms.

However, the hole left behind in Flash's wake may be not a dearth but a blessing for the diversity of practices in electronic literature. Sandy Baldwin and Tiffany Zerby suggest that the relationship between works and platforms should not be governed by a dominant form:

> E-lit is always already there, in the same space as all the other components and functions of the computer and the network. The Modernist notion claims the literary would see the presentation of surreal work using Flash

as involving stratification, where the work of electronic literature does something other with the literalness of the interface. We see something different in the flat, high-speed space of contemporary digital writing and publishing: a relation of mutual inhabitation and co-parasitism. We refuse to see a dominant interface or technical system in which the artist carves out her works (Baldwin and Zerby 2014).

Indeed, as we face the new challenges of preserving, translating, and archiving the Flash era of electronic literature with a post-Flash future on the horizon, we can see that the promise of "write once" was always an illusion, a false promise that no other platform can rise to fill. Post-Flash electronic literature is found on tablets and phones, on virtual reality headsets and in installations, and even (still!) in hypertext and Flash.

References

Arah, Tom (2009), "99% Flash Player Penetration – Too Good to Be True?" *PC Pro*, February 20, http://www.pcpro.co.uk/blogs/2009/02/20/99-percent-flash-player-penetration.

Baldwin, Sandy and Tiffany Zerby (2014), "Editing Electronic Literature Scholarship in the Global Publishing System," *Electronic Book Review*, February, http://electronicbookreview.com/thread/electropoetics/global.

Bogost, Ian and Nick Montfort (2009), *Platform Studies* (MIT Press Series Overview), http://platformstudies.com.

Geniwate (2005), "Language Rules," *Electronic Book Review*, January, http://electronicbookreview.com/thread/writingpostfeminism/programmed.

Grigar, Dene and Stuart Moulthrop (2015), *Pathfinders: Documenting the Experience of Early Digital Literature*, http://dtc-wsuv.org/wp/pathfinders/.

Kirschenbaum, Matthew (2003), "Virtuality and VRML: Software Studies after Manovich," *Electronic Book Review*, August, http://electronicbookreview.com/thread/technocapitalism/morememory.

Manovich, Lev (2002), *Generation Flash*, http://manovich.net/index.php/projects/generation-flash.

Montfort, Nick and Noah Wardrip-Fruin (2004), "Acid-Free Bits: Recommendations for Long-Lasting Electronic Literature," Electronic Literature Organization, http://eliterature.org/pad/afb.html.

Raley, Rita (2005), "Interferences: [Net.Writing] and the Practice of Codework," *Electronic Book Review*, January.

Rettberg, Scott (2014), *An Emerging Canon? A Preliminary Analysis of All References to Creative Works in Critical Writing Documented in the ELMCIP Electronic Literature Knowledge Base*, June, http://electronicbookreview.com/thread/electropoetics/exploding.

Salter, Anastasia (2015), "Convergent Devices, Dissonant Genres: Tracking the 'Future' of Electronic Literature on the iPad," *Electronic Book Review*, http://www.electronicbookreview.com/thread/electropoetics/convergent.

Sample, Mark (2013), *Electronic Literature after Flash*, http://www.samplereality.com/2013/04/10/electronic-literature-after-flash-mla14-proposal/.

Zuern, John (2007), "Letters that Matter: The Electronic Literature Collection Volume 1," *Electronic Book Review*, October, http://electronicbookreview.com/thread/electropoetics/diversified.

24

Learning as You Go: Inventing Pedagogies for Electronic Literature

Davin Heckman

In a field like electronic literature, which is both well developed and always emerging, most teachers have faced the challenge of teaching material that is regarded as "marginal" within the Humanities but relevant in the classroom. Though the scholars that circulate around the organization tend to be very interested in literary approaches, most have found themselves working in roundabout ways, slipping electronic literature into literature surveys, media studies, fine arts, and computer science classrooms.

Indeed, as Maria Engberg notes in her survey of electronic literature pedagogy in Europe, there are a range of institutional obstacles to the teaching of electronic literature, and these obstacles differ depending on national, institutional, and disciplinary contexts. Citing Jörgen Schäfer's experience teaching eliterature in Germany, Engberg points to the various places where electronic literature can fit into a broader curriculum: "1) literary studies; 2) communications or media studies; 3) art and design schools or creative writing programs; and 4) computer science departments."[1] In response to the scant attention to electronic literature in German

[1]Maria Engberg (2014), "Electronic Literature Pedagogies," in Sandy Baldwin and Scott Rettberg (eds.), *Electronic Literature as a Model of Creativity and Innovation in Practice: A Report from the HERA Joint Research Project*, 73, Morgantown, WV: West Virginia University Press.

academic settings, Schäfer's recommendation is "to 'reanimate' the so-called *Allgemeine Literaturwissenschaft* (or 'general study of literature') of the 1970s and 1980s in German universities."[2] The conclusion reflected broadly across the various approaches in Engberg's survey is that the electronic literature teacher must be open to a variety of approaches and opportunities, and must draw upon the community of international researchers, scholars, and institutions to support the work of teaching electronic literature.

While it might be daunting to participate in a field of practice that has very few established institutional homes, the capacity to teach electronic literature in dialog with English, Media Studies, Fine Arts, Computer Science, Rhetoric, Performance Studies, and other disciplines adds value to existing curricula by opening up insights into technology through considerations of medium, form, language, poetics, narrative, semiotics, design, culture, etc. Beyond the *Electronic Literature Collection*[3] and the activities of the Electronic Literature Organization (ELO), there is no central, universally acknowledged institution that is synonymous with electronic literature. In the American national context, literature's inertia coheres in the Modern Language Association, the Norton Anthology, and a number of high-profile programs that compete for top honors in English (Harvard, Yale, Berkeley, Stanford, Princeton, Cornell, etc.). For the scholar of electronic literature, there are not "programs," but an international network of practitioners and programs that are friendly to this work. Often, intrepid individuals are doing original scholarship with support from a committees and colleagues that are open to consider experimental works. For instance, a professor from the University of Ghana, Kwabena Opoku-Agyemang, is in the process of documenting an entire field of contemporary conceptual electronic poetry in Ghana.[4] Reham Hosny, a professor from Minia University in Egypt is developing a database of works in Arabic and has organized the first conference on Arabic Electronic Literature at King Khalid University.[5] Both Hosny and Opoku-Agyemang have presented their work at ELO conferences and have contributed to the ELMCIP Knowledge Base, and worked with Sandy Baldwin at WVU and RIT (the three also co-edited Hyperrhiz 16: Globalizing Electronic Literature). I hold them up here not simply to highlight their contributions to the field, but because in a field such as this, the best work does not come from centralized "high profile"

---

[2]Ibid.
[3]*Electronic Literature Collection*, Vols. 1–3, accessed December 5, 2016, http://collection. eliterature.org.
[4]Kwabena Opoku-Agyemang (forthcoming), "Magpie Poetry": The My Book of #GHcoats Project and African Conceptual Poetry," in Joseph Tabbi (ed.), *The Bloomsbury Handbook of Electronic Literature*, New York, NY: Bloomsbury Publishing.
[5]*Arabic E-Lit* (accessed January 3, 2017), https://arabicelit.wordpress.com.

programs, but from a distributed network of scholars that are largely excited about new work.

As an emerging global field that generates expressions via increasingly decentralized media, the scholar of electronic literature must be mobile, flexible, and sensitive. The emergent character of this work is a benefit: while the professor can initiate the practice of reading works of electronic literature as *literature*, can provide institutional cover for the validity of this work, and can require documented outcomes of research and practice, it is often the students that engage with the transmedia landscape who bring the work to class, form the research questions, and produce novel results. So, far from being at a disadvantage, the para-institutional nature of electronic literature curriculum is that which can keep the classroom nimble, dynamic, and fun.

The aspiring e-literature professor should consider a patient strategy of compiling research that speaks to the specific institutional context that one operates in, seeking areas in which electronic works complement or complicate existing curricula in a meaningful way, and work diligently to create places in the curriculum that can include electronic literature as a standalone subject or part of a dynamic portfolio of rhetorical, computational, and/or aesthetic practices that make sense within a broader educational setting.

In the end, the most convincing argument for *teaching* electronic literature is its effectiveness as a pedagogical tool. And the most convincing argument for *studying* electronic literature is the potential for knowledge production. Does electronic literature improve one's appreciation and understanding of the dominant codes of meaning in the twenty-first century? Can electronic literature open up deeper appreciation and understanding of culture and history? Can electronic literature be used to develop student writing? Can electronic literature increase our awareness of and competence with digital technologies? Can electronic literature improve student engagement in the learning process? And, most importantly, can electronic literature open up a critical perspective on society during a period of radical historical upheaval?

Basic Strategies

The obvious place to begin when discussing any new learning experience is to first and foremost begin with an encounter with material. For students who have no prior experience with digital arts or literature, an unprimed encounter with a new text in a novel format offers ample opportunities for thinking about the work. And since there are powerful generational differentials in play regarding platforms, media usage, and user experience, even a naïve reading of a work of electronic literature can provide a rich learning experience.

Selecting a single work or a handful of works for a "cold" introduction and providing opportunities to navigate/experience the text in the fullest context available is often an eye-opening experience for students and teachers alike. The experience of the work as a phenomenon often opens up questions that lead into rich terrain, and the raw read through of a work in a classroom setting can sustain theoretically rich discussions. Basic phenomenological questions (like, What is this? Why would someone make this? What is the point?) lead readily into meditations on form, genre, intention, interpretation, politics, and poetics, and allow students to foreground their own intuitive understandings of the works in question and generate critical comments.

The key, however, is to select works that reward exploration and play. Selecting something with a big "wow" factor can provide an easy preface to deeper exploration of the broader practices.[6] While the scholarship on these works probe the depth and sensitivity of practice that form them, they are good introductions because they are accessible. Such works touch on familiar cultural forms, communicate in strongly visual languages, are relatively intuitive to navigate, and have a disarming charm that draws many into conversation/controversy over the value and place of these works within broader schemes like film, literature, gaming, etc. The goal is to enjoy the first experience of the electronic literature and to establish interest before digging deeper into the field.

After introducing examples, the next step is to establish a basic definition of "electronic literature" and work through a variety of approaches to electronic literature. A good working definition, which has its roots in N. Katherine Hayles' foundational "Electronic Literature: What is it?," is the one offered by the ELO:

Electronic literature, or e-lit, refers to works with important literary aspects that take advantage of the capabilities and contexts provided by the stand-alone or networked computer. Within the broad category of electronic literature are several forms and threads of practice, some of which are:

- Hypertext fiction and poetry, on and off the Web
- Kinetic poetry presented in Flash and using other platforms

---

[6]Depending on the context, works by Jason Nelson, Donna Leishman, Christine Wilks, Alan Bigelow, J. R. Carpenter, Nick Montfort, Serge Bouchardon, Rui Torres, and/or Stephanie Strickland can offer some strong starting points. While none of these writers can be characterized as "simple," many offer works that are rewarding for the naïve reader of electronic literature. However, the *Electronic Literature Collections* are filled with strong examples of electronic literature that can satisfy a wide range of readers.

- Computer art installations which ask viewers to read them or otherwise have literary aspects
- Conversational characters, also known as chatterbots
- Interactive fiction
- Novels that take the form of emails, SMS messages, or blogs
- Poems and stories that are generated by computers, either interactively or based on parameters given at the beginning
- Collaborative writing projects that allow readers to contribute to the text of a work
- Literary performances online that develop new ways of writing

This little passage identifies a conceptual definition that marks "electronic" more than it marks "literature," is itself a powerful discussion starter, and leads through a range of questions about quality, cultural attitudes, formal practices, print traditions, etc. Second, it identifies a rough bundle of forms that scholars have identified in the field. If one wishes for a deeper or more sustained discussion of the definition and forms, Hayles' 2007 essay provides a very thorough discussion of the field that can quickly build an awareness of the origins of the field.[7]

For theoretical reasons, I am a strong advocate of moving from the "estrangement" produced by the cold encounter and subsequent whirlwind tour of the field into zones of familiarity. Once the raw phenomenological response is registered, the return to the familiar offers students the chance to assert some order over a sprawling and often mystifying field. The quickest way to get people thinking theoretically about works is to draw out more deliberately formed responses that rely upon the critical experience of the student (as provided by personal research and structured curriculum). If your course is focused on a cluster of practices (for instance "computer generated texts," "glitches, errors, and accidents," "poetry: oral, print, digital," or "electronic gaming"), the texts selected will be more limited than a general course on electronic literature, but materials can be arranged in terms of their chronological order (first to last), by genre (hypertext, digital poetry, generative works, database-driven works, literary games, etc.), by comparisons to the extant knowledge of the audience (for example, as compared to genres of print literature, gaming, cinema, fine art, interface design, etc.). In any case, the goal is to present a variety of approaches and practices.

---

[7] N. Katherine Hayles (2007), "Electronic Literature: What is it?" Electronic Literature Organization, http://eliterature.org/pad/elp.html.

Once students feel reasonably comfortable talking about electronic literature, an extremely productive teaching strategy is to turn them loose on the field as a field in upheaval. Beyond the *Electronic Literature Collection*, a number of international database projects are busily trying to document the field of practice as it emerges. Through databases like the ELMCIP Knowledge Base, NT2, LIKUMED, Po.Ex, I <3 e-poetry, the Electronic Literature Directory, Hermeneia, ADELTA, and others, students can explore the field from a variety of perspectives. The SYNAPSE project will make all participants in the Consortium for Electronic Literature (CELL) searchable through a common interface, creating ample opportunities for budding researchers to gather, tag, and critique works of electronic literature.[8]

Having surveyed the various established resources, the next step, of course, is to invite them to find (or make!) their own works of electronic literature, develop arguments that establish similarities and differences from recognized practices and works, and to document these works in their research. Several of the CELL partners encourage user contributions of bibliographic data, descriptive content, and critical responses. A key benefit of studying electronic literature is the strong potential for meaningful research to contribute to the field. A number of professors in the field have seen students publish their contributions in databases and journals and/or to participate in creative projects. As a teacher, always emphasize the unsettled nature of the field, encourage students to take positions and ask questions, and view the occasion of research, writing, and argument as an occasion to contribute to the field of humanistic discourse during a period of upheaval.

Reading Works of Electronic Literature

If your goal as a professor is to provoke interactions with digital works that will contribute to the collective understanding of the class, there are some theoretical and methodological approaches that can be used to bootstrap readers of electronic literature into deeper engagement with the field. If literacy is a prerequisite for the appreciation of print literature, commensurate "reading" strategies must be applied to the work of electronic literature to deepen one's understanding of the work. Just as the tradition of literary criticism has revealed that this deeper understanding can be supported by different kinds of depth (historical/cultural depth, linguistic depth, hermeneutic depth, etc.) and that there can be a variety of productive "serious" readings that are nevertheless limited, we can accept that there are

---

[8]*CELL: Consortium on Electronic Literature* (accessed December 5, 2016), http://cellproject.net.

a plurality of serious approaches to electronic literature that often produce readings that are simultaneously accurate with respect to their domain of analysis while being in tension with alternative approaches. The reading strategies a student might choose to adopt will likely be determined by their own competencies and the curricular demands, but it is absolutely important that critical readers of electronic literature take seriously the tensions that the work contains, even if they are unable to provide a full account of the work.

Ground zero for analysis is Media Specific Analysis. An approach that has precursors in phenomenology, cultural studies, and media ecology, Media Specific Analysis (MSA) as it relates to literary criticism is most clearly articulated by N. Katherine Hayles, whose work *Writing Machines* "performs" MSA through both careful reading and provocative design.[9] The basic gist of MSA is that one must not simply take the medium of transmission for granted, the scholar must consider the way the content of the text interacts with its existence as a material object. A number of writers (Johanna Drucker, David Jhave Johnston, John Cayley, and others) have identified various ways in which the text matters.[10] This media-reflexivity is critical to the definition of electronic literature that is employed above.

My preferred articulation of this question, the definition offered by Serge Bouchardon and Davin Heckman in "Digital Manipulability and Digital Literature," parses the digital work into three layers: content, form, and technical design.[11] This tripartite model asks readers to consider what the work is about, what cultural form it employs (everything from tropes to genres, from styles to art forms), and how it is constructed as a technical object.[12] Bouchardon and Heckman note that these three categories exist in tension with each other in the literary work: What the work is about is often related to the genre of its expression. How the work is expressed is often as technical as it is aesthetic. The technicality of the work contributes

---

[9]N. Katherine Hayles (2012), *Writing Machines*, Cambridge, MA: MIT University Press.
[10]An excellent overview can be found in N. Katherine Hayles and Jessica Pressman (eds.) (2013), *Comparative Textual Media*, Minneapolis, MN: University of Minnesota Press.
[11]Serge Bouchardon and Davin Heckman (2012), "Digital Manipulability and Digital Literature," *Electronic Book Review* (accessed December 5, 2016), http://electronicbookreview.com/thread/electropoetics/heuristic.
[12]This threefold approach also resonates strongly with Stiegler's discussion of individualism and the productive dynamism of the psychic, social, and technical. This model of cultural production, long present, but rarely discussed, in the history of literary criticism gains greater visibility in this approach. In light of competing anxieties over social media and digital culture, it seems that teaching electronic literature not only revitalizes individual interest in literature, but that it re-establishes the critical urgency of Literature itself. See Bernard Stiegler (2009), *Acting Out*, trans. David Barison, Daniel Ross, and Patrick Crogan, Stanford, CA: Stanford University Press.

to the "message" of its expression.[13] In teaching students to write about electronic literature, I ask them to describe its content, its formal aesthetics, and its technical specifications. From here, it is only a matter of time before the careful reader notices the degree to which these layers are entangled with each other. A key question of "reading" in the twenty-first century is the role of the machine as an "interpreter" of the text, from the question of translating source code into output to the larger question of macroanalytic readings of human behavior. While the approach I outline is fairly formulaic as a writing prompt, the tensions lead into provocative research questions, and, in the best cases, explosive essays on digital culture itself.[14]

Making Works of Electronic Literature

While I do not consider myself an e-lit "author," I almost always offer electronic literature students the opportunity the opportunity to engage in a practice-based research option: (1) Start with a critical objective. (2) Follow with a selection of relevant examples of creative work. (3) Read the critical material that addresses the specific objective or mechanism. (4) Attempt to fulfill the critical objective through experimentation. And, (5) write a formal response to document the process.

It is often useful to establish some sort of constraint within which the student must explore the form. The approach that I often use is to ask students to draft a narrative text for the medium that they are most comfortable with (often print or video) and then to prototype the project without access to their preferred tool. An alternate approach is to write supplementary materials for a central text (a movie, book, or poem) that does not exist. The goal is to tell a story or create a sensation without recourse to the typical tools of expression, probing the limitations and strengths of other media and to reflect upon the technical specificities, cultural codes, and content-level associations that accompany our modes of expression. While I leave

---

[13]For "close readings" of code, it is useful to expose students to some simple programming exercises to highlight the difference between what the machine "reads" and what the human "reads." There are a number of works which make these differences visible as objects of consideration, and there is an entire community of scholars that are engaged in "Critical Code Studies," an approach that pays scrupulous attention to code. For more information on Critical Code Studies, see *Humanities and Critical Code Studies Lab*, http://haccslab.com.
[14]I typically ask my students to start with this three-layered description, then to identify the points of tension. As an added constraint (because I am often thinking of a database as a potential home for student writing or for condensed writing as building blocks for larger papers), I ask students to complete this task in 500 words of less, with multiple drafts to eliminate obvious or repetitive statements and to pack the micro essay with insights that move to describing the piece's tension.

the door wide open to explore "analog" media as well, one might prescribe an array of tools if specific skills are required by the curriculum. Other approaches involve more specific constraints: copy and adapt source code to create a transformative work, find an imaginative use for a tool or platform that you use every day, construct a collaborative project or participate in a netprov performance, etc.[15] And, of course, students who are steeped in digital design and programming tools would be prime candidates for developing more complex works. Regardless of one's ability or experience, the key is to experiment.

Resources

An obvious way to extend one's teaching in electronic literature is to review syllabi and lesson plans from scholars working in the field. The fastest way to find these is to visit the ELMCIP Knowledge Base, which is a treasure trove of teaching resources.[16] Rita Raley's "Electronic Literature" (Fall 2009) is a good example of a first-year writing course designed to satisfy general studies requirements,[17] while Jessica Pressman's "Digital Literature" (Fall 2010) is an upper-level survey course in electronic literature.[18] Mark Sample's "Electronic Literature" (2015) is a massive online course, open to the public.[19] The syllabus for John Cayley's "Writing Material Differences" (Spring 2012) explores "the material poetics" of writing within a transcultural context that considers calligraphy, print, and digital texts with a strong emphasis on Chinese writing.[20] Nick Montfort's Comparative Media course, "The Word Made Digital" (Fall 2009), explores "non-narrative" forms of digital writing in the context of games, electronic

---

[15]Network Based Improvisational Performance, or Netprov, is an approach that relies upon group participation. Due to its open nature and the looseness of improvisational practice, it often relies upon platforms and tools that require little technical introduction. Netprov performances are often staged on Twitter, Instagram, Facebook, and other social media platforms. See Rob Wittig (2011), "Networked Improv Narrative (Netprov) and the Story of Grace, Wit & Charm," Master's Thesis and Creative Project, Norway: University of Bergen, http://robwit.net/?project=114.

[16]"Teacher Resources," ELMCIP Knowledge Base (accessed December 5, 2016), http://www.elmcip.net/teaching_resource.

[17]Rita Raley, "Electronic Literature," ENGL 146EL, Syllabus (UCSB, Fall 2009), http://transcriptions-2008.english.ucsb.edu/curriculum/courses/overview.asp?CourseID=315.

[18]Jessica Pressman, "Digital Literature," English 391a, Syllabus (Yale, Fall 2010), https://anthology.clmcip.net/materials/syllabi/Pressman-2010-US.pdf.

[19]Mark Sample, "Electronic Literature," Archived course (edX/Davidson, 2015), https://www.edx.org/course/electronic-literature-davidsonx-d004x.

[20]John Cayley, "Writing Material Differences," LITR 1230J (Brown, Spring 2012), https://wiki.brown.edu/confluence/display/wdm/wmd+-+course+syllabus+-+Spring+12.

literature, digital arts, online content, and code.[21] Talan Memmott's course, "Rhetoric and New Media," (Spring 2010) is focused on analysis and application of digital rhetoric.[22] While Aya Karpinska's "Electronic Writing" (Spring 2008) is a "project-oriented workshop to explore techniques for effective and innovative use of text in digital media."[23] Lisa Swanstrom's "New Cyborg Theory" (Spring 2011) is a graduate course in science fiction that incorporates electronic literature to enhance a print-heavy reading list.[24] In fact, a visit to the ELMCIP Knowledge Base's list of teaching resources includes over forty syllabi, over a dozen exercises and lesson plans, plus numerous additional resources (to which I hope you will add your own!), that can help the prospective electronic literature professor build a plan that will suit the needs of students.[25]

For those who wish to participate in more participatory approaches to experimental pedagogy, UnderAcademy College and Meanwhile Netprov Studios offer opportunities for immersive play in the creation and analysis of digital texts. UnderAcademy features courses and seminars taught by leading artists and scholars in the field, typically around absurd provocations and prompts, and culminating in significant creative outputs.[26] Similarly, Meanwhile Netprov Studios frequently opens its network-based improvisation performances to public participation, in many cases enlisting entire classes to participate in the creative practice.[27]

Other resources include the network of databases represented by the Consortium for Electronic Literature.[28] These databases provide free access to comprehensive information about works of electronic literature, scholarship in the field, artists' websites, and other resources. Soon, these databases will be linked under a common search engine, SYNAPSE, providing teachers and students with access to primary and secondary sources with which one can build syllabi, construct reading lists, and build research projects. More exciting, perhaps, is the possibility of contributing to partner databases. Many of these databases invite user contributions,

---

[21]Nick Montfort, "The Word Made Digital" CMS 609J (MIT, Fall 2009), http://www.elmcip. net/teaching-resource/word-made-digital-cms-609j-fall-2009.

[22]Talan Memmott, "Rhetoric and New Media," EN1306 (Blekinge Institute of Technology, Spring 2010), http://www.elmcip.net/sites/default/files/files/attachments/teaching/rhetoric_and_ new_media_11.pdf.

[23]Aya Karpinska, "Electronic Writing," LITR 0210D, Syllabus (Brown, Spring 2008), http:// www.technekai.com/ewriting/wiki/index.php?n=Main.Syllabus.

[24]Lisa Swantsrom, "New Cyborg Theory," LIT 6932, Syllabus (Florida Atlantic University, Spring 2011), http://newcyborgtheory.wordpress.com.

[25]"Teacher Resources," ELMCIP Knowledge Base.

[26]UnderAcademy College (accessed January 3, 2017), https://underacademycollege.wordpress. com.

[27]Meanwhile Netprov Studios (accessed January 3, 2017), http://meanwhilenetprov.com.

[28]CELL: Consortium on Electronic Literature.

both by individuals and by institutions, allowing students to contribute their research to the scholarly community.

Additional resources, many of which can be found through the SYNAPSE search tool, include artist websites, journals, videos, curated exhibits, and digital repositories that are available online.

Choose Your Own Adventure

While I recognize that my overview of pedagogy is going to be hampered by my limited experience relative to the ever-expanding universe of electronic literary practices, I hope that in identifying basic approaches alongside a growing catalog of resources that you will have (or be able to find) everything you need to teach a course in electronic literature. The most important feature of the electronic literature community is the enthusiasm of its members—from the authors who have invented and re-invented literary practices to the scholars who have greeted such work with curiosity and enthusiasm, from the pioneering teachers who integrate emerging practices into established disciplines to intrepid students who bring new works into critical consideration. So, yes, I invite you to engage with this enthusiastic community. But, more importantly, I invite you to become that community— to make your own way through the field of electronic literature.

References

Bouchardon, Serge and Davin Heckman (2012), "Digital Manipulability and Digital Literature," *Electronic Book Review*, accessed December 5, 2016, http://electronicbookreview.com/thread/electropoetics/heuristic.

CELL: Consortium on Electronic Literature, accessed December 5, 2016, http://cellproject.net.

Electronic Literature Collection, Vols. 1–3 (accessed December 5, 2016), http://collection.eliterature.org.

Engberg, Maria (2014), "Electronic Literature Pedagogies," in Sandy Baldwin and Scott Rettberg (eds.), *Electronic Literature as a Model of Creativity and Innovation in Practice: A Report from the HERA Joint Research Project*, 71–86, Morgantown, WV: West Virginia University Press.

Hayles, N. Katherine (2007), "Electronic Literature: What is it?" Electronic Literature Organization. Last modified 2007 (accessed December 6, 2016), http://eliterature.org/pad/elp.html.

Hayles, N. Katherine (2012), *Writing Machines*, Cambridge, MA: MIT University Press.

Hayles, N. Katherine and Jessica Pressman (eds.) (2013), *Comparative Textual Media*, Minneapolis, MN: University of Minnesota Press.

Humanities and Critical Code Studies Lab (accessed December 6, 2016), http://haccslab.com.

Meanwhile Netprov Studios (accessed January 3, 2017), http://meanwhilenetprov.com.

Opoku-Agyemang, Kwabena (forthcoming), "'Magpie Poetry': The My Book of #GHcoats Project and African Conceptual Poetry," in Joseph Tabbi (ed.), *The Bloomsbury Handbook of Electronic Literature*, New York, NY: Bloomsbury Publishing.

Stiegler, Bernard (2009), *Acting Out*, trans. David Barison, Daniel Ross, and Patrick Crogan, Stanford, CA: Stanford University Press.

UnderAcademy College (accessed January 3, 2017), https://underacademycollege.wordpress.com.

Wittig, Rob (2011), "Networked Improv Narrative (Netprov) and the Story of Grace, Wit & Charm," Master's Thesis and Creative Project, Norway: University of Bergen (accessed December 5, 2016), http://robwit.net/?project=114.

Teaching Resources

Cayley, John. Syllabus. "Writing Material Differences," LITR 1230J, Spring 2012, Brown University, https://wiki.brown.edu/confluence/display/wdm/wmd+-+course+syllabus+-+Spring+12.

Karpinska, Aya. Syllabus. "Electronic Writing," LITR 0210D, Spring 2008, Brown University, http://www.technekai.com/ewriting/wiki/index.php?n=Main. Syllabus.

Memmott, Talan. Syllabus. "Rhetoric and New Media," EN1306, Spring 2010, Blekinge Institute of Technology, http://www.elmcip.net/sites/default/files/files/attachments/teaching/rhetoric_and_new_media_11.pdf.

Montfort, Nick. Syllabus. "The Word Made Digital," CMS 609J, Fall 2009, MIT, http://www.elmcip.net/teaching-resource/word-made-digital-cms-609j-fall-2009.

Pressman, Jessica. Syllabus. "Digital Literature," English 391a, Fall 2010, Yale, https://anthology.elmcip.net/materials/syllabi/Pressman-2010-US.pdf.

Raley, Rita. Syllabus. "Electronic Literature," ENGL 146EL, Fall 2009, UCSB, http://transcriptions-2008.english.ucsb.edu/curriculum/courses/overview.asp?CourseID=315.

Sample, Mark. Archived course. "Electronic Literature," edX/Davidson, 2015, https://www.edx.org/course/electronic-literature-davidsonx-d004x.

Swanstrom, Lisa. Syllabus. "New Cyborg Theory," LIT 6932, Spring 2011, Florida Atlantic University, http://newcyborgtheory.wordpress.com.

"Teacher Resources," *ELMCIP Knowledge Base* (accessed December 5, 2016), http://www.elmcip.net/teaching_resource.

SECTION IV

Artist Interventions

25

My cODEwORk ARTicle

Michael J. Maguire

My cODEwORk ARTicle asks what does the term codework mean to me, to someone who has tried to understand and engage with (& in) it? Is it poetry or art, does it really matter to me as a digital artist what label is assigned to my practice. In this confined flat space it could prove problematic to strive to reach any definitive answer concerning praxes. Thus, I tried to pick up the concept of codework and shake it a little to see if anything useful, illuminating, informative, or even entertaining drops out as a result. Do we consider it poetry, categorize it as textual art? Let me shake (down) codework in four connected directions, and see if I can find (in) out anything from academic use or even entertainment value for the answer.

a mODE of cODEwork (mis)understood as Art
Codework ART (mis)understood as Study.
a mODE of cODEwork (mis)understood as Comedy
Codework ART (mis)understood as Technological Writing

First, any writing about codework will always offer the opportunity to eulogize the work and works of Mez, (Netwurker/ Mary Anne Breeze), John Caley, Rita Raley, Alan Sondheim,[1] and several significant others. The practice, knowledge, and persistence of an activity called codework, its very existence, is owed to the insight and artistry of those four individuals. Dig into discussions from early exchanges on list-servs and later open deliberations on EBR.[2] Ted Warnell, Carl Banks, Florian Cramer, and Talan

---

[1]http://collection.eliterature.org/1/works/sondheim__internet_text.html.
[2]http://www.electronicbookreview.com/tags/codework.

Memmott are also among many others who have engaged with what might obtusely be argued to be an activity partly responsible for the gestation of specific types of intellectual framing of many contemporary born-digital creative practices and artifacts. Codework has certainly been influential in terms of born-digital poetry, promoting a programmatic paradigm, which has undoubtedly inspired at some level artists and poets who make modern digital poetry. Anyone who has fixed bugs in commercial codebases for a living might, however, rebut such arty neuro-flatulence by declaring that all code involves work, that's what its primary *function* is, what goes in must come out, it works, that code in a computer code context is ultimately karmic ... radiohead karmic ... he talks in maths, like a detuned radio ... electric, microelectronic, motherboards, always about numbers, the scientific, the logical appeals ... deductive, reductive, take the pencil, don't develop a million dollar pen, anecdotal, experiential, this thing won't do what you're telling it to do, write it, run it, compile it, break it, fix it, fix it better, better again don't break it, all your codebases are belong to us, it may well just be a cultural information thing à la Floridi's fourth revolution.[3] Nick Montfort[4] makes a strong case for programming as much as a social activity, engaged in by enthusiasts, as controlled computation culturally constituted solely by professional labor. Code and codemonkeys conceived as geek central is just that; a conception. Often behind conceptions lie assumptions and more often initial ignorance. The joy of learning, the joy of code that works, the joy of code work balanced only by the bliss of ignorance the enthusiasm before it ... breaks ... like a wave ... Breaks ... to wait for input ... breaks ... for respite ... breaks ... hearts ... breaks ... patience ... breaks ... and breaks again.

The interrogation of one's own art practice, the requirement for some fathomable progressive consideration of the motivation for first order technological creativity, understood perhaps as drive, as urge for new ways, new elements, nutrients that feed into a desire for computer-mediated experimental expressiveness, and which are therefore surely and admittedly sadly possessed of some narcissistic import. (un)Even technological refraction of a once clear-minded artistic intent to examine the freedoms promised by digital utopianism become blurred as code and language combine within and below surfaces and internal illocutionary assumptions. A history of codework can be drawn, as venom from a wound, by extracting meaning out of the collection of curated articles in the ELMCIP Knowledge Base[5]

---

[3]Luciano Floridi (2014), *The 4th Revolution: How the Infosphere Is Reshaping Human Reality*, Oxford: Oxford University Press.
[4]Nick Montfort (2015), *Remediating the Social University of Edinburgh ELMCIP*, Edinburgh: University of Edinburgh Press.
[5]https://elmcip.net/.

and elsewhere, it can be electronically or mechanically drawn like lazy lace curtains across the open windows of the creative mind by viewing, using, and exploring works by the aforementioned acknowledged, if not renowned, practitioners. It can be drawn, carthorse like, across the open landscape of our consciousness, drawn slowly, yet with a kind of flamboyant artistic absorption, between obscure rotating points within some Ciceronian hermeneutic circle. By buying into such acts of deliberate and controlled artistic literary endeavor, bye bye-ing convention, not adhering to a series of even our own expectations, nor casting ourselves against performative impulses to play with both sides of the rules of code, the affordances of language, and some ludological strain of combinational creative artistry. Is it a modern magic which you or I again, descending, might then simply call codework for the want of a more complex term. Acts of the apostrophes weighed and weighted on a byte-based biblical digital scale, a dimensionally strange combinational creative <iframe>that consciously sequestered away the child nodes in the night. If there's a fire is in our head, we head out, we write, express, connect with something *natural*, we create, today our natural world resides, resituated within a networked, interactive narrative home. Our digitally ubiquitous environment means that out right is tethered to a line break, a closing tag rejected yet included, some syntactical abyss that stares right back until you fix that flaw. Literacy, whether visual, cultural, like numeracy, is eventually acquired, sweltering, drown, in such degrees of digital literacy, behaviors not bounded by Asimov, but unbounded by the laws of Moore,[6] Metcalfe,[7] and Parkinson.[8] These three sides to the story of codework suggest a surface story that embraces personal, technological, and the aforementioned creative expression. Obviously, any such suggestions become themselves limited in their depth of understanding, the definite and definitive articles have their place, but only alongside or nested within. But The code is not The code when it sits on surfaces and lends itself to easy reach and shorter grasp, this digital dance of codework demands more complex steps and that the dancer submerges herself in deeper connotation and wilder submission.

Q: wot do u get if u stitch 2gether standardized literary conventions [think: the monumental output of bill Shakespeare = the staccato pulsings of emily dickinson] with coded poetics steeped in digitally-drenched communication?
A: mezangelle.

(http://bit.ly/rfFpDH)

---

[6]http://www.mooreslaw.org/.
[7]https://www.cs.umd.edu/~golbeck/downloads/Web20-SW-JWS-webVersion.pdf.
[8]https://www.collinsdictionary.com/dictionary/english/parkinsons-law.

Looking at the landscape of text, inscription above, from the back of Mez's print book[9] I see that the above, the surface and the interior, the interior meaning, the suggestion, those potential combinations that originally begat recipes for creative computer based celebration and speculation. Calling it a creole is a myopic injustice, rather it is a linguist mountain, range of possibility within what Anne Balsamo has called the "technological imagination,"[10] an opportunity to ascend and do some *dreaming within the machine*, en route, that adventure embarked upon becomes conducive in terms of the opportunities that chance itself affords, the very nature of random combinational creativity, the so very wrong of occasionally getting it so very very right, and now getting it beyond previous rights. The concept of poetics as coded becomes more than attractive as it gets less self-referential and mustering an academic will to studiously surrender to the ambiguity of others that reveals, is again its own contradiction evidencing itself. A computer-based counterpoint itself guilty of expression, judging any anchored point where, ambiguity and obscurity touches such *Lexia to Perplexia*[11] itself, the plastic (green)screen of technological phenomena beyond superficial comprehension, such *great work* wot makes us think, becoming a greater work that makes you think and feel, striving to find the greatest, offering thinking, feeling, a sense of wonder, purpose, relevance, connection, code.

When first publically presenting my then work in progress, "cAMEltext. Net" alongside Steve Gibson's "*Grand theft Bicycle*"[12] and Christine Wilks' "Underbelly"[13] at the first transliteracy conference in the UK in 2008/9, I presented, discussed, and contextualized my making of the work by introducing my own nascent concept of: "Dreaming within the machine" as some speculative approach or modus operandi, some method of creating works of digital literature, which focused on the imaginative creative flow and an almost autonomic relationship with the enabling technological tools. Cameltext began as an artistic intention to engage with the inherently ecstatic Persian poetry of Rumi while attempting to carry multiple meanings close to the surface of dearly departed derivative drafts of work. By surface I refer to that easily interpretable layer of meaning which operates round and about deliberate acts of interpretation, reflection, analysis, etc. Not sign and signified more point, and pointed, point and pointy, in the same vein

---

[9]Mez Breeze, *HUMAN READABLE MESSAGES [MEZANGELLE 2003–2001]* (Traumen. at, 2011).

[10]Anne Marie Balsamo (2011), *Designing Culture: The Technological Imagination at Work*, Durham NC: Duke University Press.

[11]http://collection.eliterature.org/1/works/memmott__lexia_to_perplexia.html.

[12]http://grandtheftbicycle.com/.

[13]http://crissxross.net/elit/underbelly.html.

as John Lilis "anoint anointy."[14] My goal was to carry two meanings on the surface of the text, as with the descriptive term cAMEltext.Net which itself carries both the terms cameltext.net[15] and AMEN on its surface, the second by deliberate use of dispersed capital letters. Camelcase consists of a mixed case writing system beloved of programmers, where normal capitalization conventions are not respected and the camelcase is used to highlight keywords. My draft work drew to itself three bounding parameters, camelcase, and the case of the camel as "a horse designed by committee," a colloquial commentary upon collective or collaborative practices having the potential to be overtly flawed by virtue of their inclusivity and necessity for placation as cooperation. Finally, the quote from the Koran that one must: *Trust in God but tie up your camel.*

Again, as a faithful contradiction or an exemplary paradox of faith, in this instance some maybe creative god will protect you if you have a practical awareness of your own ineptitude. Then the ludological nature of my own work, representing the traversal of the poem as a game, with macro and micro goals which need to be achieved rather than merely read, strived to instantiate that framing triumvirate. Whether it was achieved in: "I AM the song THAT heard itSELF sing" is a matter more for other critics than for its maker.

Codework understood simply as some study of intellectual, cultural, born-digital, writing experiment can wrongly suggest a central critical path can become a simple vector of (in)accessibility to any later corporeal mentioned joy of codework, as a field of overcast endeavor, complete with hovering intellectual electronic kerstal above it, acts of deliberate interpretation and codification versus acts of spontaneous or instantaneous judgment, carefully carved in code, running revelations as a celebration of the prior foundational binaries that gave rise to preceding forms of inscription, whether blood on velum, grooves in stone, ink on paper, or green text on black screens, Codework seen as JavaScripted Jazz, a cultural artistic phenomena very much of our present.

They were the best of lines and the worst of lines, like exhausting long form literature, goatboy like, lazed lines, an infinitesimal infinity inversion on those stones, in those stones, found in that field, meaning something, if only she could read or speak Ogham, might it light a lantern upon Cuirithir and Liadain ? Instead it was the 28K light of ten men made mutant in Williams defender.[16]

[14]Carl Reiner, Steve Martin, and George Gipe, *The Man With Two Brains*, dir. Carl Reiner, perf. Steve Martin, Kathleen Turner, and David Warner.
[15]http://digitalvitalism.com/RW05.html.
[16]Digital Dickens. Unpublished email correspondence.

The moist sweaty pulsating joy of codework, jocular jouissance, playing with syntax, fumbling around in and out, out and in with language, tickling those traditional, normal, or accepted processes involved in writing and generating poetic expression, such salubrious creative acts may well mirror two fundamental approaches to writing comedy: begin with serious intention and enlist standard conventions, only post draft does the comedy writer look for the angle that helps to tilt his prose, dialog, or content towards that comic 15 degrees which yields the laughs, susceptible sometimes to unacceptable somnolence waiting for the serendipitous intervention of an aristophanetic comedy goddess, (in)complete with tyche somewhere in her name. The alternative is sitting down and approaching a codework project with that 15-degree comedic tilt already in place in the first instance. Combining code and an almost standard writing syntax, delivers an idiomatic, even alphanumeric palette of expression from which the maker of codework can chose, does she attempt to adopt a codeworker (netwurker) mindset and allow a freeflow of mixed (informed and practiced) expression from within that idiomatic mode, or draft, design, and refine, in the manner of the second comedy approach, post-structural outline, employ some wordplay fundamentals and perhaps magically morph the not so initially coded work into a kind of codework mode. Again, like the existence of God it may be very much a matter of personal veracity. Think different ... think syntactically ... think hexadecimally ... think underneath the box, and feel the difference at a deeper sense of completion should you be blessed enough to reach that point.

Shaking Codework as "technological writing, suggests in turn a from of digital writing, computer based or (re)mediated writing, locating it somewhere on a screen or in a space within creative writing practice." A start to sounding out, the resonance of the concept of autopoesis coming to loudly dominate western workshop conventions, the idea of an individual voice seems to have propelled much of that pedagogical instruction and the field in general. Talent and craft and their role if any, in specific questions about codework-like creations, obviously arise from my own previously stated position as a writer who subscribes to the age-old concepts of vitalism, that while there is focus on dissecting the surface of craft, creating stages, both waypoints and prosceniums, and I welcome any renewed attention to the role of the subconscious and indeed the separate and certainly diverse idea of consciousness itself. The possibility of a superconscious influence on the entire writing process, the creative impulses and energies beyond the surface layers surely find their natural bedfellow adjacent and elbowing next to the concept of codework. While poets work a poem, and perhaps all writing is rewriting, there is always active the unconscious processes that make significant, yet mostly unacknowledged, contribution to the entire model of creative writing. Code, computer code, impels its author to acknowledge not only meaning and connotation but also function and interpretation at a surface level, your voice is again karmic like code itself. But all of these explorations and speculations, while experience based, are open to individual understanding, epistemological framing, intellectual alignment, and obviously misunderstanding, by author, reader, maker, critic or passer by.

But misunderstanding is key here, whether it is the *Massage in brothal* by *the police* or the *Exe.termination* itself offered by the previously alluded to *Lexia to Perplexia*, it still offers a nonthreatening context for engagement, codework in some way highlights that potential, as a piece of information that subverts the standard expectation of the reader viewer and perhaps asks more difficult questions than a complex temporal forking path. Instead, a chasm of complex multiplicity sits like a loose thread waiting to be pulled, exponential potential lures through simple-seeming superficial signs. Come play this way, engage and immerse. I am as ever optimistic that Codework and its digital literature siblings will eventually find a home within the Irish Creative Writing firmament. Up until now it has been conspicuous by its absence in Ireland's creative writing literature, simply put; the field in Ireland has to date been oblivious of its existence and thus leaves an actual vacuum in relation to its study and growth. It is a fact that life tends towards complexity, organisms evolve, computers and our networked culture afford huge creative potential. The opportunity afforded by engagement and even misunderstanding of codework grows daily, and will eventually, in my humble view, come to find its place in the pantheon of creative cultural practices in Ireland and elsewhere.

26

Locative Narrative

Jeremy Hight

The moment of greatest cartographic awareness in human history is arguably happening right now. Maps are active on smartphones and in cars. New ways of tagging maps, annotating spaces, and even storing and searching information are on these maps. Cartography is no longer static maps and globes but a more malleable and layered entity. Our sense of space is also entering new areas of resonance and contextualization in this more active mapping and sense of location. An area to emerge and expand in this cartographic period is locative narrative (geospatial narrative) but it began before this ubiquitous mapping moment by several years.

A street is in fact a punctuated space. A sidewalk is as well. A city is a more complexly punctuated collective amalgamation of spaces small to large (like sentence to paragraph to full text). A sentence is mediated by length, vocabulary, punctuation, and detail in the same way a street is mediated by stop signs, stop lights, and length and detail. City planning is, in a sense, placement of spatial comma and semicolon to mediate space, flow, and information. A city is a scripted space by said design. The larger definition in animation of "narrative" is something moved over a duration and thus something changed and this was inferred to the viewer. Therefore, our physical spaces are already narrative in a sense. There also is in any space a connotative and denotative read.

Locative narrative has emerged as a range of works that place narrative along physical space (geolocate) be it on maps, on malleable maps like Google maps, or, more commonly, in the physical world itself. Martin Rieser in his 2009 essay "Locative Media and Spatial Narratives" describes how "The emergent field of 'locative' media art explores the convergence of computer data and location using portable media."[1] He continues with how "the predominant uses of mapping and spatial information necessarily lead us to a radical reassessment of the nature of representation." Rieser elucidates how locative narrative is a location aware progression of both narrative form and technological functionality and usage for creative exploration. It is both a radical move into a malleable and content aware cartography and a logical progression.

Ann Galloway and Matthew Ward in their 2005 essay "Locative Media as Socialising and Spatialising Practices: Learning from Archaeology" early on discusses the general range of locative works as "everything from mobile games, place based storytelling, spatial annotation and networked performances to device-specific applications."[2] The emergent field at that time was a range of location aware works with a subset being locative narratives. The field has since expanded out into augmented and mixed reality and films with location-specific components. The works under the umbrella of locative narrative explore the inherent social and functional tensions in a space as well as the narrative potential of space, data, and one's physical engagement in a location.

Spaces are never of only present itself. There are ghosts. Past is a ghost. Erasure is a ghost. Forgetting over time is a ghost. Suppression of controversial information is another ghost and perhaps the most complex of all. A writer can take a room and based on details chosen describe paint as dark or bright as possible, as full of portent and subtext as possible or as light and of the passing moment. Prose has long held these elements in its basic tools and architecture. What happens when this is applied to the world itself and physical location(s)? What is possible when a writer writes with the physical world as details, metaphors, subtext, or meta-text? The possibilities are nearly endless and have become a huge thread within the emergence and resonance of locative narratives. It is important, however, to first look back at key precedents that have helped this field form and emerge.

Locative narrative has roots that go back in many directions. One is in forms of writing playing with the space within a text. A text is its own sort of cartography as it is laid out for navigation in the larger space of presentation and thus has a kind of architectural internal logic as well as physicality and spatiality. Typography also is a sense of almost girders as individual letters are bent thick shapes clustered then into words, sentences, and paragraphs. A traditional book thus is a building or even a city in a sense; there are rooms within the hole, spaces, internal elements, parts to

---

[1]Martin Rieser, "Locative Media and Spatial Narratives," *NeMe*, last modified May 28, 2009, http://www.neme.org/texts/locative-media-and-spatial-narratives.

[2]Ann Galloway and Matthew Ward, "Locative Media as Socialising and Spatialising Practices: Learning from Archaeology (Draft)," forthcoming *Leonardo Electronic Almanac*, MIT Press, last modified 2005, http://www.purselipsquarejaw.org/papers/galloway_ward_draft.pdf.

whole and how it all "moves" upon reading. Some experimental texts have moved these elements even further into new areas and directions that are key lineages to locative narrative.

George Perec's *Life: A User's Manual* was published in 1978. It has been called a postmodern masterpiece as well one of Oulipo. It may well be these things but it moves beyond the border lines drawn by such naming. It can be argued that placing a deeply resonant work of varied elements in one specific field is reductive and constricts the ways the work can be viewed and experienced to the semantics and semiotics of a cluster of related works and definitions. This book is many things. It is the rooms of a single building instead of chapters. It is the map of these rooms. It is a chess board and the reader moves along like a pawn on a board not page by page in a narrative in a book. It also is a puzzle. There are to this day new theories and images trying to "solve" the puzzle. It is also beautiful writing telling stories by listing elements in each room. It shows that narrative and space can be malleable and can be read on several levels and contexts at once.

Mildorad Pavic's *Dictionary of the Khazars* came out in 1984 and is completely fictional as far as the Khazars and is not a dictionary but a "novel" constructed as a fictional encyclopedia. The reader can read it in its entirety or can read one paragraph entry and have read the work. This is Pavic's intention and design. The text is now a space that has let go of authorial control and is instead a space open to navigate. Pavic has made a space that is experientially navigated and is many spaces within the larger space of book. It is essentially a landscape. This work elucidates how text can be experienced in many paths within the larger ideas and meta resonances of an author and small to large experienced can be considered reading/experiencing the work.

These two works together are a direct vein into locative narrative as text is shown to be a space, an experience parts to a whole and not just nonlinear or linear but to move through as though in a city, mountainside, building hallway, or seaside. An interesting third bit of text is actually something made to be read primarily by young teens. *Choose Your Own Adventure* books were created by Edward Packard in 1978 for an audience generally 10–14 years of age after telling his daughters variations of bedtime stories of the same character in different adventures. It initially was rejected many times but became a phenomenon as it opened up narrative to a kind of game with many paths inside a printed book. The instructed along the path gave options of pages to turn to and thus forking paths often leading to either death or a new section and possible forking paths to come. It was a lot like the play of video games that had a certain end point and several ways to "lose" before reaching that end point in a narrative and space. Text was game play and a space of many paths and experiential lengths. This form has come to influence many experimental writers and inform locative narrative as geospatial placement of narrative creates not only an experiential interface of text and narrative, but also has ties to game narrative and space as having options and being multidirectional.

Locative narrative also has clear and strong ties to previous art movements and works dealing with physical space and its resonance and even annotation in nondigital forms. Land art took the age-old question of the white space (exhibition space and white walls such as museum and gallery) and the problemitization of work taken to the white space and pushed it out altogether. There are positive and negative semiotic reads just as there as negative positive epiphanies (flash of realization of idea vs. flash of realization of failure of impossibility of idea or its negative effects). A negative semiotic read is a kind of extended rhizomatic (shared root system) symbolism and association as opposed to more familiar positive or general semiotics. The age-old negative semiotic of the gallery or museum space is the antiseptic nature of the white walls, the erasure of all things outside this hierarchically charged space. It can be tied back to the negative semiotic read of a hillside village in feudalistic times. The low-lying area of crops and serfs looking up the hillside see the king's castle and its high walls as exclusionary. It is a rarified space with rare passage. The positive of art being accepted in galleries in museums, of course, is the work being seen as having merit and appeal. The negative, however, is that this space is excluding all other works, other places, and things outside the tastes of the gatekeepers. It also is seen a rarified space of commerce above aesthetics.

Land art argued in many works that the physical world was an exhibition space, was material for art, and was not the natural sublime (the quality of that sunset beyond words and that crushing beauty of something greater than the human realm) but simply of itself. Nature was to be seen at times sublime, but also humble, open, malleable to a degree, open to many paths and contexts. Robert Smithson had long been exhibiting in galleries when he moved toward spaces outside and works like his 1970 classic work *Spiral Jetty*. Smithson made a work of physical location and materials, location, and aesthetics. The jetty is an elemental form both seen in nature and in created works. The spiral is an artwork and it is location specific. Gone is the white space and its positive and negative semiotics. It is a location-specific and aware work and is an annotation of a physical space like we take for granted now on smartphones.

GPS has roots going back to 1956 but became a working system in 1995. A triangulated "line of sight" is fixed between navigational satellites utilizing latitude and longitude. The public signal has slightly less accuracy than the military and first was used in airplanes, by fishermen, hikers, and geocaching (easter egg-like hunts for signals placed in physical locations) enthusiasts. The system gridded and guided smart bombs in the First Iraq War. Early GPS art includes the important early work *The Telepresent* by Stephen J. Wilson in 1997. The work was a decorated box to be passed on like a present to others in physical space. It used GPS to track its location on a shared map.

What it saw with its camera was shared to others and determined by who it had been given to. Wilson brought together tenets of land art (out of the gallery into the physical world), video art, telepresence, human behavior psychology (he was very interested in who it would be given to and why), and autonomy (it had no wires or specific protected ownership when passed around). It opened up many doors and directions to come.

In 1999 Teri Rueb first created *Trace*. The work utilized GPS to trigger sounds and poems along hiking trails in Yoho national park in British Columbia. The songs, sounds, and poems were memorials along the trail and ruminations on death, life, mortality, and the space in between. The work was geospatial augmentation and an experiential interface. The audio forms an overlay on specific locations along the trails and a transformative space of elegy and memorial. This moves from *The Telepresent*'s groundbreaking sense of space and signal into something also of text space overlay in physical space while holding onto the sense of aesthetics and power of location found in *Spiral Jetty*.

Trace opened up a new area of geolocative writing and art as a kind of space to place memorials along physical space. Physical space became a kind of curated as well as augmented space in the same way that flowers are placed at the sites of fatalities along streets and roads. This was also a crucial early move into the possibilities of mixed space and mixed reality with technology and screen space tied to physical space.

34 North 118 West began in 1999 and was completed in 2001 and first shown in 2002. The project utilized the public GPS signal at first with an external GPS unit (like hikers used at the time), then a GPS card on an old 100 dollar laptop. The project began as an exploration of the connections between global positioning satellite data and the way raw materials moved across physical space in the early twentieth century along railroads. This opened up research initially into cartography as well. The map is an overlay of data and system of functionality above landscape just as railroads have long veined the landscape with a system of timing, architecture, data/ material, and a specific dialect of its own iconography. GPS also has a grid space laid out across physical space with its own iconography, system functions, and methods of working with parcels of data (smart bombs in the First Iraq War, for example).

34 North 118 West became the first locative narrative. The project laid narrative as a kind of skin across the landscape taking elements of detail, metaphor, and subtext from the keys of type to a kind of imagined hybrid keyboard of writing with text and the physical world itself. The place could reveal a deeper metaphor in say a dry river bed, could take the facade of an older building as that meta reference in a story along with the text itself. Utilizing the Global Positioning Satellite system's ability to place data by latitude and longitude, the project also allowed place to have a "voice."

34 North 118 West created a kind of "narrative archeology" by taking historical research of a hundred plus years of a four-block area of Los Angeles and culling historical data. The research found that waves of people had moved through while the buildings essentially were the same from the early twentieth century. The city was shown to be iterations, layers almost like sedimentary layers of earth over time. The cliched perception/negative semiotic of Los Angeles is of a perpetual present continually botoxing and erasing its past. The city has let go of many historical landmarks like Schwab's pharmacy (legendary for stars being discovered in the mid-twentieth century) and the beheaded legendary restaurant (literally top section cut off and placed on a mini mall across town while rest was leveled) the Brown Derby.

A common generalist conception of archeology is the researcher digging in the soil and finding at different depths different layers of artifacts and thus time and history (striations of a sort like sediment layers themselves or tree rings in measure). The verticality of the dig becomes time and the artifacts become placed spatially giving a basic sense of space tied to information. *34 North* laid out short prose/prose poetry made of historical data *horizontally*. To move through a city space and have information trigger from the past while moving along streets, sidewalks, and alleys was clearly another type of archeological "dig."

34 North placed colored squares on a map (an early augmented reality aspect) along a map of a four-block area of Los Angeles. The map moved a cursor as the person or persons (up to six people at a time with connected headphones) moved along streets, parking lots, or alleys. The pink squares indicated areas of augmentation with narrativized historical data from research. The map indicated the date of the audio as well. The effect was of place being given voice. The narratives were indeed a kind of skin laid across physical space, but were also a way for place to voice layers of past and even things lost or suppressed over time. There was not the need to read books far away or lose voices as all could now be signal, could be tied to place itself, channels even of voices of past where they occurred.

Locative narrative has traveled the path of many things once deemed "avant garde." The history of creativity and technology echoes the same kind of sine wave of emergence and experimentation. This is by far not a negative. Avant garde has an early meaning of basically being that first line emerging in battle. A new tool comes along and is played with and explored. Later tendrils and threads form of emergent paths and areas of exploration. Later fields and sub-fields begin to gel of aspects of play and contextualization. The field becomes documented and studied and a feedback loop begins of looking both backward and forward over time. The work of early pioneers becomes either lexicon, reference point to move from or both.

Another interesting aspect of locative narrative is its context in the literary discussion of form and experimental vs. traditional. Robert Frost is a great point of reference. His work is seen by many as having the clarity and familiarity seen as precepts of "traditional" writing. At the same time he late in his life spoke of how he imagined a place and its spaces and iconography along with a character not himself that navigated this space. This Robert Frost is in fact a proto avatar and his writings of a kind of second life or internet space, a simulation. These aspects are the bedrock of much "experimental" writing. Locative narrative has become a range of forms and works and many have the clarity and familiar structures of literary fiction, poetry, short shorts, and even Oulipo in its listing of things like list poetry. Locative narrative at the same time is a form taking older forms into newer areas of structure, form, presentation, malleability, and play. These aspects are of experimentation.

Locative narrative has and will continue to move in new tributaries and hybrids. Locative narrative film making, locative psychology, locative journalism, locative comedy, QR code narratives, augmented reality narratives, map augmentation narratives, and mixed-reality game storytelling are just some of the areas to emerge thus far under the geo-locative umbrella. The initial magma of play with a tool has cooled into fields within fields and an ever-expanding lexicon.

27

Come Play Netprov!: Recipes for an Evolving Practice

Rob Wittig and Mark C. Marino

Netprov is internet improv. Omnivorous in their reappropriation, netprov artists hijack whatever media are being used currently for private, friend-group, and public communications to enact fictional characters, spur-of-the-moment satires, and carefully planned speculative narratives. Shaped, planned, and nurtured, a netprov is facilitated by netrunners, but they are not alone. In our (Mark and Rob's) evolving terminology, "netrunners" plan and produce netprovs, "featured players" collaborate with the netrunners and carefully craft their characters and participation, and "players" from far and wide drop in and play for a short or long time, leaving their mark on the shape of the collaborative creation.

Make a Twitter account for your high-school self and attend school with everyone who's ever lived! Pretend to go without technology for a week and share the hell out of it via social media! Live-tweet an imaginary TV show with fellow super-fans! These are some of our recent netprov invitations to "play and go deep" as our motto goes.

The irresistible creative impulse to take a brand new form of written/pictorial communication designed for practical use and perform a silly, serious, satirical character voice within it is something we (Mark and Rob) share with each other, and with other creators (such as Laurence Sterne and Tina Fey) throughout time. Netprov brings literary and theatrical traditions into electronic network. Digital, netprov-like projects have been going on since the beginnings of the internet and proliferate today.

Writing netprovs with others is an exhilarating creative challenge and a ton of fun. We'd love you to try it—with us or with your friends! To that end we're going to define the form, talk a little about what netprov players bring to the common playspace from their different fields, and share the recipes for a few of our most recent netprovs.

Characteristics of Netprov

The basic catalog of formal elements of netprov that Rob outlined in his master's work at the University of Bergen is useful:

> Netprov is networked improv narrative. Netprov creates stories that are networked, collaborative and improvised in real time.
>
> Netprov uses multiple media simultaneously. Netprov is collaborative and incorporates participatory contributions from readers. Netprov is experienced as a performance as it is published; it is read later as a literary archive. During the performance, netprov projects incorporate breaking news. Netprov projects use actors to physically enact characters in images, videos and live performance. Some writer/actors portray the characters they create. Netprov is often parodic and satirical. Some netprov projects require writer/actors and readers to travel to certain locations to seek information, perform actions, and report their activities. Netprov is designed for episodic and incomplete reading.

Although netprov is always fiction, not all netprovs are what might be called "story based." Some netprovs are character-based tableaus, like an improv theater skit, in which the plot events serve mostly to heighten the character studies. Others are language arts games with no necessary time sequence. Netprov narratives with a strong story are holographic in structure, which means that the basic fiction and the overarching narrative must be constantly re-told so that readers dropping in at any time can be brought up to speed and invited to play effectively.

What World Do You Come From?

Netprov lies at the intersection of literature, theater and performance, mass media (film and television), games (in particular, Alternate Reality Games [ARGs], in which players physically enact roles and compete in real life), avant-garde visual arts (in galleries and museums), and born-digital internet, personal media, and social media practices. What has become clear to us is the importance of the methods and attitudes people trained in these various fields bring with them to the netprov space.

When Is a Netprov Finished?

Our interest in methods is not merely theoretical. Methods from different worlds impact practical decisions about netprovs in progress. For example, say we've just finished a netprov such as *I Work For the Web*, in which players are encouraged to become self-aware about the daily media practices they perform that generate income for corporations (browsing, "likes," "favorites," etc.) and complain about them like a job. What becomes of the netprov once the initial play period is over? From a traditional, literary point of view the text has been composed and the archive is ready to be read as a finished work. From a theater point of view, one performance has been given and an unlimited number of repeat performances of the same scenario can now be produced. From a film and TV point of view, the first iteration is viewed as Season One, and it's time to plan Season Two, which tracks the same characters from exactly the point in the story where the first iteration left off. From a game point of view, multiple versions of the same scenario can be re-staged simultaneously, at any time, with new groups of players. From a visual arts point of view the same scenario can be re-enacted, with site- and time-specific variations, upon the next offer of patronage from a gallery or museum. Finally, from the point of view of digital culture any of these options can be accommodated and new options added, such as the continuation of the characters' story in real time, with a real-time lag since the last performance, a porting of the story to a new media platform, or the migration of the same characters to an entirely different netprov scenario.

How Should a Netprov Relate to its Audience?

Attitudes toward audience from the tributary worlds of netprov vary widely. As a sketch for a future in-depth study, we will outline some of the differences.

The relationship between authors and readers in the literary world still bears the marks of print-era hierarchy and one-way communication. Authors are on a pedestal. Authors lead, and readers follow. Best-selling authors are a brand representing a reliable mental experience that is always different in particulars, but not necessarily new in form and strategy. Genre fiction authors are expected to stay quite close to their models. Avant-garde authors are expected to lead into new territory, but stay only a step ahead of their readers. Theater also has its mainstream and avant-garde, but in general, giving an engaging experience to the audience—whether they knew they wanted it or not—is the common goal. Mass media, because of production costs, is focused on audience metrics and is compelled to follow public tastes wherever they lead, even as creators try to shape those tastes.

The game world's attitude toward audience couldn't be more different from the literary world's. In the game world, the audience is on the pedestal, and the creators are humble servants. Game creators are obsessed with reaching out a friendly hand to their audience, pleasing their users and keeping their players happily on task. The pleasure of play guides creative decisions. The visual art world, because of its unique history (i.e., the most radical Modernism became the mainstream, which it did not in any other of the arts) embodies the opposite extreme. Visual artists, as a useful overgeneralization, are often still treated as high Romantic geniuses and do not reach out to their audience. If museum goers do not understand the work, it is considered the audience's fault, rather than a lack of good communication on the part of the creator, as it would be considered in other fields.

Born-digital audiences are no longer audiences at all. They are content creators. (Let's not use the cringeworthy web moniker "prosumers.") They donate creative energy and marketing data to savvy companies with every move they make, since every gesture can be tabulated. Facebook, for example, is simply a display mechanism for user-created narrative. Digital audiences think of themselves as drifting from pleasure to pleasure at the same time as they complain about their lost time. Digital audiences already fulfill the dreamy, then-impossible ideal of theorists such as Roland Barthes. Everyone is an author; no one is a reader.

How to reconcile these different attitudes? In our current experimental netprov formula, the netrunners lead with the basic concept and overarching narrative in the literary mode, collaboration is organized in the film/TV production mode, the featured players aim to please in the theater mode, and players participate in a born-digital social media mode. Very recently we've created new roles in the audience-friendly game mode: Player-Care Coordinator and Featured-Player-Care Coordinator, whose job it is to reach out and encourage players, providing feedback as a reward to good creativity.

Rob: How do netprovs work? Mark, should we demonstrate for them?
Mark: Oh, you're writing all of this down now—everything I say?
Rob: Well, it seemed like a good moment to switch to dialogue.
Mark: Beleaguering Boogers.
Rob: Focus.
Mark: Oh, right. The projects we're about to mention can be found archived at meanwhilenetprov.com—the hub of the hubbub of our netprov activity.

An Unreal Reality Show: SpeidiShow: SpeidiShow, a Netprov (2013)

Rob: Why don't we start with SpeidiShow?

Mark: Okay, maybe you should explain the idea behind this idea.

Rob: Sure, but first, tell folks who Speidi are.

Mark: Um, I think they know. Spencer Pratt and Heidi Montag. I mean, they're Reality TV stars. *The Hills? The Hills, New Beginnings? Celebrity Big Brother UK? I'm a Celebrity, Get me out of Here!*

Rob: Okay, good. So they were already familiar with the pseudo-real world of popular performance. For future readers who have no idea what Reality TV is ... it's a genre of television programming featuring performers who are "real people" who are just living their lives or participating in challenges and contests.

Mark: Wait, Rob, "real people?" Scare quotes?

Rob: Well, they are real people, but they are performing in segments that are at minimum staged if not also full-out scripted.

Mark: You're blowing my mind!

Rob: Anyway, Spencer and Heidi are Reality TV stars.

Mark: Indeed, and we'd already worked with them before on Tempspence (aka Reality, being @SpencerPratt), a netprov in which we pretended to be an obscure British poet who had found Spencer's phone, and hence his Twitter account.

Rob: So in the role of Tempspence, as his fans dubbed him, we proceeded to play Surrealist-inspired poetry games with his followers and narrate the poet's fictional romantic life.

Mark: Yes, that netprov went on for about three weeks, and tempted Spencer to play again with us that same year. So we pitched SpeidiShow.

Rob: The basic premise built on something I'd been doing with my students, where we'd all pretend we were watching some television show that doesn't exist and would react to the events on the show, live tweeting a show that doesn't exist. So we brought that model into SpeidiShow, creating a fictional Reality show of Reality shows for Spencer and Heidi.

Mark: And by that we mean that the fictional "SpeidiShow" takes on a different format each week. One week they're giving marital advice, which funny because on *The Hills* and for a time after played out a dysfunctional relationship—while in real life, they're a very close couple. The next week, they'd engaged in competitive yoga.

Rob: The "Show" aired once a week on Thursdays, and during that hour, we and our featured players, choosing a variety of roles, pretended we were either on it, producing it, or watching it.

Mark: Meanwhile, during the week, we'd tweet the stories of the
various characters involved in the show, including Spencer and Heidi,
getting tangled in behind-the-scenes drama as they prepped that
week's episode.

Rob: The show was wild. I remember our haunted Big Brother house
episode in which the ghost of Gertrude Stein made an appearance.

Mark: And then there was ... the Aura Baby.

Rob: Oh, my, yes.

Mark: So we created this idea of an "aura baby" that is the progeny of
two people's auras. Heidi desperately wanted an aura baby. Spencer
needed some convincing. During the netprov, Twitter followers
witnessed a public conception.

Rob: Right The Immacu-Twit Conception. *New York Magazine* picked
up the story and seemed to enjoy the joke, speculating about what to
give at an Aura Baby shower. In our Speidi project, the tabloids and
more serious press acted as another participant in the netprov.

Mark: I guess our Speidi work made us think a lot about celebrity, but
more specifically this class of Reality celebrity who is trapped in a
purgatory of perpetually publicly performing.

Rob: Precisely! Because Reality celebrities are amplified versions of all
of us in social media.

Rob: Right, we kept returning to puns about "keeping it real."

Mark: At the same time we were making it up—exaggerating the antics
that Spencer and Heidi had begun but taking it to an absurd level,
like chasing down leads with Winnie the Pooh.

Rob: Or engaging in espionage as they hunted for the disappearing bees.

Mark: Some of Speidi's followers got it. But others on Twitter seemed to
prefer the simpler storylines of Bad Spencer and Superficial Heidi.

Rob: Meh.

#1wknotech (2014, 2015)

Rob: A lot of our netprovs come out of our reflections on contemporary
anxieties. For examples, I teach a drawing class, and I noticed that
my students' sketchbooks were full of scenes of their friends, sitting
together in lounges and dorm rooms, hanging out together by staring
at their smartphones. And you notice the same thing, right, Mark?

Mark: We're getting a lot of Likes right now on that Facebook post.

Rob: Exactly. But remember your story, about hiking?

Mark: Oh, right. So I like to go hiking—you know, totally out of the
hubbub of the city—off away from where I can even get cellphone
coverage—and I have a REALLY good carrier. Anyway, so I hike up

this mountain and get to this peak with dropdead gorgeous views. Valleys. Ocean. Trees. You can see it all, and I think—Yes, this is peaceful. I am unplugged. I've got to take a selfie to capture this and then post it online to share it with my friends.

Rob: Exactly, so in that spirit, we came up with #1wknotech or One Week No Tech, a thought experiment in which we asked participants to *imagine* (emphasis here) giving up technology for one whole week and then to live tweet every moment of that experience.

Mark: Yes, so students were pretty freaked out when we first proposed it, but then, they mellowed out when they realized that it meant they got to use their mobile devices even more.

Rob: The goal wasn't really to make them more technology obsessed but to play out the paradox, the simultaneous and opposing pulls away from this technology that is doing damage to our lives and yet constantly back toward the world of friends and likes and online social approval.

Mark: Some of our favorite pieces were the memes we had students create. In one, a student of mine stands staring down at a deodorant stick laying flat in her hand, a poor replacement for her smartphone.

Rob: I think this netprov taught us the power of a good hashtag and how light a netprov could be. Basically, the whole premise is in the hashtag—the irony of a week without technology that has a hashtag for reporting back.

Mark: Also, it was an easily repeatable netprov, and one that students enjoyed. Students expanded the netprov by interpreting "technology" in different ways. While some gave up digital technology, others gave up utensils, shoes, et cetera.

Rob: Yes, it was a terrific example of students helping us explore an idea.

I Work for the Web (2015)

Mark: Some of our projects emerge out of our daily goofing around on social media.

Rob: Actually, most of our project emerge that way.

Mark: Good point.

Rob: So, one day, I saw Mark posting strange—well, stranger than usual—things on Facebook, claiming he was clocking in for his job working for Facebook, and his job was to post and Like things. He'd address his boss as Zuck, meaning Facebook-founder Mark Zuckerberg.

Mark: I Liked my job. A LOT!

Rob: That little act laid bare the whole playbor economy and reminded me of something serious: by posting content on Twitter and Facebook, we were working as unpaid content providers for them.

Mark: And from that little seedling grew I Work for the Web, which begins as the misguided viral marketing campaign of a fictional media telecom giant, Rockehearst Omnipresent Bundlers (R.O.B.), who own and therefore control the internet. In its poorly conceived marketing campaign, the company attempts to get folks to imagine how great it would be if their job WAS Liking and posting on the internet.

Rob: Only the campaign backfired because it reminds many web users that they are indeed already working for the internet for free. So, some internet workers decide to protest.

Mark: In response to the foolhardy corporate campaign, a union movement is born, called the International Web and Facetwit Workers— which fortunately also spells out IWFW—helping us keep to just one hashtag. We encouraged the participants to choose a side or to flip flop.

Rob: During I Work for the Web, we also introduced Nighthawks, named for the Ed Hopper painting, a fictional cafe/bar where the workers of the web could gather to drink, complain, and try to unionize—at least till Andrew Rockehearst sent in the Pinkertons— or rather, the Pingertons. In fact, the opening prompt for I Work for the Web was What Happened at Night Hawks? The conflicting accounts started us off.

Mark: That netprov was an example of one that had easy one-time fun—

Rob: Yes, remember the web waffler? The account from one of your students who served up waffles in the form of ## to anyone who wanted them?

Mark Very tasty. Or someone could play in "story mode," we could call it, participating in the overarching story arc as the struggle to unionize grew to a crucial make-or-break vote for or against the union cast, of course, with Like thumbs up or down!

Rob: Oh, but remember, the vote was preceded by a hilarious walkout, people taking pictures of their fingers walking away from their keyboards.

All-Time High (2015)

Rob: Now All-Time High was different for lots of reasons, first because it was primarily driven by performance poets Claire Donato and Jeff T. Johnson.

Mark: Performance poets. I like that. They are also the wit and wiles behind Special America.

Rob: They had the idea that it would be it would be awesome to imagine going back to high school—what a nightmare, rt?—and to go back with everyone.

Mark: Everyone who has ever lived!

Rob: What if everyone was back in high school, including you?

Mark: Wait, do I *have* to go back to high school to play this?

Rob: Yes, Mark. See, people have such visceral feelings about high school, and this netprov gave people a chance to relive and re-explore old wounds.

Mark: At the same time, bringing back the dead allowed for some historical remixing. Sappho, the moody fragmented poet, interacting with Napoleon, the perpetually frustrated teenage tyrant with the— what would you call it?

Rob: Napoleon complex?

Mark: *Exactement!* This netprov featured a four-week Twitter narrative, punctuated by four big synchronous live events.

Rob: They were iconic high school and high school movie moments: The Big Game, The Big Test, The Big Dance, and Graduation. The live events gave a chance to do some synchronous play. But we were in for bigger surprises.

Mark: Yes, it was during this netprov that players changed their characters dramatically. One Twitter account kept cycling through characters based on roles the actor Johnny Depp had played. Another drug dealer, actually played by the same featured player, Mike Russo, also underwent a series of transformations, changing the display name and icon as the character evolved over time.

Rob: Yes, and Claire followed suit with her characters. That opened up Twitter again for us because it showed yet another way of using Twitter. One account could transform into different characters or different versions of the same character by changing handles or profile pictures. You know, that example demonstrates the way netprovs do not just use existing platforms, like the cuckoo bird laying eggs in other birds' nests, but transform them, find untapped affordances, like

Mark: The cuckoo bird egg rolling into the side of the nest and discovering a cuckoo-bird-egg shaped pocket.

Rob: Right.

Mark: Really?

Rob: I think so. Yes, netprov players don't merely adhere to and play off existing social conventions of networked platforms— conventions that have existed for all of two seconds in the grand scheme of things. Beyond crowdpleasing and fanserving, engaging the masses or pleasing the elites, a netprov-er surfs the waves of whimsy. So why not play?

28

A Collective Imaginary: A Published Conversation

Kate Pullinger and Kate Armstrong

When Kate Pullinger and Kate Armstrong were approached about collaborating to produce a written piece on the subject of social platforms within the context of electronic literature for this publication, they decided to take the social aspect of the subject and apply it to the process of configuring a response. This dialog, which unfolded within a shared document over a period of three months at the end of 2014, is the result.

PULLINGER: Sometimes I feel that the word "community" no longer has any real meaning when it comes to talking about the world beyond our own local neighborhoods. I find it horribly easy to disassociate from communities that don't revolve around people I interact with frequently face-to-face. What does the word "community" mean to you now?

ARMSTRONG: I think there was a heyday of the word "community" that was super-powered by the dot com era, when startups in that early moment were tapping into emergent communities of interest and it was possible to see how this was going to cause a culture shift. I think this is still the legacy of what community means when people talk about it in a digital context. But at the same time, perhaps strangely, "community" doesn't always often mean *knowing anyone* in a community. Of course, this is hilarious because the whole thing is predicated on connecting to people. I think we're at a point when that connection both is—and feels—more akin to being a node in a massive interconnected web than having any direct or improved contact with individuals or even specific groups of individuals.

PULLINGER: I agree. However, there are still times when the crowd comes together to respond jointly in a way that preserves or even augments the role of the individual in a collective endeavor. *Letter to an Unknown Soldier* is an example of that. This project aimed to create a digital war memorial by asking everyone—and we meant everyone, though the project was UK-based—to write a letter to the statue of the Unknown Soldier that stands in Paddington train station in London, England. The statue, a work of great beauty in my opinion, is of a World War I soldier in full trench infantryman uniform, with his trench coat and his boots and a big knitted scarf, reading a letter he's just torn out of an envelope. We asked people to respond to a simple question: if you could say whatever you want to say to that soldier, what would you say? The project was open to submissions for five weeks in the run-up to the August 4, 2014 centenary of Britain's declaration of war against Germany. More than 22,000 people wrote letters to the soldier. Through this, we achieved an extraordinary snapshot of what people were thinking about that war, and war in general, during the summer of 2014. Once the project really took off (when the British Prime Minister wrote a letter three days after we launched, we realized it was going to be big) we had a few approaches from publishers. From the beginning we had thought that one outcome might be a book of some kind. In the end, the publisher we went with was William Collins, the history imprint at Harper Collins UK, who published a lovely small hardcover edition of the book with 138 of the most interesting letters from the project in it. It came out a few days before Remembrance Day. We held a launch party for all the letter writers in London later that month, and it was one of the most wonderful book events I've ever been to. More than 100 of the writers came, and they were mostly people who had never been published, and who did not think of themselves as writers. The print book was like a sort of talisman for them—I did this, and look, it's real. I found that so interesting and also moving.

ARMSTRONG: Your project surfaces something fresh that is also old-school, which is the form of the letter. People connected through *Letter to an Unknown Soldier*—to themselves, to the work itself, and to each other—through the national historical lens.

FIGURE 1 *The Unknown Soldier, Platform One of Paddington Station, London, England; Photo Credit: Dom Agius.*

To bring it back to social platforms, historically people would write letters to each other and would not always meet very often face-to-face. But when they would meet they would have a new connection that had been enabled through the written word. One of the strange things about the internet is that it creates a context where once in a while someone pops into focus from within this vast but vague web of connection. At those times it is possible to get a sense of someone in a way that is outside how you would ever have known them in real life. In this way I see community functioning like a random generator that throws out a demi-connection once in a while, and in doing that, creates a real connection. It is mostly dissociated from face-to-face interactions, but when the two realms happen to overlay, you can accelerate a friendship from knowing someone as basically a celebrity to knowing them as an actual friend. I don't mean that the person is a celebrity; I mean that in our current culture you sometimes enter a relationship knowing the same amount about a person as you might know about some random celebrity. And that makes it more isolating, if you never see them face-to-face, but also accelerates a connection when you do see them face-to-face, so that you don't have to waste a lot of time understanding who that person is or what their projects are. You can leapfrog into real discussion.

PULLINGER: That's so true, and so clearly put. In fact, this accelerated friendship is one of the things I enjoy most about the connections we make online. But when it comes to your own art and writing practice, is community—however you might define it—an important factor?

ARMSTRONG: My work centers on narrative forms that use dynamic information sources as part of their material structure. So if looking to the question of whether this dynamic information is ever emerging from, or informed by, a community, I think the answer is likely no. I think it's information that is generated by the activity of people, but in my work it doesn't usually matter how those people connect to each other—only how their activity connects to the work. That might sound psychotic, I don't know. But I'm more interested that a wave of activity is generated by people when they use the network and that this activity is part of the contemporary condition. But at the same time, the logic of what is happening within a stream of information comes from what it is, who is making it, where it comes from, what it means.

PULLINGER: One of the things I enjoy most about your work is the feeling that you are making sense of the endless streams of information and activity, that you have mastered the art of stepping back, listening in, and somehow pulling out, or pushing in, narrative—a narrative layer emerging from the chaos. This fulfills a fundamental desire of mine—of most people I think—to make sense of the world by creating story. Is that a valid interpretation of your work?

ARMSTRONG: I like that description. And to add to it, wanting to try to make sense of the relationship between story and world. What happens when they are attached in functional, technical ways? How can these things be linked in new ways and across contexts? I think community is part of this. I am wondering if the mechanisms of community operate differently from context to context for your projects and what that means for you in your work.

PULLINGER: It's been a source of frustration to me that the distance between these two communities of practice has been, and remains, so separate within my creative practice—literary fiction and digital fiction.

My recent novel, *Landing Gear*, grew up out of my digital fiction project, *Flight Paths*, and was, in part, a deliberate attempt to see if I could pull these two realms closer together. But I think the simple truth is that there are different types of readers, and different modes of reading, and that it remains difficult to draw readers from one mode to another—from the book to the screen, even if the book itself is being read on a screen. I still feel that we are at the very beginning of thinking about the potential for electronic literature and that in twenty years time there will be multiple modes of reading—from long-form prose narratives like the novel to highly networked, responsive, multimodal forms of the book. But we aren't there yet, and sometimes I feel that my attempts to draw the two readerships closer together are, well, maybe not psychotic, but a bit pointless at this stage.

ARMSTRONG: I agree, we are just at the beginning. And even right now there is a surge in micro-encounters with text. Even given the dominant position of the image, networked culture depends on text. I am thinking of the experimentation that is happening when writers intervene inside existing platforms, like Mark C. Marino and Rob Wittig's recent works[1] with Twitter as a narrative platform. My own work, *Why Some Dolls Are Bad*, was an early experiment with Facebook in 2007. One of the things that strikes me with this kind of work is that it blasts apart the distinction between "books" and "other reading."

FIGURE 2 *From* Space Video, *Kate Armstrong.*

---

[1]Such as Mark C. Marino and Rob Wittig (2013), *OccupyMLA*.

PULLINGER: Yes, and the other factor blasting apart these distinctions is the rise of mobile as the most important device—the single screen—for all consumption of content, from books to movies, particularly among young people. But to return to our discussion of the networks of activity that the internet affords, can you talk a bit more about that in relation to your work *Space Video*?

ARMSTRONG: *Space Video* happened because my collaborator Michael Tippett and I noticed that there were shared aesthetic qualities of video imagery that accompanies very disparate cultural and scientific phenomena including guided meditation, hypnosis, undersea and space exploration by NASA, motivational speaking, PowerPoint backgrounds, science fiction, psychedelic drug culture, electronic music, popular spirituality, and computer effects. There is a volume of video imagery sloshing around on YouTube that connects to these subjects in one way or another. These shared aesthetic qualities are familiar to us as a culture: a preponderance of things such as pyramids, philodendrons, eyes in the palms of smooth blue hands, heads that open to the universe in a flourish of lurid Photoshopping, slow pans of imaginary stars, op art, celestial storms, infrared yoga positions. The text fragments I've written for *Space Video* blend in with videos that people are uploading to YouTube in real time, so the whole thing becomes like a big, groovy film.

I think the role of the social here has two dimensions: one is the way the work is about popular or collective imagination in relation to these subjects, and how there is a questionable universality to what humans think of when they think of outer or inner space. The other is the way that YouTube is an unprecedented platform for apprehending collective activity—especially when it comes to the creation of undisciplined, trippy videos that stand as placeholders for what is ultimately nonvisual activity. There, we can really see the internet functioning as a collective imaginary.

PULLINGER: This is key, and I see your work as a vivid exploration of what this means. I think in my own work I've only really seen this in action once, in *Letter to an Unknown Soldier*. Somehow we pulled off the feat of producing a coherent work of art through collective participation on a huge scale, an act of the collective imaginary.

ARMSTRONG: But most of your other digital work is deeply engaged with participatory activity that happens through the lens of your story. This is the case for both *Inanimate Alice* and *Flight Paths*, isn't that fair to say? We had talked at an earlier point about authorial voice, and you'd said that in most of your digital fiction work the authorial voice is "a collective voice conjured by project collaborators." Maybe this is the juncture where authorial voice can be seen as a kind of constraint that can be used to form generative systems. Maybe this is what we can arrive at—that through networked literature one of the things we can do is form creative constraints that elicit and shape social participation.

29

Addressing Torture in Iraq through Critical Digital Media Art—*Hearts and Minds: The Interrogations Project*

Roderick Coover, Scott Rettberg, Daria Tsoupikova, and Arthur Nishimoto

Hearts and Minds: the Interrogations Project is an interactive virtual reality narrative artwork developed by an interdisciplinary team including humanists, social scientists, artists, and computer scientists from four different universities. The project, originally made in the CAVE2™ virtual reality theatre environment[1] at the Electronic Visualization Laboratory at the University of Illinois at Chicago, attempts to extend and make accessible difficult narratives of war and torture based on actual accounts from soldiers involved. *Hearts and Minds* uses VR as a narrative platform to represent a complex contemporary issue and to provide a platform for discussion and debate of military interrogation methods and their effects on detainees, soldiers, and society. We have published on this project previously in computer science and technical venues,[2] as well as digital arts venues.[3]

---

[1]See Febretti et al. (2013) for description of CAVE2™.
[2]See Tsoupikova et al. (2015) (SIGGRAPH) and (2016) (SIGGRAPH Asia).
[3]See Tsoupikova et al. (2015) (ISEA).

Our contribution to this volume focuses on the work from an artistic and narrative perspective and on how the work functions as a digital humanities project: one which brings important documentary material addressing an important contemporary problem to contemporary new media environments for critical engagement.

Hearts and Minds makes use of the CAVE2™ environment for a multisensory artwork addressing a complex contemporary problem: as American soldiers are returning from wars in Iraq and Afghanistan, it is becoming increasingly clear that some of them participated in interrogation practices and acts of abusive violence with detainees for which they were not properly trained or psychologically prepared. This project addresses a period of recent American history in which torture was both officially sanctioned and informally institutionalized. *Hearts and Minds* is intended to provide a window into both this institutionalization of torture and its effects on the young men and women who served as its instruments, few of whom joined the military believing they would become torturers. Many American soldiers are returning home with post-traumatic stress disorder (PTSD). American soldiers and citizens are left with many unresolved questions about the moral calculus of using torture as an interrogation strategy in American military operations. By giving voice to and in some ways situating the viewer in the perspective of soldiers who engaged in acts of abusive violence, *Hearts and Minds* further encourages citizens to consider carefully our complicity in acts done in our name.

Hearts and Minds bridges art, computer science, and social science research. Artist Roderick Coover (Temple University) and writer Scott Rettberg (University of Bergen) worked with the research scholars John Tsukayama and Jeffrey Stevenson Murer (St Andrews University) to distill central themes and stories from the significant and extensive research project—based on hundreds of hours of original interviews with veterans—carried out by Tsukayama (Tsukayama 2014). Coover and Rettberg worked with artist and virtual reality researcher Daria Tsoupikova (University of Illinois at Chicago) and computer scientist Arthur Nishimoto (University of Illinois at Chicago) to bring the script to fruition in the CAVE2 at the Electronic Visualization Laboratory at the University of Illinois at Chicago and subsequently in other media environments.

Tsukayama's interviews include revelations of a highly sensitive nature, including narratives of participation in acts of abusive violence that entailed violations of human rights. The interviewees granted Tsukayama the right to use their stories in his dissertation and in subsequent research outcomes derived from it, provided that their identities remained anonymous. The tapes of recorded interviews were destroyed after transcription, except for short samples to prove their authenticity, and Tsukayama did not retain any personal contact information for the soldiers he interviewed. The text was condensed into an accessible and coherent set of stories that would preserve the accuracy of the testimonies while voice actors would perform the roles of veterans, further assuring their anonymity.

Hearts and Minds as Creative Digital Humanities

We present this work here, in an electronic literature publication, and in a digital humanities research context, in part to argue that work of this kind should be considered in the broader context of the digital humanities. This is not an uncontroversial position. Some would argue that the scope of the digital humanities should be limited to the application of digital tools to traditional humanities subjects. While digital humanities includes applications such as digital editions, text encoding, various applications of computational linguistics, data-mining, visualization, and different applications of GIS and 3D modeling in disciplines such as literary studies, philology, history, archeology, and philosophy, digital humanities are not typically concerned with digital art, nor with contemporary geopolitical or social concerns. Indeed, while we have been engaged and fascinated with the growth and increasing institutional power of the digital humanities in the past decade, it is surprising how little attention the digital humanities per se has paid to digital culture and in particular how the contemporary products of electronic literature and digital art somehow seem to fall outside the frame of "digital humanities" in many contexts. Just as the digital culture of the present will be lacking if it is not engaged with and contextualized by the humanities, digital humanities will be deeply impoverished if it fails to engage with digital art and electronic literature but instead defines itself as a purely retrospective endeavor focused only on using the technologies of the present to consider the cultures of the past.

Hearts and Minds: The Interrogations Project is an artwork and narrative, but one that also functions as a digital humanities project that might serve as a model for future collaborations that bring together digital methods and technologies, social science, arts, and the humanities. Interdisciplinarity is an element of most digital humanities (DH) projects. While anyone working in DH knows that while the word "interdisciplinary" looks good on a grant application, in truth interdisciplinarity is difficult to achieve, and is often uncomfortably situated once it happens. Consider how the work of the digital humanities is divided and valued: how we must balance between technological development, "grunt work" such as gathering, cleaning, and filtering data, with analyzing and writing up that data. When multiple researchers and multiple disciplines are involved, there is always a question of the division of labor and how credit will be apportioned and perhaps even more fundamentally what terms and discourses will be applied to the given project: whose language will we speak? Sometimes DH projects are interdisciplinary only in the sense that tools and technicians are employed to tackle research questions that are fundamentally situated in the discipline of one principal investigator: the technologists serve the humanist and provide tools to address a particular research question or challenge. A project like *Hearts and Minds* models a different type of "all-in" collaboration, which while difficult is worth pursuing: we entered into the project thinking of it not purely as an art project, and not purely as a narrative project, and not purely as social science research, and not purely as technological research, but from the beginning as all of those together. This has entailed both collaboration and negotiation from the impetus of the project to the present, both between the individual actors involved and the disciplines in which we are institutionally situated.

The *Hearts and Minds* project developed as a result of cross-disciplinary relationships—friendships—as much as anything else. Rettberg knew digital artist and CAVE researcher Daria Tsoupikova from her brief stay with the Electronic Literature research group in Bergen as an intracountry Fulbright lecturer several years before we began the project. They had stayed in touch and planned to work with each other on a future project, and when he had the opportunity to take a sabbatical in Chicago during the spring semester of 2015, she was able to arrange some time for us to work on a project in the CAVE2. Filmmaker Roderick Coover and Rettberg had collaborated on a number of projects film and new media projects together over the preceding several years and he asked Coover to join in developing the new CAVE project. Rettberg and Coover began bouncing around themes and ideas that might work well in the immersive 3D theatre environment of the CAVE2. Our projects have typically centered thematically on contemporary social, political, and environmental challenges. Coover mentioned a conversation he had had with a friend—collective violence researcher Jeffrey Murer—about John Tsukayama's dissertation research on prisoner torture in Iraq. Every collaborator offered a different set of skills and a different disciplinary background to the shared effort.

Project Development and the AudioVisual Approach: Roderick Coover

The visual environment of the work includes 3D modeling and panoramic photography. The project presents the audience with an environment that begins in a reflective temple space with four doors opening to ordinary American domestic spaces: a boy's bedroom, a family room, a suburban back yard, a kitchen. The user navigates the environment. The virtual scene is continuously updated according to the user's orientation. Certain objects in the room have ambient audio and visual cues which encourage the user to trigger them. Once the given object is triggered, the walls of the room fall away and the audience members find themselves in abstracted desert landscapes—poignant and surreal landscapes of memory. The modified panoramic images which surround the audience at this point—originally photographed at US Army bases in the American West and at Pinochet's prison camps in Chile—reference both battlefield environments and metaphorically suggest a space of interiority. Perhaps most importantly, for the audience these environments function as a "listening space" in which they can hear, focus on, encounter, and confront some disturbing true stories told in American voices.

Hearts and Minds employs creative visual methods as a means to make challenging research accessible and meaningful on differing levels. The project grew out of a series of conversations about the research I had with John Tsukayama and Jeffrey Murer, mostly on Skype. We first addressed questions of how visual methods might contribute to John's research and its reception. While we shared interests in the potential of visual media to give voice and emotion to the data, we also both feared that a more conventional documentary approach risked sensationalizing the material. We were in agreement in a quest to create the space to engage the stories without excessive dramatization.

My approach drew on a combination of methods. One of these methods is drawn from interpretive and visual anthropology—an area in which I have extensive training and experience. Interpretive anthropology offers methods to engage subjective and illusive materials, those of the poetics and rhetorics of language, of performance, of sensation, and of creative expression. Emphases on motifs, objects, metaphors, and other turns of phrases are designed to help ground subjective accounts and provide points of translation. This was very valuable in my work with John. For example, we discussed ways that a common object like a folding chair look on differing meanings for the soldiers and gave meaning to their stories. Away from the home comforts of lounge chairs and sofas, the hard folding chair is immediately a sign of displacement. In the stories, the chairs become tools of interrogation, and in some cases, tools of violence and torture.

As we talked further, this attention to objects in the imagination helped shape the form of the project. Computer games were an important part of the soldiers' experiences. They are important in how young men envisioned the war experience in advance of enlistment; they were—and still are—broadly used as recruitment tools; the interface in some weaponry has close parallels to those of games; and soldiers describe playing games as a form of relaxation away from the battlefield. The games are also places of escape. Our choice to use a gaming format therefore was apt in a number of ways. It evokes the surreality of home and away, and of engagement in real and imagined worlds. It suggests landscapes in which violence is enacted, but it also mirrors back that violence: the worlds of memories, like those of games, resemble lived experience but also have deformations, disjunctions, and displacements. The form further questions the relationship between play (or indeed industrialized play) and human actions.

The relationship between visual references and language is valuable to pursue in this context. In this case a curious method arises from some other collaborations with writer Scott Rettberg in which we explore combinatory forms, such as in our works *Three Rails Live* and *Toxicity: A Climate Change Narrative*. Those works use code to shuffle images and language. They draw together stories based on scientific study of contemporary environmental conditions with evocative visual environments. The database structure provides a useful way of working through material. An object, such as a folding chair, may have direct references in one scene as an object of torture and placed in another, it returns that references with others with which it might be joined, such as those of a folding chair as an object of travel or an object of ceremony. One begins to describe a web of significations. Attention to the text is required to point to inherent and apt references, to avoid overly elaborated and illusory connections.

A second area of concern was how to conjure from the stories landscapes of memory, and how to place these stories within such landscapes. One aspect of this challenge is that the stories were being told after the war when the soldiers had returned and become veterans. Further, the soldiers' motivations for telling their stories often seemed to involve a difficulty in reconcile differing worlds, the awkwardness of returning home to find that the familiar had become strange. Meanwhile, the landscapes of their memory were frequently incomplete, abstract, and altered. Daria Tsoupikova's 3D modeling artistry in building the home settings would help express levels of defamiliarization in the home environment, while computer scientist Arthur Nishimoto's skills in interaction design and in creating the movements into the memory landscapes could articulate the conditions of travel and translation that are inherent in entering into the world's' stories, memories, and unnamable anxieties. While later this work would involve the extensive visual construction of the memory landscapes and work with actors to bring the stories alive, the next part of the narrative lay with Scott, in condensing John's research into manageable stories.

Project Development and the Writer's Approach: Scott Rettberg

When Rod and I first talked about the materials, I wasn't immediately convinced that we could do justice to the material, and to these soldiers' stories, in a CAVE 3D environment. The type of atmospheres and interactions that we can produce in these visualization environments are typically game-like and somewhat cartoonish. I was worried that we would risk exploiting the material, trivializing it by putting it into an inappropriate context. However, when we had a Skype conversation about the material and its potential representation in a digital artwork, as well as the limitations of the CAVE VR environment, John and Jeffrey convinced me the project was worth pursuing in this form. They had reached out to Rod because they both felt that the stories the soldiers had told should be heard in other contexts than conventional academic research publications, and they were excited about the possibility of art functioning as a medium to communicate the issues involved to audiences that the research might otherwise not reach. They also felt that a VR environment might situate the audience in a different way than a documentary or fictional film might, by immersing the audience more directly. When we met on Skype, as humanists: writer, filmmaker, and social scientists, we were able to reach a kind of shared consensus and understanding of what was at stake. John writes,

> When Jeffrey Murer told me about Rod Coover's interest in creating a multimedia experience for users to gain insight into some of the experiences revealed in the Detainee Interaction Study, I was immediately intrigued. In working with them and Scott Rettberg I developed a sense that they would honor the trust the veterans gave me that their stories would be treated respectfully and shared with others.

After I read through John's dissertation and the interviews, we had another conversation and at this point the conversation shifted from considering the project as social science research, and as factual testimony, towards considering it from literary and artistic perspectives. John's dissertation traced an arc, a set of patterns and stages in the development of different soldier's perspectives, attitudes, and embodied experiences of participating in or observing acts of battlefield torture. As I began to think about translating the research and how to stay true to its intent, those stages would become a story arc represented through the different rooms that the user encounters in the work. We also considered metaphor. In many of the interviews, the soldiers kept returning to the idea of "home"— in both the battlefield and after they had returned to civilian society, "home" had been on their minds. When they were at war, they felt a sharp disjunction between the reality they faced and the things they were doing in Iraq with their idea of who they were or had been at home. And when they returned from Iraq, "home" was also central to the way they described their experiences. After their return home, they had become estranged from civil society, they had come to feel displaced and unsettled in everyday life. They could no longer feel "at home" in themselves. Out of this discussion, we arrived at the idea that homes, domestic environments and the objects within them, should be the central environmental metaphor of the piece.

Metaphors were also key to the way that the stories are triggered through the user interface. The mundane everyday objects that trigger the stories serve as visual metaphors or icons related to the stories connected to them. This is in keeping with accounts of how victims of PTSD experience the ordinary world as a middle ground between the present and the traumatic past. The sight of everyday objects can trigger buried memories and traumas. We also discussed how to portray this transition "between worlds"—when each object is triggered, the walls of the environment fall away and the environment changes. The 3D domestic environment changes and the user is surrounded by landscapes surrounded by surreal 2D panoramas, surreal landscapes meant to suggest both the battlefield and more strongly perhaps a kind of interiority. Further visual metaphors and cues, such as metal folding chairs, mentioned often in the stories of interrogations, or a child's tricycle, were layered into the panoramic environments. We might pause for moment here to note that in this development stage of the process the conceptual work that we were doing was deeply informed by our background as humanities researchers. Our discussions of how the project should be structured were shaped by not only by our experiences as writer, filmmaker, artist, but also by our research and understanding of how metaphor functions in poetry, in cinema, and in visual art.

To stay true to the voices in the interviews, we decided to change very little of the soldiers' testimonies in their interviews with John in the script. Outside smoothing some transitions, I changed very little with the soldiers' monologues. I decided to put the fragments of testimonies into four voices, composites representative of types roughly characterized in the thesis, but the stories they told were essentially lifted verbatim from the interviews. In this sense the writing involved in the project is not about the creation of story from whole cloth—it is instead a matter of selecting fragments from a large pool of material and providing an architecture for them to fit together and make sense. The writing (or translation) involved is much more about distilling the stories in a way that language is condensed, representative of more than what is actually said. With a background in writing fiction, I sometimes struggle with this in writing for film and media art: my impulse as a writer is to represent as much of a world as possible through the written word. But in writing for media art, one needs to think much more like a minimalist poet, distilling experience rather than using language alone to model a world. While in a novel the written word stands alone on the page, in electronic literature, film, or media art, it is one channel among several. In this case, the visual environment, the human voice, the user's movement and interaction all play signification roles in our experience of the work. Much of the work involved is in balancing and harmonizing these channels so that they don't compete but instead serve each other symbiotically.

Collaboration in the CAVE

As we took the project from the script to realization, each of us played distinctive but separate roles in the project. Once we made the decision about what type of environment we wanted to create, we also made the decision to develop the project in Unity, a popular platform used to develop many contemporary commercial and independent games. One of the advantages of the CAVE2 compared with some earlier projection CAVE environments is that it can support a wide variety of development platforms in both Linux and Windows, as opposed to a platform that is necessarily custom-developed for the particular space. For CAVE artworks this is an important development, as it means works are now transportable from one contemporary 3D visualization environment to another, and importantly to other platforms as well. Although there is a history and an interesting corpus of electronic literature and digital art developed for CAVEs, it has been a great frustration for many working in these environments that because they were typically custom-designed for one specific CAVE, they were often written about more than they were actually seen by audiences.

Developing work for CAVEs was sort of the opposite of work made for the web in this sense: while work on the network was published everywhere on the network at accessible all over the world at the same time, work in CAVES could only be seen in one place by one audience at one time. This new model of works that are portable to other CAVEs and other devices is an important development and may well bring more artists to CAVEs in the future. The Electronic Visualization Laboratory (EVL) was generous in enabling us to have a good chunk of dedicated time in the CAVE2, a facility that is more often occupied by engineers and scientists doing things like examining 3D models of protein chains. But the EVL, the lab that developed the first CAVE, had a long history of collaborations between artists and scientists that stretches back to the 1970s.

Our roles in this project were fairly clearly defined, which made the relatively swift modular development of the various parts of the project feasible. Daria Tsoupikova began to work on the 3D room models, and also brought Arthur Nishimoto, a computer scientist and Unity developer, into the project to begin work on scripting and interaction design. Meanwhile in Philadelphia, Rod was working on the panoramas and with voice actors. Scott, who had been refining the text, scheduled a time to join Daria and Arthur in the CAVE to discuss the structure and designs and they communicated with Rod virtually through Skype. Once a critical mass of the components were together, we all met in Chicago in the CAVE and test out a prototype, the first of the rooms.

Meeting together for four intensive days, we rapidly prototyped the model for the project and tested out various ideas of interaction design, the use of visual and sound cues, and how movement and audience interaction would function in the space itself. There are iterative contextual shifts involved designing a project like this on paper, in a recording studio, in the Unity platform on the computer screen, to the actualized environment of the CAVE itself. The project didn't move as a finished entity from the screen into the CAVE but in a cycle of testing in the CAVE. We worked physically in the space of the visualization environment, taking notes and identifying problems and ideas, rebuilding and testing again. Working in a CAVE environment was advantageous to collectively experiencing and sorting through the materials. After Rod returned to Philadelphia, we continued this cycle for a number of weeks in Chicago while Rod continued to develop visual elements of the piece.

During the final stages of the project's initial development we shifted from thinking of the work primarily as a playable interactive work, and instead as an interactive performance work. The last part of our development work in Chicago included two performance events in June and July 2014. We asked performance artist Mark Jeffrey to join us in presenting the project. In the CAVE2 what the audience sees is focalized on the perspective of one person, whose movements are tracked in the space—the interactor literally moves physically through the virtual environment and, using a wand, also triggers the interactive events. Seeing a performer encounter the work and make specific decisions about his own movement in response to the digital artwork also changed our perspective on it. While the 3D screens and spoken voice are essential to the CAVE experience, it is also a theater-in-the-round performance, as our attention as an audience is split between the virtual and the physical. We watch and listen to the materials of the digital work, but we also watch the focalized performer. It is also a particularly important aspect of this piece that the members of the audience are also watching with the others in the audience. It is a collective encounter with some disturbing material that reflects back on our society, our complicity in what is done in our name. The fact that we are watching it together with others magnifies some of its effects, and emphasizes our shared responsibilities. The discussions that we share after screening the work are perhaps its most important aspect of the work.

Following successful installations around the world, we then returned to the concept of the playable interactive work as an educational tool and potentially one that could be used to by veteran's groups, human rights organizations, and others to build discourse. In public performances there were always researchers, artists, and invited scholars to discuss the work. Further, public exhibition allowed users to share experience afterwards through conversation. In preparing to release the playable object, the foremost lesson from performances was that such an objection would need context. To do so, we added own reflections on the work through short essays; we invited Jeffrey Murer to add a commentary on his experience, and we solicited additional commentaries from differing fields. Thus, from the core research the project results in a public experience in artistic and scholarly venues, including immersive CAVE environments, and a work for personal devices that can be used by individuals, organizations, students, and educators.

Conclusion

Systematic abuse is difficult to stop without listening those who lived within it—both the believers and objecters who confront the memories of carrying out the tasks a nation asked of them. *Hearts and Minds: The Interrogations Project* puts us uncomfortably in the shoes of those who have tortured in their country's name and have come back home, in many ways just as broken as the victims of torture themselves. After the revelations of Abu Ghraib, after the US Senate report on CIA torture, and after attempts during the Obama administration to remove torture from approved lexicon of the US military and intelligence apparatus, it may seem unnecessary to ask audiences to return to the memory of this historical period, and instead to dismiss it as a mistake which, once acknowledged, can be dismissed and forgotten as a relic of another time. One would hope that the lessons have already been learned. Instead, in 2016, we found that discussions of torture had returned to the public sphere. The Republican candidate for president not only refused to condemn torture—he actually made torture of terrorism suspects one the main planks of his platform. A surprising proportion of the US population remains receptive to using torture as an interrogation method, in spite of the fact that all available evidence indicates that it is not effective in its stated purpose of extracting useful evidence. It seems the lessons of these episodes have not yet been absorbed into the popular consciousness. There is still much work to be done to communicate the effects that torture has on the people, and the societies, who choose to inflict it on others.

The arts and humanities serve many functions to society, and from time to time—particularly recently it seems—we are called upon to justify the existence of humanities disciplines within university environments that are driving by increasingly utilitarian approaches to education. As humanities researchers we quite naturally resent this interrogation of the practices, research, and pedagogy that we have committed our professional lives to. We come back with the response that one of the roles of the humanities is to serve as an archive, as a part of academia that preserves our cultural memory. Projects such as *Hearts and Minds* ask us to think of that act of preserving memory not only from a comfortable distance, but also in ways that are engaging very directly with the recent past and in the present, functioning as critical digital media as we collectively address our situation within a challenged sociopolitical reality.

References

Febretti, A. et al. (2013), "CAVE2: A Hybrid Reality Environment for Immersive Simulation and Information Analysis," *Proc. SPIE 8649, The Engineering Reality of Virtual Reality*, 864903.

Litz, B. and W. Schlenger (2009), "PTSD in Service Members and New Veterans of the Iraq and Afghanistan Wars: A Bibliography and Critique," *PTSD Research Quarterly* 20 (1): 1–7.

Tsoupikova, D., S. Rettberg, R. Coover, and A. Nishimoto (2015), "The Battle for Hearts and Minds: Interrogation and Torture in the Age of War," *Proceedings of SIGGRAPH 2015 Posters*, ACM Press/ACM SIGGRAPH, New York, Computer Graphics Proceedings, Annual Conference Series, ACM, Article No. 12.

Tsoupikova, D., S. Rettberg, R. Coover, and A. Nishimoto (2015), "Hearts and Minds: The Residue of War," *Proceedings of the 21st International Symposium on Electronic Art: ISEA2015: Disruption*, http://isea2015.org/proceeding/submissions/ISEA2015_submission_27.pdf.

Tsoupikova, S. Rettberg, R. Coover, and A. Nishimoto (2016), "The Battle for Hearts and Minds: Interrogation and Torture in the Age of War. An Adaptation for Oculus Rift." SIGGRAPH Asia 2016.

Tsukayama, J. (2014), "By Any Means Necessary: An Interpretive Phenomenological Analysis Study of Post 9/11 American Abusive Violence in Iraq," PhD thesis, University of St Andrews, UK.

30

Poetic Playlands: Poetry, Interface, and Video Game Engines

Jason Nelson

If my digital poems could be jealous, if these interfaces could feel envy, a knee whacking desire to beat the competition, they would join together and gang-stomp the game engines. Out of the millions of users/readers my work has attracted, the poetry games dominate, drawing easily 75 percent of the interest (an emotive percentage). *game, game, game and again game; i made this. you play this. we are enemies;* and *Nothing You Have Done Deserves Such Praise* have gone pseudo-viral on the web, with articles in national newspapers, magazines, blogs, Russian MTV, Brazilian televangelists, and other odd venues around the globe. More importantly these works spread through person-to-person, forum-to-forum, message-to-message, post-to-post, meme-to-um ... -meme-maker, creating a personal sense of discovery and reader-ownership driving viral processes.

Why are the digital poetry/art games so much more compelling, more attractive, more interesting, more magnetic to a wider audience than my (or any other's) work? Some might point toward the near-crazy hand-drawn sketches and mix of strange video tales and universal themes. But many of my works are equally as baffling and paranoia-inducing, just as strange and sensational, unlocked monsters, tendrils from the subconscious. Instead, I hypothesize/guess/vaguely stab it is the game interface that sparks the work's popularity. The game interface, and certainly the platform- (Mario Brothers) style engine I use, is instantly familiar and engaging to anyone who

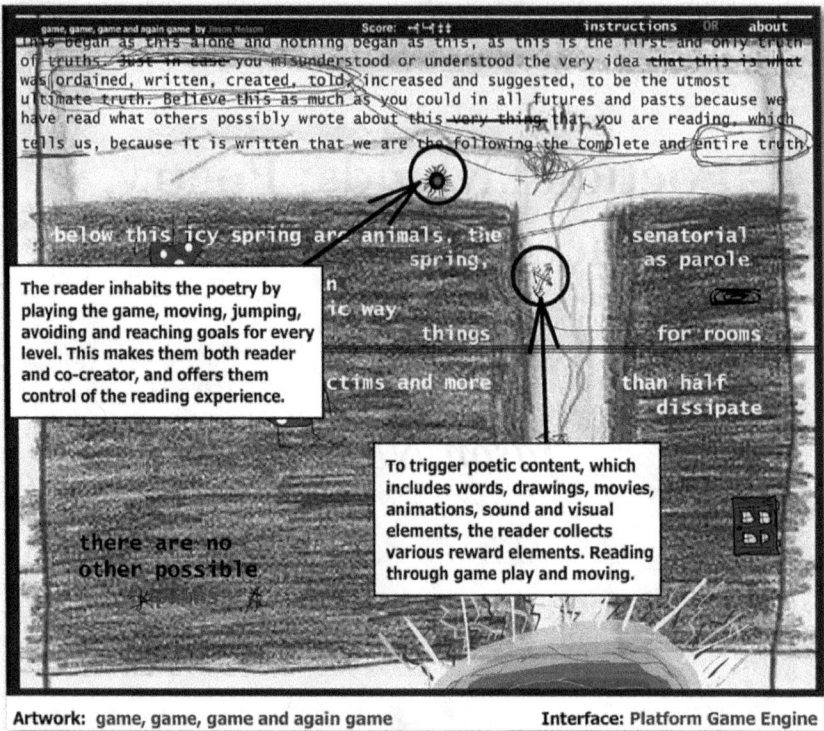

FIGURE 1 *game, game, game and again game.*

has grown up after the 1970s. Despite the immense advances in computer power and possibilities, the platform game continues to occupy a dominant and heart-spun space on the virtual gaming shelves.

This familiarity operates as a foothold, a climbing rope, a ladder, a doorway into what is otherwise a strange foreign world for most users/readers. Relatively few people read poetry (compared to cereal boxes), let alone experience net or digital art outside of a school setting—and combining the two can let them inhabit the work, to live in the giant Gerbil cage, pressing left, right, up, and mistakenly down.

The platform game engine is a linear journey with non-linear side roads, lost cemeteries of the digital settlers/pilgrims. The user/reader/player begins the game/level (typically on the screen's left) and continues, following a pathway to reach a goal (and move on to the next level or win prizes, or fall forever through a coded error). And while the poet cannot determine the exact path each player/reader will take, you can be fairly certain they will trigger specific instances should they want to move forward. This strategy might seem obvious, but it does have considerable impact on how the digital poem is constructed.

Video games are a language. They have created their own grammar, their own understanding based on rules, exploration, movement, and response. While I think certain segments or aspects of culture have been greatly impacted by video games, I don't feel that games have infiltrated fully into culture, certainly not when compared to other entertainment or communication or artwork forms. This is not to say that games cannot move beyond the same structures/game plays and ideologies that have dominated them for the last thirty years. However, there must be a greater effort made by both game makers/developers and the gaming audience to accept games on the same level as great artworks, works of literature, or music. Both groups need to learn how to "read the texts" of video games—in a more literary way, to truly play with the grammar/culture and possible new formats for video games.

To add something about online gaming worlds: the power of online gaming could be immense. The occasional news story about parents neglecting their babies to play *World of Warcraft* are a testament to that power. However, I am continually surprised at how unimaginative most of these online worlds really are. Most of the basic features in terms of fantasy or strategy are well worn clichés, and the adherence to earth-like physics is sadly uncreative. Why aren't video game creators or players demanding truly innovative gaming worlds where truly anything is possible, instead of re-treading *Dungeons and Dragons*? Let us make worlds of creatures that without the technology could never have been imagined and then given a visual/interactive birth. Then video games just might advance the culture in unexpected ways.

But what do videogames and poetry have in common? As mentioned, video games are a language, a grammar or linguistics for various texts. The sounds, the movement, the graphics, the rules (or lack of rules)—everything about a video game is a component of some kind of language. While poetry is traditionally taught as being constrained to words, I think of poetry as being based on *texts*. I mean "texts" in the broadest sense: all the elements, media, code, and artefact within a digital poetry game become a literary element/tool/device.

Therefore, the video game format, even the basic platformer game or point and click interfaces, or 3-D flying spaces I've toyed and tinkered with, are perfect grounds for a poetic playland. Additionally, many digital poems are inherently born from non-linear thinking and writing. Indeed, if technology drives poetry games, equally traditional book/page formats constrain the poem to line by line, pressed letter by bound page. But interactive technologies, and especially the game format, offer the poet the chance to make their poems not only multi-dimensional, but also interactive and multi-temporal. Prior to building haphazard coding skills, I conceptualized my creative writing as a visual element translated into words or music, then reformed into print language. To have all these elements, combined with the added tools/texts of movement and interaction and dimensionality, frees up the poetry writing process in curious and wondrous ways, a meandering creek flooded free into the spidering valley.

The video game format has the additional benefit of acting as a foothold or a bridge for readers. Unfortunately, relative to other forms of communication and textual input/output, poetry readership hasn't kept pace. In my adopted homeland of Australia, a good-selling poetry book might sell a thousand copies. But my digital poetry, my games and strange interfaces, even the least popular and most difficult to understand attract tens of thousands of readers, and my most popular coax millions into playing.

The question then arises, from its long slumber half covered in shaved paper bedding, why does digital poetry appeal or at least invite so many readers? The most immediate reason is the game interface offers a familiar and inviting interface, a fun or at least interactive way for the reader to feel involved with the poetic space. The poet ceases to be an authority and instead becomes a combination of guide, artist, and theme park ride operator. Additionally, the diversity of texts within the digital poem creates, as they say in cliché-land, "something for everyone." Whereas some readers might love the combination of ambient soundtrack and responsive words, others might adore the interaction of explosions and literary allusions to religious laundromat pamphlets.

game, game, game and again game

game, game, game and again game is/was a digital poem, retro-game, an anti-design statement and a personal exploration of the artist's changing worldview lens. Much of the Western world's cultural surroundings, belief systems, and design-scapes create the built illusion of clean lines and definitive choice, cold narrow pathways of five colours, three body sizes and encapsulated philosophy. Within net/new media art the techno-filter extends these straight lines into exacting geometries and smooth bit rates, the personal as WYSIWYG (What You See Is What You Get) buttons. This game/artwork, while forever attached to these belief/design systems, attempts to re-introduce the hand-drawn, the messy and illogical, the human and personal creation into the digital via a retro-game style interface. Hovering above and attached to the poorly drawn aesthetic is a personal examination of how we/I continually switch and un-switch our dominant belief systems. Moving from levels themed for faith or real estate, for chemistry or capitalism, the user triggers corrected poetry, jittering creatures, and death and deathless noises. In addition, each level contains short videos from the artist's childhood, representing those brief young interactions that spark out eventual beliefs. *game, game, game and again game* is less a game about scoring and skill and more an awkward and disjointed atmospheric, the self-built, jumping, rolling meander of life.

Software, such as Adobe Flash or Photoshop, used to dominate the appearance and feel of images and artwork. This was partially due to the software's data processing methods (Flash loves vector, heart emoticon) and the limited range of tools and filters. The mouse also constrains the artist's ability to escape the software's overpowering aesthetic and, thus, removes or at least submerges much of the individual from many digital images. But in *gggag* I strove for a sense of anti-design within the design by relying on the hand-drawn.

Therefore this game/artwork began on paper, with charcoal and coloured pencils, each level born from the immediacy and error/curve/(e)motion of electricity-less drawing. This hand-drawn approach breaks the artwork away from both the oppressive control of software and much of design culture. In essence, this artwork/game was created as if the audience consisted of only the internal self, an offering of the personal with all its incongruities and confusions and small beauties.

Technology can be used to remove the errors and mistakes of human creation, with a prevalent emphasis on usability and glitch-free user operation. And while this artwork follows those conventions in function, it revels in the quirky, immediate corrections of the marked and corrected environment through the artwork/game. The poetic texts and animated drawings, triggered by the user/reader's movement are incomplete and in process of being edited and altered. Thus, the work itself becomes a notebook or sketchpad of personal ideas and insights into what falls away from "finished" creations.

Perhaps the most direct method of inviting the user/reader to invest the self into their experience is to offer them direct control of how the artwork is experienced. In this creation that user-inhabited space is contrasted with the creator's personal journey through belief systems. In the process of reading, the user might feel simultaneously enthralled and alienated, which mirrors the individual's experience with all group-think cultural constructs. In addition, the direct connection between action and content, between direction and poetics, encourages the user to rethink the images, drawings, and poetics in the light of the actions that opened that content.

Each of the levels within *game, game and again game* is themed/centered around a particular belief system. These systems do not inhabit religious grounds and, instead, emphasize the filters individuals use to interpret their surrounding culture-scape. The game play and poetics extend from these various belief systems. For example, within the Chemist level, which examines drugs as worldview, the syringes open surreal texts and rave-like graphics, and the game play path is largely illogical and deceiving. On the Faith level, the user has to choose between the two sides of the cross, with one choice offering a deathless death and the other continuing the path, with satirical and irreverent poetics inspired by personal events/encounters with Christianity.

Adding to the mix of personal artefact, journey, and art game are the family audio commentaries embedded into the level prize videos. These videos, shot in 8mm by my grandfather (whose last word was "gizmo"), were digitally captured by a handheld camera with the audio coming from family commentaries about the events within the videos. These remembrances are directly related to the belief systems each level explores and extend the personal, intimate thoughts into the meta-poetics of the artwork/game.

As most games have constant soundtracks and sounds/noises that clearly identify with the action or object that triggers them, this artwork/game uses sound to disjoint and translate the experiences. All levels use a different loop exploring speed, cadence, and tone to play on emotional and physical responses and to contrast or compliment particular level themes. The effects I was seeking are described by Jan Baetens and Jan Van Looy:

> From a cultural viewpoint, there are several reasons for justifying a strategic alliance between e-poetry and sound, but the strong embedding of e-poetry in the historical avant-garde is by far the most salient one. A second explanation for the text-sound link is a mechanism of psychological compensation. It is often argued that cyberculture virtualizes the body, and that this virtualization engenders different types of fear that need to be averted by an opposing mechanism of foregrounding the body.
>
> (Baetens & Van Looy 2008)

The game interface used in this artwork is more than twenty years old, dating back to early Mario Brothers and before. This familiar game play and interface was important in the creation of this artwork in that it provided an immediate and nearly invisible gameplay environment and allows the user/reader to immediately identify with the digital space. The poetry game becomes less about deciphering the conditions/rules of play and more about the contrast between the format and the poetic and drawn "content," intimate and personal, a direct extension of the artist's (my) ideologies and self.

With the gameplay being so familiar, the artwork recontextualizes some of the main features of the game. For example, the score, replaced with arrows/characters, continually spins, responsive to game play but numerically meaningless. There are unlimited lives available in the game, with the only negative consequence of encountering an enemy being pushed back to the level's starting point, accompanied by an announcement in a disembodied voice: "Come on and meet your maker" (now a slogan for a Danish fashion house). Some levels use the "warp" function to transport the creature to certain locations. This warping is taken to absurdity in the levels where the character cycles through falling, and represents the fruitless personal investigation of belief and the always altering lens.

game, game, game and again game was born from the skeleton of a retro game engine and 13 hand-drawn levels. Instead of thinking about game play and level design, my approach was to draw personal impressions of belief systems on paper and to use those drawings to create the game play of each level.

i made this. you play this. we are enemies

i made this. you play this. we are enemies is an art-game, interactive digital poem which uses game levels built on screen shots from influential community-based websites/portals. The game interface drives the poetic texts, the colliding and intersecting images, sounds, words, movements, a forever changing, reader-built poetic wonderland. Using messy hand-drawn elements, strange texts, sounds and multimedia layering, the artwork lets users play in the worlds hovering over and beneath what we browse, to exist outside/over their controlling constraints. Kiene Brillenburg-Wurth describes this layering as "together-art" and says it is "often an art of fusion: different media are not merely combined, but welded into a hybrid that rewrites older versions of the media involved" (Brillenburg-Wurth 2006: 6). In my case, the game engine and its myriad of media and interactive content is intended to create this fused artistic and poetic environment, with the arrow keys and space bar guiding you, and the occasional mouse click begging for attention. It is an ideal method of creating this collaged and layered effect, precisely because the game engine allows for triggered content to be the goal of the key/mouse-driven user/reader movement.

Each day the internet is humming with a million small interventions. From the humoresque mocking of community content sites like *Fark*, to the net gate-keepers Yahoo and Google, partisan political portals like *Huffington Post* or the open source/file sharing "pirates" of *Mininova*, the web is an easy tool/weapon for meddling/influencing and sharing/forcing/alluring your opinion on whomever clicks. Yet this digiscape is a deceiving and uneasy place, with continual streams of generic expression/content, cute dogs, and accident clips knocking against an incredible range of political/social beliefs hidden beneath the screen. Even short sequences of words, titled links, or blinking ads can reveal the strange, wondrous, and treacherous.

Timely Insert: Reactions to My Series of Game-Engine Interface Poems

An anonymous player of my newest art game *Scrape Scraperteeth* verbosely, and with great pixilated venom, described all the ways my game entirely and completely failed. Aside from poor playability, hobo-esque design, and crashing coding, he spent a considerable effort blasting me personally. Suggesting I was an art school wanker with serious mental health issues and most likely had a sordid criminal record filled with all manner of sexual deviances, he ended his diatribe with a direct threat to my skin and bones should I make another work. Others in the same forum quickly leaped to his side, wielding great textual swords of agreement. But, surprisingly, hidden in the bitter streams are islands of love, chiming comments of adoration. They tend to be quieter (as is the law of internet land), showing me love through back alley emailing or reviews and sharing my work on obscure and major sites across the net.

Since 2007, when I released *game, game, game and again game* into the gaming world (followed by six or seven more games depending on ludology), the above scenario of extreme hate or love from a polarized audience has played out many dozens of times, across all continents in a bizarrely broad range of web portals: from drug enthusiasts to adult content, in major international magazines and even elementary school syllabi. My games struck and continue striking nerves and brain stems, inspiring the sharing of something so odd, so terrible or compelling, citing notions about games that are not games, art that is not art, poetry that is, well, poetry.

In Oklahoma, where I was raised, belief systems dominate social life. For some it is Baptistic evangelism, others are ruled by Oklahoma Sooners football and everyone worships oil. So *game, game, game and again game* was built for a poetic exploration of such life-dominating notions as real estate, pharmaceuticals, or Buddhism. The hand-drawn backgrounds were created both from frustration with the ultra-clean/perfect design aesthetic of most net art and my yearnings to create a hand-made facade. One of the game's most baffling aspects are the home videos. In essence, these represent my belief system, as cheesy as it sounds, of family, with the first level's clip my mother coming out of the hospital with newly-born me in her lap.

Admittedly, a common comment about my work is simply "WTF?!" And I blunderstand why some post that critique/observation. I did not set out to share my work in popular internet forums, nor intend to play the role of "crazy dude" in gaming circles. *game, game, game and again game* and others were created as digital poems for the electronic literature community, built for galleries and academic venues. While the game was happily accepted by artists and professors, the notion of having an audience of only a few hundred (attending for free wine/cheese or because it is a university course requirement) was entirely unsatisfying. Around that time, I was helping a German PhD student (Jens Schroeder) with research into video games (involving crashing games conventions) and after a few beers where I complained about the tiny hit counts of art realms, Jens suggested I try sending my game-like creature to popular gaming blogs or culture portals.

On his advice, I sent a poorly crafted email to the generic info/tips/ contact addresses of such sites as *Kotaku, Joystiq, Destructoid, Jayisgames,* and others. Previously, as a lark, I inserted a "send me an email" note on the game's final screen. Not being used to receiving messages from anyone other than complaining students or failed eBay bid notices, I left my email unchecked over the weekend. Then on Monday morning my usual four messages were replaced by a few hundred. Within my server statistics I found all the above game blogs (and numerous cleverly named others) had reviewed my work, and indeed the game continued spreading and spreading over the next weeks and months.

As an artist it was an awakening. Here was an artwork, considered experimental in the fields of electronic art and writing (a digital poem and art-game for crusty crunk's sake), and it was being discussed, shared, blasted and praised as a game. I wasn't prepared for the extremes of player's responses. There were creative and disturbingly specific death threats, marriage proposals, including images of shaved and unshaved areas circled and labelled with detailed directions. Some people sent money, and others gave suggestions for psychiatrists. Every morning the messages kept coming, and I became addicted to checking my server statistics and vainly searching for the latest exposure.

After *game, game, game and again game*'s viral (a terrible cliché) spread tapered, I itched to make another art game. I spent months creating the zombie shooter-inspired *Alarmingly these are not Lovesick Zombies.* I explored a perpetual enemy shooter engine as a way to create an interactive sculpture generator. I crafted background videos for mini-narratives and toyed with the notions of absurd scoring goals and having levels reachable only upon losing. And sadly the game was a disaster. I had let the WWW attention camp in my head, and thus I created a game weird for the sake of being weird. Somehow the internet collective consciousness picks up on disingenuous creations and destroys them with the hammer of disinterest. If you are going to create an abstract hand-drawn poetry art-game, do it from your unique imagination (the back of your head) and not from what you think will disturb others or get the most hits. The game was met with relative web silence and sank like a narcoleptic cake-heavy synchronized swimmer. So, I decided to go back into my safe academic world and never make another game, ever, never, ever. Sniff. Sniff.

Thankfully, my gamer-hating, zombie-loving, emo-esque party lasted less than a month. What broke my brief funk was, oddly and appropriately enough, the webmaster of a series of "adult video" sites: almost a dozen sexytime clips, hot kitten, people humping websites listed *gggag* as a top link. As you might expect, the visitor count was massive, and with my work saving tons of lubricating jelly and paper towels, I was recharged. If my game could disrupt the hormone-fuelled drive of browsers, maybe my brand of artsy-crazy-poetry-game still had legs, arms, and other intertwined body parts.

I decided to thematically center my next art game on what had been preoccupying my mind for the previous year: the strange bipolar space of the gaming community and its love and hate of my craptastic creations. So in mid-2008 *i made this. you play this. we are enemies* was born. The aptly-titled game used screenshots of popular web portals, from the lumbering beasts of Yahoo/Google to fancy-pants sites like *BoingBoing* and *Metafilter*, for the level designs. I wanted to create the effect of doodle annotation, of marking up the screenshots with commentaries about the portals and what they represented. The player becomes the doodler, with each coin-like reward adding to the visuals. Oddly, introducing the idea of an intermission seemed to spur incredible numbers of emails. These missives weren't so much commentary about the use of an intermission as they were the result of a pause in the frenetic insanity. I also attempted to introduce more traditional game elements, more enemies, harder level design, and secret transporters. An unpublished version of the game included five additional levels inspired by *ESPN, Suicide Girls,* and others. Their exclusion had more to do with keeping the game size manageable than content issues.

i made this. you play this. we are enemies spread even more than *game, game, game and again game*, showing up in newspapers such as the *New York Post* and *Der Speigel*, on Russian *MTV*, and in magazines like *Wired* and others. This repeat success seemed to signal one thing: there were gamers hungry for the strange and unique, for the odd combination of poetry and art in a world dominated by clean graphics and complex game play. It even inspired some lovely copyright battles because deeper in the game are three appropriately sequenced levels utilizing Disney's main page, the RIAA (the folks that sued grandmotherly pirates) and *Mininova* (at the time a major BitTorrent destination). I received a few threatening emails from prestigious-sounding law firms demanding all sorts of madness. I would like to think my academically driven responses, hinged on satire laws, made them go away. Instead, it was most likely all those that stole the game's SWF (Shock Wave File) and placed it on gaming portals making their task seem impossible.

The follow-up, and one of my most literary theory-driven games, was *Evidence of Everything Exploding.* Continuing with the annotating doodle design approach, I chose cultural documents for the level designs, representing pivotal or interesting moments in recent human history: Bill Gate's Computer Brew letter (where he argues for charging for software), a government warning about the pre-World War I flu pandemic, the NASA moon landing document, and the patent for the pizza box among others. Moving away from the platform engine, I used a top-down shooter engine while including some of the same tricks as in previous games for pop-up narratives and other artistic content. Keeping with the theme of these documents as keys to our social puzzle, I included locked areas and required

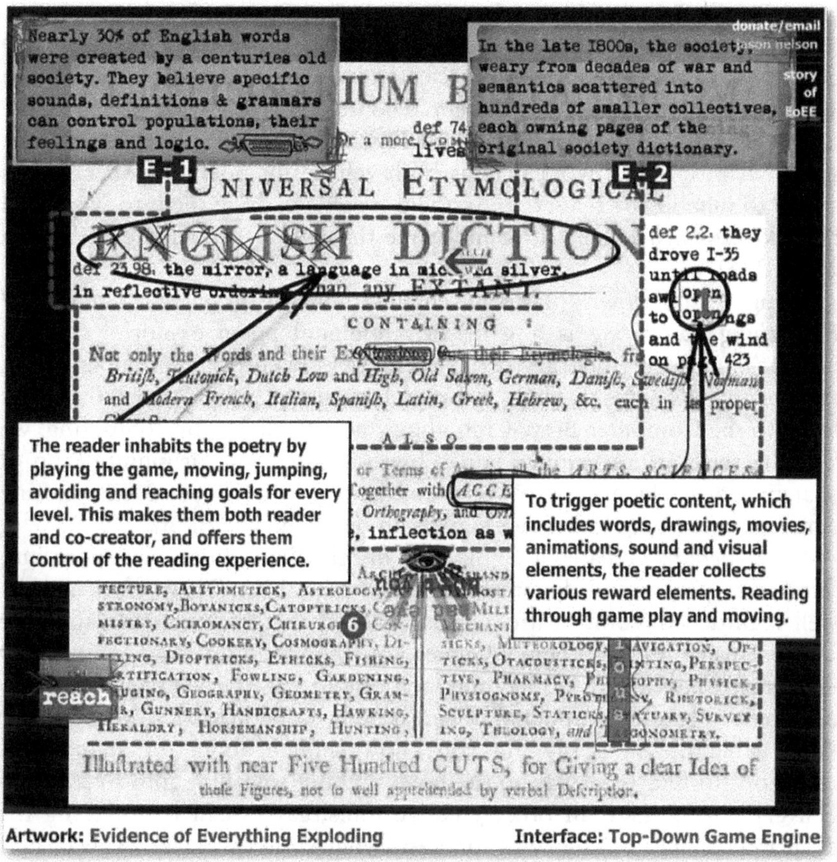

FIGURE 2 *Evidence of Everything Exploding.*

exploration for keys chased by more complex enemies. In some ways the increased difficulty of *Evidence of Everything Exploding* made for a smaller audience, as it hit that murky middle ground between proper game and art experience.

Evidence of Everything Exploding

Using documents, both historical and little-known from Bill Gates, NASA, James Joyce, Dadaism, Neil Gaiman, Fidel Castro, and others, the art game *Evidence of Everything Exploding* explores those strange moments where history does or does not turn, where unusual forces collide to create or topple storylines to build new futures. With the same hand-drawn, marked-up style, this game uses a Maze engine to guide the player through unsolvable puzzles. On each level are prophecies and stories inspired by the history and events the documents represent. The madness of the pages meets the madness of the game. And as Astrid Ensslin so beautifully describes, the surreal nature of the work acts as both vehicle for poetic wonder and as a way to inhabit the reader, to fix them breakfast, drive them to work and rethink/reconstruct their surround-scape through a surrealist lens (Ensslin 2014: 225).

Using a top–down platform engine (without gravity) *Evidence of Everything Exploding* is a game-driven digital poem exploring various historical and contemporary texts. Each level's poetic content is built from the document's sub-sub texts and curious consequences. With Bill Gates's letter to the Computer Brew Club about monetizing hobby computing, we find the seeds of an empire; James Joyce is caught in an infinite loop of changing texts; Fidel Castro's boyhood letter to the US president praising America and asking for money signals an opportunistic future.

Since then, I've created a whole herd of other less gamey excursions into interactive poetry and dynamic/generative digital art. These works, while well regarded in some realms, never reached the same massive audience as those using a game engine. Games are a common interface, a universal language. They are a ladder and a foothold for the average player to experience abstract art/poetry. When driving in Mongolia you might not know what the signs say but you know enough about the shapes and directions to find your way to the hotel without smashing into overloaded delivery trucks. Like all other creative tools, games can be anything the creator imagines, toying with, or destroying entirely, player expectations and becoming poetry. Yet there are some who argue against this notion of a game becoming a poem. Joseph Tabbi argues the boundaries between digital art, computer games, and digital literature are blurred, yet those distinctions are necessary,

> so that a literary language [can] create its own self-awareness, its own specificities, genres, and supporting networks that are needed to distinguish the literary arts from visual, oral, and computational media … Where games demand interaction and where conceptual arts bring us to a new, embodied understanding of the primacy of perception in the arts, literature does something else, something requiring continuity and development, not constant interruption through the shifting of attention from one medium to another.
>
> (Tabbi 2010: 39)

And while I agree literature does do something else, does demand attention, and continuity from the reader and writer, the two—the game and the poem—are not mutually exclusive. Additionally, the shifting attention is exactly the poetic point in many of my works: I intentionally adjust between form and media and meaning within the game environment.

Scrape Scraperteeth

My latest art game *Scrape Scraperteeth*, commissioned by the San Francisco Gallery of Modern Art, was built from the directive to make a small-scale creation, simple, and representative of my previous works. While it is not an entirely new take on the digital poetry game form, it does uniquely focus on one of the dominant events of the past few years: the real estate speculation crash. I love the notion of creating micro-games as artistic/poetic commentary on important news events or controversial topics. It might not be a complicated game but as an artistic statement I am charmed by its singular focus. Unlike many of my previous works, its primary thematic focus is a political message. Thus, it does showcase the varied usages of a game interface for digital poetry and how the platform engine can be an ideal canvas for a variety of possible poems.

FIGURE 3 Level One: the stormy cityscape of *Scrape Scraperteeth*

Scrape Scraperteeth also represents the movement of an interface from experimental form to an established approach to building digital poems. As such, I could explore other facets of the game environment previously unexplored. For example, I included textual elements that follow the hero/creature, creating a dynamic poetic layer for each layer. Additionally, the explosions for each reward use text as a visual element, creating animated concrete poems each time they are triggered. I am particularly charmed by the hand-drawn backgrounds, which, unlike my previous games, utilize and examine geography and location via photographs of the Gold Coast (Australia) skyline.

In no way does detailing my game-making experiences intend to say "look what a great artist I am" or "I'm more popular than Kangaroo Jack" (I've always wanted to type that). Indeed, I admit I am not a great game maker. My drawings are messy, and my work is difficult for those outside the net-art/digital poetry spheres to understand. Instead, my intention is to show how creating games that are truly unique with a reckless abandonand without regard to convention can actually lead to interesting artwork and also to a substantial audience. Yes, half your audience might hate you with words of violence and bitterness, but the other half will send you long, adoring notes of how your work reached some unused part of their brain, a brief crazy escape from the madness of their daily life.

The Flash Wake and The Forced Pet Metaphor

Autopsies are sometimes inconclusive, certainly when the dead remain breathing, life support pushing code into a shallow chest. Breathe, stope, breathe, stope. All extra Es.

But I've, we've, known Adobe Flash was dead/dying for at least three to four pet gerbils (depending on care and concern). So, when we arrive, collectively, at the cage one morning and our wee furry prisoner will not respond to high pitched calls or whiffs of cheese, and we ponder irrational revival methods (microwave, mouth-to gerbil mouth, surprise birthday party), the rodent's demise should not be unexpected nor feared. Gerbil maladies are rarely contagious to Javascript. Rarely.

> The dominance of Flash during that time period is an important signifier for the aesthetics of Internet culture at the time. Its ability to mix text, animation, rough video, sound, and a wide range of interactive elements within the one browser window drove the look of many works of e-lit at the time, as well as the wider Internet of the produser.
>
> (Krauth 2018: 257)

Inevitably, we ponder new creatures to "own," to reinforce our control over the immediate-scape. What makes this process so difficult is how addicted many of us were to the interface, the interaction, the multi-modal, multi-layered nature of Gerbils? Dogs, cats, miniature stretch-bears are all glorious creatures with which to play and engage. But they are unwieldy, difficult to train, and easy to un-train. Building digital poetry games from the bark of dogs sometimes makes for dramatic, yet hollow, trees, all crust and tower, crashing into power lines during even soft rains.

Since *Scrape-Scraperteeth* (my third best title), I've adventured into the long and high-grassed fields of heavy code or engines designed for the propelling of games and re-frozen foods. Prepare, cook, package, ship, leave on the street while dealing with the dangerous twin-terrors of needles and pie fillings, defrost, re-freeze. A creative process, I(T) says being semi-academic with a horror movie identity.

My future games are no longer organic pets built in the warm belly of Flash. They are erratic and stumble-drunk monsters with cybernetic limbs and processors for brains. As clumsy machines they are near immortal, alive, and scratching until the societal skill of forcing electrics through wires collapses in the maddening choice to canonize a real estate scam artist. Titles, the soft dough of a poet. Games, the risen disk, piled with the arbitrary remains of pets and practitioners. The same disease that ends an interface is the same disease that ends what's hiding in the cage.

References

Baetens, J. & Van Looy, J. (2008), "E-Poetry Between Image and Performance: A Cultural Analysis", *E-Media studies*, 1 (1). doi: 10.1349/PS1.1938-6060.A.288

Brillenburg-Wurth, K. (2006), "Multimediality, Intermediality, and Medially Complex Digital Poetry", *RiLUnE*, (5): 1–18.

Ensslin, Astrid (2014), "Womping" the metazone of the Festival Dada: Jason Nelson's evidence of everything exploding. 221–231.

Krauth, A. (2018), "Electronic Literature", Barney Warf (ed.) *The SAGE Encyclopedia of the Internet*, 257–258.

Nelson, Jason (2007), *game, game, game and again game*. Available online: *http://www.secrettechnology.com*

Nelson, Jason (2008), *i made this. you play this. we are enemies*. Available online: http://www.secrettechnology.com

Nelson, Jason (2009), *Alarmingly these are not Lovesick Zombies*. Available online: *http://www.secrettechnology.com*

Nelson, Jason (2011), *Evidence of Everything Exploding*. Available online: *http://www.secrettechnology.com*

Rettberg, Scott (2019), "Electronic Literature", Wiley.

Tabbi, J. (2010), "Electronic literature as world literature; or, the universality of writing under constraint", *Poetics Today*, 31 9 (1): 17–50.

31

A Way Is Open: Allusion, Authoring System, Identity, and Audience in Early Text-Based Electronic Literature

Judy Malloy

In tenth-century Northern France, Archdeacon Wibold created *Ludus Regularis*, an algorithm-authored game of dice in which clergy gambled for virtues (Pulskamp and Otero 2014). Centuries later, Wibold's dice-won virtues (chastity, mercy, obedience, fear, foresight, discretion, and piety, etc.) are parroted in the words that poet Emmett Williams selects for his algorithmically authored *IBM* (virgins, yes, easy, fear, death, naked, etc.). Subsequently, in the 1970s at MIT, where Wibold 's *Ludus Regularis* was probably known to mathematicians and students of chance (Kendall 1956: 2), the virtues of *Ludus Regularis* were replaced by treasures, as the authors of *Zork,* led players through the perilous Great Underground Empire in a quest to acquire nineteen treasures (Anderson et al. 1977–9). Beginning with Wibold's *Ludus Regularis*, this artist's chapter explores early text-based electronic literature and its precursors through the lens of textual, intertextual, and algorithmic allusions—whether intentional or zeitgeist inspired.

Part 1

Ludis Regularis

Hunted by bow and arrow-armed demons, crowds of people climbed the Ladder of Virtues in medieval icons and manuscripts (*Ladder*), acquiring virtues as they proceeded upwards towards heaven. The concept of the Ladder of Virtues was popularized by Saint John Climacus, but it was a tenth-century archdeacon, who, when canon law forbade clerics vice-ridden gambling, created *Ludus Regularis*, a game in which clergy could gamble for virtues.

To devise an authoring system for *Ludus Regularis*, Wibold, Archdeacon of Noyon, utilized throws of four dice. Three were cubes imprinted with groups of vowels on each of the six sides; the fourth, a tetrahedron, was imprinted with consonants on each side. Functioning to a certain extent as variables, virtues—each obtained by a combination of vowels and consonants—were grouped by ones (charity to wisdom); twos (remorse to reverence); threes (piety to exomologesis); and so forth. Once a cleric had won a virtue, it was no longer available. The cleric with the most virtues was the winner (Pulskamp and Otero 2014).

But all throws of the dice did not result in obtaining virtues. Indeed, in a recent session using John Ensley's emulator, which Richard Pulskamp and Daniel Otero provide in their comprehensive paper on *Ludus Regularis*, the first two plays resulted in no virtues, but on the third, "perseverance" was acquired.

Gentleness, liberality, wisdom, remorse, joy. The effectiveness of generative literature depends not only on how an authoring system will produce the chosen words, phrases, or lexias; but also, on the chosen words themselves.

Part 2

... what the poem amounts to, if carried out too far, is an eternal project, and, for most of us, eternity is more time than we have at our disposal for perfecting works of art ...

—EMMETT WILLIAMS

In the twentieth century, the lists of words that Fluxus poet Emmett Williams chose for *IBM*—first created without a computer in 1956, computerized ten years later, when he was asked (probably by composer James Tenney) to create a computer poem—reflect a different era, although one not necessarily without theological echoes: money, up, idiots, sex, like, quivering, evil, old, red, zulus, ticklish, kool, going, black, jesus, hotdogs, coming, perilous, action, virgins, yes, easy, fear, death, naked (Williams).

Williams' authoring system was not based on random algorithms (except possibly for "1. Choose 26 words by chance operations – or however you please") but rather was based on imposed constraints. Each letter of the alphabet was assigned a word: A = money, B = up. To begin the process, a word was chosen, "IBM," in this case. The correspondingly lettered word was then substituted, resulting in a phrase: "Red Up Going," which appears as a title. The process was then repeated as the poem expanded: "Perilous like sex, Yes Hotdogs ..." (Williams).

Part 3

The canary chirps, slightly off-key, an aria from a forgotten opera.
—Anderson et al. (1977–9). Like the altered translations that occurred in 1985 when Norman White's *hearsay* was passed (on I. P. Sharp's' ARTEX network) from Toronto to Des Moines to Sydney to Tokyo to Vienna and onwards until it returned to Toronto (White 2001), allusion is a fragile concept for working artists. A work is seen ten or twenty years ago and vaguely remembered. The work of John Cage lying in the background of mid- and late-twentieth experimental composition, not always acknowledged but often there (Kostelanetz 1988: 199). The work of Sonya Rapoport in the San Francisco Bay Area, instilling the idea of computer-mediated installation in the collective mind (Couey and Malloy 2012: 37–50). Fluxus tradition expanded and alluded to in the immense number of boxes as containers for words that comprise Jean Brown's archives (Getty). Icon-laden stamp art from Ed Higgins, echoed in the interface for my *A Party at Silver Beach* (Malloy 2003); the way video artist Joan Jonas integrated myth and life— her video *I Want to Live in the Country (And Other Romances)*—alluded to in the concluding "Song" of my *its name was Penelope*—although if I hadn't told you this, you would never know.

At MIT in 1977, whether or not with knowledge of Wibold's game, the virtues of *Ludus Regularis* were replaced by treasures, as the authors of *Zork* led players through the perilous Great Underground Empire in a quest to acquire nineteen treasures: the jewel-encrusted egg, the clockwork canary, the crystal trident of Poseidon, to name just a few.

If—despite limits on what you can carry and hostile encounters with the thief, a troll with an ax, the Cyclops, and other obstacles—all nineteen treasures are acquired, and they are all placed in a trophy case, you don't precisely get to heaven. What you get is a map that leads to *Zork II*.

Building on Gregory Yob's *Hunt the Wumpus*, Will Crowther's *Adventure*, and *Dungeons & Dragons*, *Zork* was created for the PDP-10 by Tim Anderson, Marc Blank, Bruce Daniels, and Dave Lebling. It used a sophisticated parser; incorporated MIT Culture (Montfort 2005: 95–117);

incorporated random elements; created a sprawling world model, the Great Underground Empire; and spawned the historic interactive fiction publisher, Infocom. The authoring software was Zork Interactive Language (ZIL), written with MDL.

Zork begins in an open field, west of a white house. The door to the house is boarded. Useful commands are: open, read, drop, N S E or W, climb, go down, enter, take, get, eat, move, turn on, diagnose, Odysseus (used against the Cyclops), give, say hello to, listen to, damage, echo, light, launch, attack, kill, wait, walk around, yell, smell, count, what is, wind up, pray, repent.

Along the way, readers are asked to consider issues of computer-mediated literature:

> The [windup] canary chirps, slightly off-key, an aria from a forgotten opera. From out of the greenery flies a lovely songbird. It perches on a limb just over your head and opens its beak to sing. As it does so a beautiful brass bauble drops from its mouth, bounces off the top of your head, and lands glimmering in the grass. As the canary winds down, the songbird flies away.
>
> (Anderson et al. 1977–9)

Part 4

Chaunce of the Dyse

Over the centuries, sometimes purposefully, sometimes with serendipity, in electronic literature and its precursors, narrative devices emerge, submerge, and emerge again, from a tenth-century bishop's dice-driven gambling for virtues; to allusions to the worldly Chaucerian narratives of pilgrims on their way to Canterbury in the dice-driven *Chaunce of the Dyse* (Hammond 1925; Mitchell 2009; Sergi 2011); to echoes of *Chaunce of the Dyse* in the computer-mediated output of an electronic literature-influential triangle of two men and a computer—as Lytton Strachey's nephew, Bloomsbury-bred computer scientist Christopher Strachey, and Manchester University's historic mainframe computer, aka the Manchester University Computer (MUC) (probably the Ferranti Mark 1, which was prototyped by the Manchester Mark I), and Alan Turing, the man who designed a hardwired, noise-based random number generator for the MUC—collaborate in a series of groundbreaking computer-generated love letters created with Strachey's software and Turing's hardware (Strachey 1954: 25–31).

There are fifty-six predominantly Chaucerian-allusive narratives of love, infidelity, virtue, and vice in the circa fifteenth-century manuscripts for *Chaunce of the Dyse.*

Sometimes attributed to John Lydgate, *Chaunce of the Dyse* consists of three introductory stanzas, followed by the lexias, each pictorially keyed by combinations of the throw of dice. Like *Ludus Regularis*, *Chaunce of the Dyse* is not based on the sum of the throws but rather on the fifty-six sets of combinations that throwing three six-sided dice produce.

When *Chaunce of the Dyse* was played/performed, a master of ceremonies (concealing his or her complicity with country bumpkin words—"First myn vnkunnynge and my rudenesse") read the opening ballad. Each player rolled the dice in turn and keyed the results to the corresponding text. In the process, the character of each verse was projected onto the recipient, who read the words aloud—whether to honor, merriment, or innuendo.

> ... The Chaunce of the Dyse haphazardly throws up allusions, attempting by chance to close the gap between literature and life, past and present, "game" and "earnest." This is the way the game produces, for a coterie of readers, the conditions of possibility for events to happen that confer unforeseen meanings on literary experience, respectively and prospectively.
>
> (Mitchell 2009: 63–4)

Based on then relatively contemporary works, such as Chaucer's *Troilus and Criseyde* and *The Canterbury Tales*, texts were personal, satirical, literary, character building, character destroying, embarrassing, pleasing, comedic, sardonic. Additionally, as if they were lexias in a work of generative hypertext, each lexia/node was both compactly written and intuitively linked to the other nodes. In *Chaunce of the Dyse* " ... intertextual allusions produce striking echoes across the texts," Mitchell observes (2009: 62).

It should be noted that although the texts appeared sequentially on the manuscript, their reading was determined by a random process. Contingently, when in 1995 then Xerox PARC researcher/artists, Cathy Marshall and I, provided randomly generated hypertext as one choice of reading our alternating lexias for *Forward Anywhere*, we created an interface that emphasized the difference between writing and reading the texts. Cathy writes:

> The fluidity of the process obscured the complexity of the structure. The piece is both densely interconnected, and loosely woven. The question then became, how do we express the process, make it accessible to a reader? Should we expect a reader to experience the screens in the same order in which we wrote them? Should we put the reader in front of a CRT in a darkened motel room?
>
> (Judy: black vinyl headboard) I have the lights turned out. Yellow words emerging from the black monitor ...)
>
> (Malloy and Marshall 1996)

Part 5

You are my ...Five hundred years after *Chaunce of the Dyse*, in the early 1950s, computer scientist Christopher Strachey, who was working at that time for the British National Physical Laboratory, wrote a program for the MUC in Alan Turing's Lab.

Utilizing Turing's hardwired random number generator, which improved pseudo-random results, Strachey's groundbreaking love-letter generator created a possibly endless series of letters that beginning with "You are my"; parsed and randomly inserted variables, and were signed "Yours — (adv.) M. U. C."

> Honey Dear
> My sympathetic affection beautifully attracts your affectionate enthusiasm. You are my loving adora- tion: my breathless adoration. My fellow feeling breathlessly hopes for your dear eagerness. My lovesick adoration cherishes your avid ardour.
> Yours wistfully
> M.U.C. (Strachey 1954)

Sometimes along creative ways, books fall open in interesting places, as they did for Verdi before he wrote *Nabucco* (Verdi 1942: 80–93).

Sometimes ideas are *in the air*: "So I think that what appears to be my influence is merely that I fell into a situation that other people are also falling into," John Cage once modestly observed (Kostelanetz 1988: 206).

Sometimes, an influence can be suggested, but there is no definite proof. Oxford-educated Chaucer scholar, Eleanor Prescott Hammond, wrote the classic paper on *Chaunce of the Dyse* in 1925. Her work would very probably have been known in Bloomsbury circles. And (whether consciously or not) the shifting gender identities and texts of changing ideas of love, randomly assigned in the *Chaunce of the Dyse*, echo in the 1950s in the process of Christopher Strachey's MUC Love Letters (Gaboury 2013)—and later occur and reoccur in the lives and generative poetry of the extraordinarily brilliant Fluxus couple, not-couple, couple, Dick Higgins and Alison Knowles.

Part 6

House of Dust

In New York City, after composer James Tenney presented a workshop on FORTRAN to Fluxus artists in 1967, Alison Knowles wrote the generative poem *House of Dust* (realized by Tenney), and her then husband, Dick Higgins, created and programmed *Hank and Mary, a Love Story, a Chorale.*

Into the variables for *House of Dust* (originally titled "Proposition for Emmett Williams"), Alison Knowles inserted a centered identity, evocative of her interests in natural materials, as illustrated in the variables and output below:

a house of (leaves, stone, dust, sand, wood, paper)
[place] (underwater, in dense woods, in heavy jungle undergrowth, by the sea, in green mossy terrain, among other houses, on an island)
using (all available lighting, natural light, electricity, candles)
inhabited by (friends, children and old people, people who love to read, vegetarians, horses and birds)
generating, for example:
a house of house of wood
in a metropolis
using electricity
inhabited by friends and enemies (H. Higgins 2012)

Part 7

Hank Shot Mary Dead

Created in the same time period as Knowles' "Proposition for Emmett Williams," *Hank and Mary, A Love Story, A Chorale*—written by Dick Higgins and programmed in FORTRAN IV by Higgins and James Tenney—is remarkable for the complex polyphonic ballad it produces with the permutations of only four words: "Hank Shot Mary Dead" (D. Higgins 1970).

Hank and Mary moves darkly down continuous feed computer paper, with increasingly complex repeated columns of the chorus/continuo "Hank Shot Mary Dead" playing against/with permutations such as "Mary Shot Shot Hank"; "Shot Shot Shot Shot," and finally:

Dead Dead Dead Hank
Dead Dead Dead Shot
Dead Dead Dead Mary
"Dead Dead Dead Dead" (D. Higgins 1970)

Part 8

Uncle Roger

Female narrators are unexpected in this world of men and mainframes, particularly when the historians are men. Unless the protagonist is male-created, such as Joseph Weizenbaum's innovative ELIZA/DOCTOR—which seeped so thoroughly into NPC (nonplayer character) dialog in interactive fiction—when the observer is a woman, arguably the narrative changes. For instance, in *Uncle Roger* (Malloy 1991), a male obsession with the speed of the chips (" … humming 'fast fast fast' softly to himself. I could see his black shoes and brown socks moving on the pink tiled floor") runs in the background to the touch of the hand or the intimate moment:

> Jeff and I were in the top bunk. I put my hands on his body. It was dark, and the train rocked gently on the tracks as it moved swiftly along towards San Francisco.
>
> (Malloy 1991)

Nevertheless, with a few exceptions, such as "House of Dust" and *Uncle Roger*, the early history of electronic literature is dominated by the quests of men with big machines. Moving backwards in time:

Part 9

Stochastische Texte

The idea was suggested by Stuttgart philosopher Max Bense. The input was fed into the formidable Zuse Z22 computer. To create *Stochastische Texte*, in the late 1950s, German mathematician and computer scientist (then student at Stuttgart), Theo Lutz, entered words from Kafka's *The Castle* into a program that parsed pseudo-randomly selected variables into semi-logical texts (Lutz 1959).

Arguably, a few years after Strachey's paper was published in *Encounter* in 1954, algorithmic strategies influenced by those used to create Strachey's Love Letters were used in 1959 to create *Stochastische Texte*. A primary dissimilarity between these two electronic literature precursors hinges on the textual differences between Strachey's playful romantic language and Lutz' politically charged remix of *The Castle*.

Kafka's evocative language is not evident in *Stochastische Texte*:

The tower above him here – the only one visible – the tower of a house, as was now evident, perhaps of the main building, was uniformly round, part of it graciously mantled with ivy, pierced by small windows that glittered in the sun, with a somewhat maniacal glitter, – and topped by what looked like an attic, with battlements that were irregular, broken, fumbling, as if designed by the trembling or careless hand of a child, clearly outlined against the blue.

(1969: 12)

As set forth in his *Augenblick* paper, the English versions of the words that Lutz chose were:

THE COUNT THE STRANGER THE LOOK THE CHURCH THE CASTLE THE PICTURE THE EYE THE VILLAGE THE TOWER THE FARMER THE WAY THE GUEST THE DAY THE HOUSE THE TABLE THE LABOURER OPEN SILENT STRONG GOOD NARROW NEAR NEW QUIET FAR DEEP LATE DARK FREE LARGE OLD ANGRY

In these words, there is a strength that suggests that they may have been deliberately (not randomly) chosen. What energizes this surprisingly plain list is the way Lutz chose to direct the computer to parse together two sentences on each line, using "is" as the predominate verb, and the way that the program concatenated these sentences with either a period or "and" or "or" (as if they were search terms) or "therefore." The result is effective, in part because rather than evoking the encounters with inexplicable bureaucracy that are a theme of *The Castle*—encounters that will resonate more directly in subsequent works of electronic literature and its precursors, from Perec's *The Art of Asking Your Boss for a Raise*; to the introductory language of *Moonmist* (Galley and Lawrence 1986); to the nearly insolvable puzzles of Graham Nelson's *Curses—Stochastische Texte* is strong in its activist-resonant choice of words, as if it is a response to *The Castle*, rather than a direct echo:

... A CASTLE IS FREE AND EVERY FARMER IS FAR.EVERY STRANGER IS FAR. A DAY IS LATE.EVERY HOUSE IS DARK. AN EYE IS DEEP.NOT EVERY CASTLE IS OLD. EVERY DAY IS OLD.NOT EVERY GUEST IS ANGRY: A CHURCH IS NARROW.NO HOUSE IS OPEN AND NOT EVERY CHURCH IS SILENT.NOT EVERY EYE IS ANGRY. NO LOOK IS NEW.EVERY WAY IS NEAR.NOT EVERY CASTLE IS QUIET.NO TABLE IS NARROW AND EVERY TOWER IS NEW.EVERY FARMER IS FREE. EVERY FARMER IS NEAR ...

(Lutz 1959)

In *Augenblick*, Lutz observes that:

... It seems to be very significant that it is possible to change the underlying word quantity into a "word field" using an assigned probability matrix, and to require the machine to print only those sentences where a probability exists between the subject and the predicate which exceeds a certain value. In this way it is possible to produce a text which is "meaningful" in relation to the underlying matrix.

Part 10

I AM THAT I AM

While cutting a mount for a drawing in room No. 25, I sliced through a pile of newspapers with my Stanley blade and thought of what I had said to Burroughs some six months earlier about the necessity for turning painters' techniques directly into writing. I picked up the raw words and began to piece together texts ..." (Gysin 2001b: 126)

In Paris, beginning in circa 1959, Brion Gysin and William Burroughs built on Tzara's *"How to make a Dadaist Poem"* (92: 39–41) to formalize a "cut-up" composition method. Burroughs may have used a proto-cut-up method in the writing process for *Naked Lunch*. Gysin migrated the cut-up method to computer-mediated poetry, creating the meaning-laden permutation *I AM THAT I AM* and the five-word literary theory permutation *NO POETS DON'T OWN WORDS* (both works programmed by Burroughs' lover, Ian Sommerville).

Contingently, in his innovative theory cipher *Não*—exhibited using an LED display on an electronic signboard at the Centro Cultural Cândido Mendes, Rio de Janeiro in 1984—Eduardo Kac differently interprets the combination of "no" (não) and "poets," while at the same time exploring the display and reading of digital poetry (Kac 1982–4).

As Burroughs once observed "When you cut into the present, the future leaks out" (Burroughs 1986).

Part 11

Interplay

In exploring the rhizomes of electronic literature, it should be remembered that to create many early works of electronic literature, cards were punched and then feed into a room-sized mainframe, resulting in continuous feed paper print-outs (Funkhouser 2012: 243–4). For instance, James Tenney realized Alison Knowles' *House of Dust* by programming it in Fortran IV and running it on a mainframe at the Polytechnic Institute of Brooklyn:

> The computer generated four hundred quatrains before a repetition occurred. As Knowles describes it, a foot-high stack of computer printout appeared one day on her doorstep.
>
> (H. Higgins, Introduction 2012: 195–6)

When Bill Bartlett created *Interplay* for the Computer Culture Exposition at the 1979 *Toronto Super8 Film Festival*, for many of the participating artists at I. P. Sharp terminals (in Canberra, Edmonton, Houston, New York, Toronto, Sydney, Vancouver, and Vienna), continuing dialog on computer culture emerged on continuous feed paper (Bartlett 1979).

When Roy Ascott produced the 1983 collaborative fairy tale *La Plissure du Texte*, as it traveled from artist to artist in eleven cities including Pittsburg, Vancouver, Vienna, San Francisco, and Toronto, an improvised, collaboratively authored text was printed out at many of the nodes on its journey (Ascott 1984: 24–67).

In these works, in which the artists themselves were the audience—accompanied by the sound of the printer, as the text moved from node to node—harkening back to the intertextual allusions in *Chaunce of the Dyse*—the narratives played off each other.

But when Art Com Electronic Network went online on The WELL in 1986, the nature of the environment for electronic literature changed radically, as not only did Director Carl Loeffler situate experimental writers in the midst of The WELL—including John Cage, Judy Malloy, Jim Rosenberg, and Fortner Andersen—but also sysop Fred Truck created UNIX Shell Script-based menus on which to publish electronic literature (Malloy 2016: 191–218). Some of the members of the audience were known because they responded online to ACEN works. Some were not. As the narrator's dream illustrates in *Uncle Roger*, there was a palpable feeling of the presence of the audience, even though they were not seen:

> Everything I typed on the keyboardshowed up on a large screenwhich filled the entire wall at the front of the room.Five men in tan suits were sitting around the screen,watching the words as I typed them in.
>
> (Malloy 1988)

Contingently, although the details are beyond the scope of this chapter, it should be noted that reference to the works of each other were of value in the creative practice that emerged from ACEN. And sometimes the fine line between poetic allusion and the passing back and forth of ideas that occurs in any group of artists is immaterial in the merged creative process.

References

Anderson, Tim, Marc Blank, Bruce Daniels, and Dave Lebling (1977–9), *Zork*, Cambridge, MA: MIT/PDP-10.

Ascott, Roy (1984), "Art and Telematics," in Heidi Grundmann (ed.), *Art Telecommunication*, 24–67, Vancouver: Western Front; Vienna: BLIX.

Bartlett, Bill (1979), *Interplay*, Toronto Super8 Film Festival.

Buchloh, Benjamin H. D. (2012), "The Book of the Future: Alison Knowles' 'House of Dust'," in Hannah B. Higgins and Douglas Kahn (eds.), *Mainframe Experimentalism: Early Computing and the Foundations of the Digital Arts*, 200–8, Berkeley, CA: University of California Press.

Burroughs, William (1986), "Origin and Theory of the Tape Cut Ups," *Breakthrough in Grey Room*, Sub Rosa Records.

Coover, Robert (1999), "Literary Hypertext: The Passing of the Golden Age," *Digital Arts and Culture*, October 29, Atlanta, GA, www.nickm.com/vox/golden_age.html (accessed July 4, 2018).

Couey, Anna and Judy Malloy (2012), "A Conversation with Sonya Rapoport (on the Interactive Conference on Arts Wire)," in Terri Cohn (ed.), *Pairing of Polarities*, 37–50, Berkeley, CA: Heyday Press.

Chaunce of the Dyse, MS Fairfax: 16; MS Bodley: 638.

Crowther, Will and Don Woods (1976–7), *ADVENTURE*, reconstructed by Donald Ekman, David M. Baggett, and Graham Nelson, www.web-adventures.org/cgi-bin/webfrotz?s=Adventure (accessed December 23, 2016).

Funkhouser, Christopher (2012), "First-Generation Poetry Generators: Establishing Foundations in Form," in Hannah B. Higgins and Douglas Kahn (eds.), *Mainframe Experimentalism: Early Computing and the Foundations of the Digital Arts*, 243–65, Berkeley, CA: University of California Press.

Gaboury, Nick (2013), "A Queer History of Computing: Part Three," *Rhizome*, April 9, www.rhizome.org/editorial/2013/apr/9/queer-history-computing-part-three/ (accessed July 4, 2018).

Galley, Stu and Jim Lawrence (1986), *Moonmist*, Cambridge, MA: Infocom.

Getty Research Institute (n.d.), *Jean Brown Papers, 1916–1995, Series VI. Art Objects, 1958–1986*, www.archives2.getty.edu:8082/xtf/view?docId=ead/890164/890164.xml;chunk.id=ref2994;brand=default (accessed December 23, 2016).

Gysin, Brion (2001), "Cut-ups: A Project for Disastrous Success," *Back in No Time: The Brion Gysin Reader*, 125–32, Middletown, CT: Wesleyan University Press.

Gysin, Brion (2001), "*I AM THAT I AM*," in Jason Weiss (ed.), *Back in No Time: The Brion Gysin Reader*, 81–8, Middletown, CT: Wesleyan University Press.

Hammond, E. P. (1925), "The Chance of the Dice," *Englische Studien*, 59: 1–16.

Higgins, Dick (1970), *Computers for the Arts*, Somerville, MA: Abyss Publications. [Contains a printout and partial program for *Hank and Mary* and a printout and documentation for *House of Dust*, under its original title *Proposition no. 2 for Emmett Williams*.]

Higgins, Hannah B. (2012), "An Introduction to Alison Knowles's The House of Dust," in Hannah B. Higgins and Douglas Kahn (eds.), *Mainframe Experimentalism: Early Computing and the Foundations of the Digital Arts*, 195–9, Berkeley, CA: University of California Press.

Jerz, Dennis J. (2007), "Somewhere Nearby is Colossal Cave: Examining Will Crowther's Original 'Adventure' in Code and in Kentucky," *Digital Humanities Quarterly* 1 (2), www.digitalhumanities.org/dhq/vol/001/2/000009/000009.html (accessed December 23, 2016).

Jonas, Joan (1976), *I Want to Live in the Country (And Other Romances)* [Video].

Kac, Eduardo (1982–4), *Não*, www.ekac.org/nao.html (accessed December 23, 2016).

Kafka, Franz (1969), *The Castle*, New York, NY: Modern Library.

Kendall, M. G. (1956), "Studies in the History of Probability and Statistics: II. The Beginnings of a Probability Calculus," *Biometrika* 43 (1/2): 1–14.

Kostelanetz, Richard (1988), *Conversing with Cage*, 206, New York, NY: Limelight Editions.

Ladder of Divine Ascent. Saint Catherine's Monastery, Sinai Peninsula, Egypt. This icon depicts St. John Climacus, leading monks up the ladder.

Lutz, Theo (1959), "Stochastische Texte," *Augenblick* 4: 3–9, www.stuttgarter-schule.de/lutz_schule_en.htm (accessed July 4, 2018).

Malloy, Judy (1988), "Terminals," *Uncle Roger*.

Malloy, Judy (1991), "Uncle Roger, an Online Narrabase," in Roy Ascott and Carl Eugene Loeffler (eds.), *Connectivity: Art and Interactive Telecommunications*, Leonardo 24 (2): 195–202.

Malloy, Judy (2003), *A Party at Silver Beach*, 2nd edn., 2012, www.well.com/user/jmalloy/weddingparty/begin.html(accessed July 4, 2018).

Malloy, Judy (2016), "Art Com Electronic Network: A Conversation with Fred Truck and Anna Couey," in Judy Malloy (ed.), *Social Media Archeology and Poetic*, 191–218, Cambridge, MA: MIT Press.

Malloy, Judy and Cathy Marshall (1996), *Forward Anywhere*, Cambridge, MA: Eastgate.

Mitchell, J. Allan (2009), "Consolations of Pandarus: *The Testament of Love* and *the Chaunce of the Dyse*," in J. Allan Mitchell, *Ethics and Eventfulness in Middle English Literature*, 61–8, New York, NY: Palgrave Macmillan.

Montfort, Nick (2005), "Zork and Other Mainframe Works," in Nick Montfort, *Twisty Little Passages: An Approach to Interactive Fiction*, 95–117, Cambridge, MA: MIT Press.

Nelson, Graham (1993), *Curses*, http://www.highprogrammer.com/alan/games/video/ifmaps/curses.pdf.

Perec, George (2011), *The Art of Asking Your Boss for a Raise*, trans. David Bellos, London: Verso.

Pulskamp, Richard and Daniel Otero (2014), "Wibold's Ludus Regularis, a 10th Century Board Game," *Convergence*, June, accessed on the Mathematical Association of America website, www.maa.org/publications/periodicals/convergence/wibolds-ludus-regularis-a-10th-century-board-game (accessed December 23, 2016).

Sergi, Matthew (2011), "Interactive Readership in the Chance of the Dice," paper presented at *Reading the Middle Ages*, March 26, Berkeley, CA: University of California.

Strachey, Christopher (1954), "The Thinking Machines," *Encounter* 3: 25–31.

Tzara, Tristan (1992), "How to Make a Dadaist Poem," in Tristan Tzara, *Seven Dada Manifestos and Lampisteries*, 39–41, London: Calder Publications; New York, NY: Riverrun Press.

Verdi, Giuseppe (1942), "An Autobiographical Sketch," in Franz Werfel and Paul Stefan (eds.), *Giuseppe Verdi: The Man in his Letters*, trans. Edward Downes, 80–93, New York, NY: L. B. Fischer.

Weizenbaum, Joseph (1966), "ELIZA – A Computer Program for the Study of Natural Language Communication between Man and Machine," *Communications of the ACM* 9 (1): 35–45.

White, Norman (2001), *Hearsay. Telematic Connections: The Virtual Embrace*, curated by Steve Dietz, Curator, Walker Art Center, telematic.walkerart.org/timeline/timeline_white.html (accessed December 23, 2016).

Williams, Emmett (n.d.), *A Valentine for Noel*, Stuttgart: Edition Hansjörg Mayer. [*A Valentine for Noel* is also the source of the words quoted in the section heading.]

Yob, Gregory (1976), "Hunt the Wumpus," *The Best of Creative Computing*, v. 1, 247. [*Hunt the Wumpus* was created in BASIC in circa 1974.]

INDEX OF NAMES

INDEX OF TERMS AND WORK